# Green Building Project Management

# Green Building Project Management

Guest Editor

**Xingwei Li**

MDPI

Basel • Beijing • Wuhan • Barcelona • Belgrade • Novi Sad • Cluj • Manchester

*Guest Editor*
Xingwei Li
College of Architecture and
Urban-Rural Planning
Sichuan Agricultural
University
Chengdu
China

*Editorial Office*
MDPI AG
Grosspeteranlage 5
4052 Basel, Switzerland

This is a reprint of the Special Issue, published open access by the journal *Buildings* (ISSN 2075-5309), freely accessible at: https://www.mdpi.com/journal/buildings/special_issues/greenbuildings.

For citation purposes, cite each article independently as indicated on the article page online and as indicated below:

| Lastname, A.A.; Lastname, B.B. Article Title. *Journal Name* **Year**, *Volume Number*, Page Range. |
|---|

**ISBN 978-3-7258-6958-9 (Hbk)**
**ISBN 978-3-7258-6959-6 (PDF)**
**https://doi.org/10.3390/books978-3-7258-6959-6**

# Contents

# About the Editor

**Xingwei Li**

Xingwei Li is a Professor at Sichuan Agricultural University and a master's supervisor, specializing in construction management and environmental sustainability. His research focuses on green building materials, construction and demolition waste management, sustainable urban development, and the environmental management of construction enterprises. Recognized as one of the World's Top 2% Scientists by Stanford University and Elsevier in both 2023 and 2024, Prof. Li has an h-index exceeding 20, an i10-index surpassing 30, and a Field-Weighted Citation Impact (FWCI) over 3. He has published more than 60 peer-reviewed research papers, all indexed in SSCI, SCI, A&HCI, or EI. Prof. Li has received competitive research funding from the National Natural Science Foundation of China (NSFC) and the Natural Science Foundation of Sichuan Province. He has extensive editorial experience in leading international journals. He served as Associate Editor of Management Decision from 2023 to 2024, and as a member of the Editorial Board of Humanities & Social Sciences Communications from 2021 to 2025. He is currently serving on the Editorial Boards of several reputable journals, including *HighTech and Innovation Journal* (EI, since 2020), *Buildings* (SCI/EI, since 2023), and *Scientific Reports* (SCI, since 2024).

# Preface

This Reprint assembles twelve peer-reviewed articles that examine the transition toward sustainability in the building sector through complementary analytical lenses, including evolutionary game theory, behavioral modeling, supply chain analysis, and digital life cycle management. Collectively, the contributions address how policy instruments, stakeholder strategies, market preferences, and technological tools interact to shape green building development and construction and demolition waste resource utilization.

A central theme concerns government governance and strategic interaction in green building and recycling projects. Several studies employ evolutionary game models to analyze how policy design influences the behavior of governments, construction enterprises, suppliers, remanufacturers, financial institutions, and third-party intermediaries. The findings clarify the differential effects of static and dynamic rewards and penalties in preventing greenwashing, the role of tax incentives and infrastructure support in promoting collaborative innovation, and the influence of ecological compensation mechanisms on government green development behavior in cross-regional waste disposal. Additional research in this Reprint also examines cooperation and financing arrangements within green building practice. Studies on design and construction consortia and green patent pledge financing highlight how contractual design, risk sharing, and financial mechanisms influence the stability of collaboration and the capacity of small and medium-sized enterprises to pursue green transformation.

Market demand constitutes another important dimension. Drawing on established behavioral theories, several articles analyze green home buying intentions and willingness to pay for green real estate, emphasizing the roles of perceived value, behavioral control, pricing considerations, and social norms. Related research extends this perspective to the adoption of innovative energy facilities in tertiary sector buildings, connecting consumer preferences and policy incentives with technology uptake decisions.

Circular economy challenges are addressed primarily through construction and demolition waste management. Contributions in this area explore supply chain coordination and stakeholder interaction under government intervention, as well as value cocreation among firms and the public. Together, they deepen understanding of how market mechanisms and policy instruments affect recycling performance and industry development.

Technological approaches are represented by work on data-driven evaluation methods for green interior decoration and integrated digital platforms for life cycle management of building envelope components, both aimed at strengthening circularity and improving decision support. This Reprint brings together governance analysis, behavioral research, circular economy modeling, and digital innovation within a shared sustainability framework for green building project management.

**Xingwei Li**
*Guest Editor*

buildings

MDPI

*Article*

# Considering Consumers' Green Preferences and Government Subsidies in the Decision Making of the Construction and Demolition Waste Recycling Supply Chain: A Stackelberg Game Approach

**Yating Han †, Haoxuan Zheng †, Yicheng Huang and Xingwei Li ***

College of Architecture and Urban-Rural Planning, Sichuan Agricultural University, Chengdu 611830, China; 201908440@stu.sicau.edu.cn (Y.H.); 201908323@stu.sicau.edu.cn (H.Z.); huangyicheng@stu.sicau.edu.cn (Y.H.)

* Correspondence: xwl@sicau.edu.cn

† These authors contributed equally to this work.

**Abstract:** Resource utilization of construction and demolition waste (CDW) is regarded to be an important means of achieving the sustainable development of the economy and the environment. However, previous research has not fully considered the green degree of products in the demand function of CDW remanufactured products. This study aimed to clarify how consumers' green preferences and government subsidies affect decision making in the supply chain. First, a CDW resource utilization supply chain model composed of building materials manufacturers and retailers was constructed using consumer behavior theory. Second, the optimal decision making of members under conditions of decentralized and centralized decision making was analyzed using the Stackelberg game solution. Finally, the validity of the model and conclusions were verified by numerical simulation. The main conclusions are as follows. Government subsidies have a different impact on the pricing of new building materials products and CDW remanufactured products. Under decentralized decision making, the optimal profit of the CDW resource utilization supply chain with government subsidies is higher. However, under centralized decision making, the optimal profit is also related to consumers' green preferences. According to consumers' green preferences, choosing different decision-making models can not only improve the total profit of the CDW resource utilization supply chain, but also improve the reuse rate of CDW.

**Keywords:** consumer green preference; government subsidies; closed-loop supply chain for construction and demolition waste (CDW); Stackelberg game; product green degree

## 1. Introduction

As a result of the rapid development of the world economy and the rising living standards of the global population, urbanization and industrialization are accelerating. An increasing amount of construction and demolition waste (CDW) is generated in the demolition, retrofitting, and new construction of buildings and municipal infrastructure projects [1–4]. Statistics show that more than 10 billion tons of CDW are generated every year, of which about 700 million tons are generated in the United States [5], more than 800 million tons are generated in the European Union [6], and about 230 million tons are generated in China [7].

Engineering construction units, as CDW producers, often follow the principle of prioritizing interests, and will accumulate garbage and send it to landfills. If these wastes are not properly disposed of, they will wreak havoc on the environment, adversely affect arable land, and threaten people's physical and mental health. Therefore, determining how to balance economic development and environmental protection is particularly important, and the effective implementation of CDW management has become a priority action item for the global sustainable development goals [8].

In terms of technology, many materials in CDW can be reused as renewable resources after being sorted, shredded, or removed. For example, CDW can be used as an eco-friendly alternative in the production of eco-efficient cement [9]. In terms of management, construction materials are wasted due to a lack of basic technical and labor knowledge, poor material handling, poor material quality, and other reasons. This wastage can be prevented by strict and regular monitoring of workers, accurate estimation of material quantities, and proper storage [10]. In addition, the European Union, Japan, the United States, etc. have promulgated relevant laws and regulations, ordering manufacturers to be responsible for the entire life cycle of products, and requiring them to recycle and dispose of their products. In addition, an increasing number of enterprises have implemented closed-loop supply chain management [11].

Various issues related to CDW management have also attracted extensive attention from the academic community. The "3Rs" principle has long been proposed in Western countries, and studies have shown that "reducing, reusing, and recycling" of CDW is a hierarchical and well-planned planning strategy [12–14]. Several scholars have also reviewed methods that can quantify CDW generation [15,16]. These studies have promoted the development of CDW management and proposed new directions for subsequent topics. At present, in the research on the supply chain of CDW resource utilization, scholars often believe that consumer demand is only related to the product price, and the green degree of the product has not yet been considered. Based on this, the pricing of green new products and green remanufactured products has become a key decision for supply chain enterprises.

The continuous deterioration in the global ecological environment has awakened people's awareness of environmental protection. Due to the enhancement of consumers' awareness of environmental protection, environmentally friendly products are increasingly favored by consumers [17]. In order to better adapt to the needs of consumers, enterprises in the closed-loop supply chain have gradually shifted from the traditional extensive management model to the consumer-centered intensive management model. This intensive management model can meet the needs of green consumers and maximize corporate profits. However, most enterprises believe that, the greener the product, the better the product meets the needs of consumers, and the higher the profit they will receive. However, this will eventually result in the price of green products exceeding consumer expectations, causing manufacturers to use green materials to produce products that will not be able to be sold. This hurts profits of both manufacturers and retailers. In addition, because their product demand is affected by consumers' green preferences, manufacturers have to invest in the research and development of green products to improve their competitiveness [18,19]. However, the implementation of green technology innovation inevitably increases the production cost of enterprises; as a result, some enterprises may be reluctant to develop green technology [20]. Based on this, many scholars have conducted research and found that the government can adjust the subsidy rate and supervision probability, which can control the distribution mode of corporate profits, and thereby affect the decision-making behavior of stakeholders in the supply chain [21–23]. Therefore, major global economies have successively issued a series of laws, regulations, and subsidy policies in order to promote the green production of enterprises and achieve sustainable development. For example, the United States enacted the Solid Waste Disposal Act as early as 1965 [24], becoming the first country to legally determine waste utilization. In 2021, China proposed that, during the "14th Five-Year Plan" period, it will thoroughly implement relevant laws and regulations, vigorously promote waste reduction, resource utilization, and the harmless disposal of bulk solid waste, and promote the comprehensive utilization of the resources industry to achieve new development [25].

Based on the current problems faced by the construction industry, it is important to focus the impact of consumer green preferences and government subsidies on CDW resource utilization supply chain decisions for the development of recycling and remanufacturing. With the aim of solving problems, and by combining real, down-to-earth, and key breakthroughs [26], this study introduced the product's green degree into the consumer demand function, which complements the decision-making research on the CDW resource utilization supply chain. Moreover, the contributions and implications of this paper are as follows: (1) Considering the influence of consumers' green preferences on the pricing of CDW remanufactured products, this paper enriches the theoretical research on consumer behavior in the remanufacturing closed-loop supply chain. (2) At the same time, this paper provides effective management advice for enterprises and governments. In summary, this study focuses on answering the following three questions: First, how will government subsidies affect the behavioral decisions and profits of building materials manufacturers and retailers in the CDW resource utilization supply chain? Second, how will consumers' green preferences affect the behavioral decisions and profits of building materials manufacturers and retailers in the CDW resource utilization supply chain? Third, in order to make the CDW resource utilization supply chain more profitable, should decentralized or centralized decisions be chosen?

Based on consumer behavior theory, this study established a CDW resource utilization supply chain model that considers consumers' green preferences and government subsidies. This paper compares the decisions of building materials manufacturers and retailers under decentralized and centralized decision-making. At the same time, it provides effective management suggestions for enterprises and governments regarding the CDW resource utilization supply chain.

The remainder of this paper is structured as follows. Section 2 provides a relevant literature review; Section 3 presents the problem description and model assumptions; Section 4 considers consumers' green preferences, and presents the construction of four game models of decentralized and centralized decision making between building materials manufacturers and retailers, with or without government subsidies; Section 5 presents model analysis and discusses the impact of relevant parameters on the decision making of building materials manufacturers and retailers; Section 6 presents the numerical simulation, and intuitively obtains the change in the profit of the CDW resource utilization supply chain under different models; finally, Section 7 summarizes the conclusions of this paper.

## 2. Literature Review

This study aimed to solve the CDW management problem from the perspective of closed-loop supply chain operation, based on consumers' green preferences and government subsidies, to study the relevant decisions of the CDW resource utilization supply chain. Therefore, this paper reviews the relevant literature in terms of three aspects: the CDW closed-loop supply chain, consumer behavior theory, and the closed-loop supply chain under government subsidies, as shown in Table 1.

**Table 1.** Research related to construction and demolition waste, consumer behavior theory, and the supply chain under government subsidies.

| Research Topics | Dimensions | Source Papers |
|---|---|---|
| CDW management | The vast majority of CDW can technically be converted into new building materials. | [27] |
| | The impact of CDW management stakeholders' behavior and economic incentives on CDW management | [28–30] |
| | Research on CDW management from the perspective of supply chain operation | [31] |
| | Decision-making behavior of recycling units and CDW remanufacturers under the evolutionary game | [32] |
| | The decision and profit of the CDW resource utilization supply chain under the Stackelberg model | [33–35] |
| Consumer behavior theory | On the new theory of consumer behavior | [36] |
| | Consumer green preferences improve green innovation | [37,38] |
| | Consumer environmental awareness and coordination in the closed-loop supply chain | [39] |
| | Dual-channel green supply chain pricing policy considering consumers' green preferences | [40] |

**Table 1.** *Cont.*

| Research Topics | Dimensions | Source Papers |
|---|---|---|
| Supply chain under government subsidies | The dominance of the government in recycling of CDW | [41] |
| | Government subsidies stimulate recycling of CDW | [42,43] |
| | Research on the supply chain of government subsidized green products | [44,45] |
| | Research on the supply chain of government subsidized remanufactured products | [46–48] |
| | Research on consumer preferences and government subsidies on decisions and profits of supply chain members | [49,50] |

### 2.1. CDW Management

In order to alleviate the negative impact of CDW on the environment and promote the stable development of the CDW recycling industry, the scientific management and effective utilization of CDW has become the focus of extensive attention of experts and scholars. Regarding CDW recycling research, some scholars noted, from a technical point of view, that the vast majority of CDW can be converted into new building materials through proper recycling treatment, thereby contributing to the sustainability of the construction industry [27]. Using the technical feasibility of CDW recycling as a guarantee, some scholars have found from a management perspective that the attitudes and behaviors of CDW management stakeholders, and economic incentives, have a significant impact on CDW management [28–30]. To promote the better formation and operation of the CDW resource utilization supply chain, there is an urgent need to study CDW management from the perspective of supply chain operation, to clarify the decision-making process of stakeholders in the supply chain [31]. Therefore, some scholars have constructed evolutionary game models and found that business leaders, the government supervision rate, the government cost subsidy rate, and the recycling unit's effort profit coefficient will affect the decision-making behavior of recycling units and CDW remanufacturers [32]. Other scholars have found, by solving the Stackelberg model, that factors such as the learning effect, reference effect, information sharing, and fairness concerns, can affect the decision making and profit of the CDW resource utilization supply chain [33–35].

The above review proves that it is important to closely study CDW management by constructing a game model. The existing research mainly involves the different decision making of supply chain members, such as the government, construction waste recyclers, recyclers, and retailers, but has not fully considered the behavior of consumers in the supply chain. Based on this, from the perspective of consumer behavior, this study examined how consumers' green preferences and government subsidies affect the decisions of manufacturers and retailers in the CDW resource utilization supply chain. In addition, this study examined the decision making of the CDW resource utilization supply chain when both new products and remanufactured products have a green degree.

### 2.2. Consumer Behavior Theory

Consumer behavior theory shows that consumers' purchasing decisions are not only closely related to the product itself, but are also influenced by consumer preferences [36]. In reality, consumers' purchasing decision behavior is often affected by a variety of factors, and different consumers have different consumption preferences for various factors. Studies have found that, as a result of the increasingly serious problem of environmental pollution, consumers are more willing to buy low-carbon, energy-saving, and environmentally friendly products. This behavior will change the product attributes of producers, prompt manufacturers to enhance their green innovation capabilities to produce green products, and promote the development of green industries [37,38]. Therefore, the green preferences of consumers have attracted great attention from the government and enterprises. Some scholars believe that the profits of retailers and manufacturers in the supply chain are related to consumers' green preferences. For example, Xu et al. [39] found that the product environmental protection level, recycling rate, and corporate profits at each node were positively correlated with consumers' green preferences, but negatively correlated with the green investment coefficient. Abbey et al. [40] empirically found that green consumers, and consumers who perceived remanufactured products as green, generally found remanufactured products to be more attractive.

The above research results suggest that consumers' green preferences have an impact on the pricing decisions of stakeholders in the supply chain and the green degree of products. However, most studies on consumers' green preferences focus on green manufacturers and do not consider the impact on green remanufacturers' decision making. Moreover, consumer green preferences have not been considered in the CDW resource utilization supply chain. Therefore, further research on consumers' preferences for green products based on consumer behavior theory will play an important role in the development of CDW resource utilization supply chain.

### 2.3. Closed-Loop Supply Chain under Government Subsidies

The government, as a supervisor and regulator, plays a leading role in the process of recycling CDW [41]. Questionnaires have found that government subsidies can effectively stimulate the recycling of CDW [42,43]. In recent years, a large amount of research has examined closed-loop supply chain management under government subsidy policies.

First, research on subsidizing green products has shown that, when the government subsidizes the manufacturer, the increase in the government subsidy will improve the green degree of the product, but the government subsidy is not always beneficial to the green supply chain and the manufacturer [44]. In addition, scholars have studied how governments use consumer subsidies to promote the development of green technologies, and how policy adjustments interact with industrial production decisions over time [45].

Second, the research on subsidized remanufactured products generally shows that government subsidies effectively stimulate the demand for remanufactured products and promote the development of the remanufacturing industry [46,47]. For example, Huang [48] et al. studied the impact of government subsidies on channel members' pricing decisions and recycling mode choices, and found that government subsidies have a positive effect on remanufacturers' and collectors' willingness to remanufacture.

Aware of the importance of consumers' green preferences and government subsidies, some scholars have considered the impact of supply chain members' related decisions and profits under the combined effect of the two. For example, Barman et al. [49] explored pricing strategies, green strategies, and compared optimal decision making for both centralized and decentralized models. Yu et al. [50] found that the improvement in consumers' environmental awareness and good government subsidy policies will motivate manufacturers to produce more green products and generate profits for manufacturers. However, whether these conclusions can be obtained in the construction waste resource utilization supply chain remains to be confirmed.

To summarize, there are still some deficiencies in the research on the supply chain of CDW resource utilization. First, few scholars have integrated the product's green degree into the closed-loop supply chain of CDW. Second, no scholars have considered both government subsidies and consumer green preferences in the construction waste resource utilization supply chain. In view of this, in this study, Stackelberg game theory was used to solve for and compare the optimal pricing and revenue of building materials manufacturers and retailers under decentralized and centralized decision making, taking the product green degree, consumers' green preferences, government subsidies, and other factors into consideration.

## 3. Problem Description and Associated Assumptions

Consider a closed-loop supply chain model consisting of building materials manufacturers, retailers, and consumers. Among these, building material manufacturers are responsible for the production of new products and recycling construction waste from consumers, and then remanufacturing the recovered construction waste as raw materials.

The retailer is responsible for selling new and remanufactured products to consumers at different prices, and does not take part in the recycling of the products during the process. The government can choose whether to subsidize CDW remanufactured products. The manufacturer's decision problem is to determine the product green degree and wholesale price of the product to maximize its own profit, and the retailer's decision problem is to determine the retail price of green products to maximize its own benefit. The formed game model is shown in Figure 1.

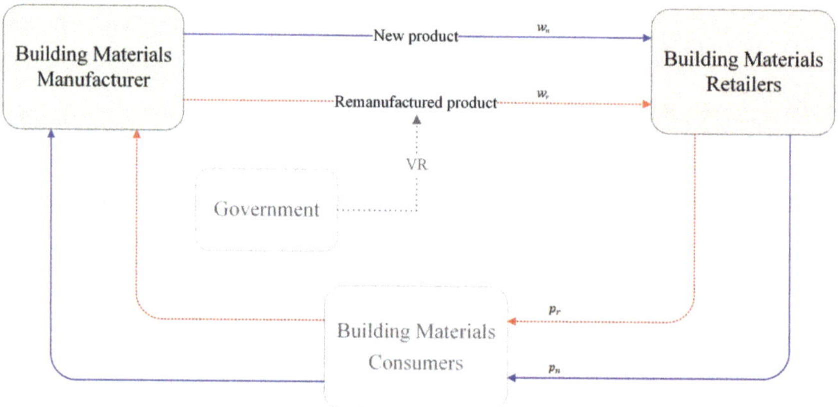


**Figure 1.** Game model of building materials manufacturer and building materials retailers. Note: VR is the government subsidy for remanufacturing; $w_n$ is the wholesale prices for new products of building materials; $w_r$ is the wholesale price for CDW remanufactured products; $p_n$ is the retail price of new products of building materials; $p_r$ is the retail price of CDW remanufactured products.

The basic definitions of the parameters used in the study are shown in Table 2.

**Table 2.** Model parameters.

| Parameter | Definition |
|---|---|
| $\gamma$ | Consumer green preference coefficient. $0 \leq \gamma \leq 1$ |
| $v$ | Unit price subsidy granted by the government for CDW remanufactured products. $v \geq 0$ |
| G | Recycling of CDW. |
| $h$ | Number of consumers with active recycling awareness. $h > 0$ |
| $k$ | Sensitivity of consumer recycling attitudes to b changes. $k > 0$ |
| $\eta$ | Research and development effort coefficient of building materials manufacturers. $\eta > 0$ |
| $b$ | Unit CDW recycling costs. |
| $c_n$ | Manufacturing costs of new products of building materials. |
| $c_r$ | Manufacturing costs of CDW remanufactured products. |
| $\delta$ | Discount rate of value of remanufactured products relative to new products. $(0 < \delta < 1)$ |
| $q_n$ | Demand for new products of building materials. $q_n \geq 0$ |
| $q_r$ | Demand for CDW remanufactured products. $q_r \geq 0$ |
| $\Pi_X^Y$ | The profit of $X$ under the $Y$ model. $X = m, r$ represent building materials manufacturers and retailers, respectively; $Y$ = N, VR represent CDW remanufacturing, without and with government subsidies, respectively. |
| $p_n$ | Retail price of new products of building materials (decision variables). |
| $p_r$ | Retail price of CDW remanufactured products (decision variables). |

**Table 2.** *Cont.*

| Parameter | Definition |
|---|---|
| $\mathcal{W}_n$ | Wholesale price of new products of building materials (decision variables). |
| $\mathcal{W}_r$ | Wholesale price of CDW remanufactured products (decision variables). |
| $g$ | Product green degree (decision variables). |

### 3.1. Model Assumptions

(1) Manufacturers of building materials produce two products: new products of building materials and remanufactured products of CDW. The unit production cost of new products is $c_n$, the unit production cost of CDW remanufactured products is $c_r$, the unit cost of recycling CDW is $b$, and $c_n > c_r + b > 0$.

(2) Due to the different raw materials used for production, the two products are substitutes for each other and there are differences in quality, which is in line with reality. A retailer of building materials wholesales new products from manufacturers at a unit price $\mathcal{W}_n$ and sells them to consumers at a unit price $p_n$, and wholesales remanufactured products at a unit price $\mathcal{W}_r$, and sells the remanufactured products at a unit price $p_r$ to consumers. It follows that $\mathcal{W}_n > \mathcal{W}_r > 0$, $p_n > p_r > 0$, and $p_n > \mathcal{W}_n > 0$, $P_r > \mathcal{W}_r > 0$.

(3) In order to stimulate consumer purchases, building materials manufacturers need to invest in green product research and development by purchasing new equipment or technological innovation. It is assumed that there is a quadratic function between the manufacturer's green product development cost and product green degree ($g$) [51,52]; that is, $\frac{1}{2}\eta g^2$ [53], where $\eta > 0$ is the green product development cost coefficient.

(4) The recycling quantity of CDW is a linearly increasing function of recycling price, namely, $G = h + kb$, where h represents the number of consumers with active recycling awareness in the supply chain, k represents the sensitivity of consumers' recycling attitude to changes in b, and $k > 0, h > 0$ [54,55].

(5) There is a Stackelberg game relationship between building materials manufacturers and retailers, in which the manufacturer is the game leader and the retailer is the game follower. Both the manufacturer and the retailer are risk neutral and information symmetric.

### 3.2. Demand Function

Referring to the research in [56], it is assumed that market demand is only affected by the product value and green degree. Consumers compare and analyze the utility of the two products to decide which product to buy. $a$ is the value of the new product of building materials considered by consumers, and $a$ obeys a uniform distribution in the interval [0,1]. $\delta a$ is the value of products remanufactured from construction waste that consumers consider, and $\delta$ is the discount rate of remanufactured products relative to new products $(0 < \delta < 1)$. The market demand for green products is negatively correlated with the price, and positively correlated with the green degree of the product. It can be found that the effect functions of consumers on green new products and green remanufactured products are $U_n = a - p_n + \gamma g$ and $U_r = \delta a - p_r + \gamma g$, respectively, where $\gamma$ represents the green preference coefficient of consumers, and $g$ represents the product green degree.

When $U_n > max\{U_r, 0\}$, that is $\frac{p_n - p_r}{1-\delta} < a < 1$, the demand function of new products of building materials is shown in Equation (1):

$$q_n = \int_{\frac{p_n - p_r}{1-\delta}}^{1} da = 1 - \frac{p_n - p_r}{1 - \delta} \tag{1}$$

Similarly, when $U_r > max\{U_n, 0\}$, that is $\frac{p_r - \gamma g}{\delta} < a < \frac{p_n - p_r}{1-\delta}$, the demand function for CDW remanufactured products is shown in Equation (2):

$$q_r = \int_{\frac{p_r - \gamma g}{\delta}}^{\frac{p_n - p_r}{1-\delta}} da = \frac{p_n - p_r}{1-\delta} - \frac{p_r - \gamma g}{\delta} = \frac{\delta p_n - p_r + \gamma g(1-\delta)}{(1-\delta)\delta} \quad (2)$$

## 4. Model Building and Solving

### 4.1. Without Government Subsidies

#### 4.1.1. Decentralized Decision Model (N Model)

Under the decentralized decision-making model (N model), the closed-loop supply chain Stackelberg model composed of building materials manufacturers and retailers is a master–slave game model, in which the manufacturer is the leader and the retailer is the follower. Manufacturers and retailers are independent of each other, and their decision-making goal is to maximize their own profits. The manufacturer first decides the wholesale price of new and remanufactured products ($W_n$, $W_r$) and the green degree of the product ($g$), and then the retailer decides the retail price of new and remanufactured products ($p_n, p_r$) based on the manufacturer's decision.

Profit function for building materials manufacturers:

$$\max_{(W_n,\ W_r)} \Pi_m^N = (W_n - c_n)q_n + (W_r - c_r)q_r + (\mu - b)G - \frac{1}{2}\eta g^2 \quad (3)$$

Profit function for building materials retailer:

$$\max_{(p_n, p_r)} \Pi_r^N = (p_n - W_n)q_n + (p_r - W_r)q_r \quad (4)$$

Since $\Pi_r^N$ is a concave function about $p_n$ and $p_r$, we can obtain:

$$p_n^{N*}(W_n, W_r, g) = \frac{1}{2}(1 + g\gamma + W_n) \quad (5)$$

$$p_r^{N*}(W_n, W_r, g) = \frac{1}{2}(g\gamma + \delta + W_r) \quad (6)$$

We can substitute $p_n^{N*}(W_n, W_r, g)$, $p_r^{N*}(W_n, W_r, g)$ into the manufacturer's profit function $\Pi_m^N$, and find the first-order and second-order derivatives of $W_n, W_r$, and $g$ respectively, so as to obtain the Hessian matrix of the manufacturer's profit $\Pi_m^N$ with respect to $W_n$, $W_r$, and $g$.

$$H = \begin{bmatrix} -\frac{1}{1-\delta} & \frac{1}{1-\delta} & 0 \\ \frac{1}{1-\delta} & -\frac{1}{(1-\delta)\delta} & \frac{\gamma}{2\delta} \\ 0 & \frac{\gamma}{2\delta} & -\eta \end{bmatrix} = \frac{\gamma^2 - 4\delta\eta}{4(1-\delta)\delta^2}$$

**Lemma 1.** *When $\gamma^2 - 4\delta\eta < 0$, there is $(W_n, W_r, g)$ such that $\Pi_m^N$ achieves the maximum value.*

Prove: The first-order sequential principal subform of the Hessian matrix $H$ is $-\frac{1}{1-\delta} < 0$, the second-order sequence main subform is $\begin{bmatrix} -\frac{1}{1-\delta} & \frac{1}{1-\delta} \\ \frac{1}{1-\delta} & -\frac{1}{(1-\delta)\delta} \end{bmatrix} = \frac{1}{(1-\delta)\delta} > 0$, and the main subform of the third-order sequence is $H = \frac{\gamma^2 - 4\delta\eta}{4(1-\delta)\delta^2}$; hence, when $\gamma^2 - 4\delta\eta < 0$, the third-order sequential main subform $H < 0$. Therefore, the Hessian matrix $H$ is a negative definite matrix, that is, there is $(W_n, W_r, g)$ so that $\Pi_m^N$ takes the maximum value.

When the condition $\gamma^2 - 4\delta\eta < 0$ is satisfied, $\Pi_m^N$ is a concave function about $\mathcal{W}_n$, $\mathcal{W}_r$, and $g$, so the optimal decision of the manufacturer can be obtained:

$$\mathcal{W}_n^{N*} = \frac{\gamma^2 - \gamma^2\delta - 4\delta\eta + (\gamma^2 - 4\delta\eta)c_n + \gamma^2 c_r}{2(\gamma^2 - 4\delta\eta)} \tag{7}$$

$$\mathcal{W}_r^{N*} = \frac{-2\delta^2\eta + (\gamma^2 - 2\delta\eta)c_r}{\gamma^2 - 4\delta\eta} \tag{8}$$

$$g^{N*} = \frac{\gamma(-\delta + c_r)}{\gamma^2 - 4\delta\eta} \tag{9}$$

By substituting $\mathcal{W}_n^{N*}$, $\mathcal{W}_r^{N*}$, $g^{N*}$ into $p_n^{N*}(\mathcal{W}_n, \mathcal{W}_r, g)$, $p_r^{N*}(\mathcal{W}_n, \mathcal{W}_r, g)$, the optimal decision of the retailer can be obtained as:

$$p_n^{N*} = \frac{1}{4}(c_n + \frac{3(\gamma^2 - \gamma^2\delta - 4\delta\eta + \gamma^2 c_r)}{\gamma^2 - 4\delta\eta}) \tag{10}$$

$$p_r^{N*} = \frac{-3\delta^2\eta + (\gamma^2 - \delta\eta)c_r}{\gamma^2 - 4\delta\eta} \tag{11}$$

By substituting the optimal pricing into Equations (3) and (4), $\Pi_m^{N*}$ and $\Pi_r^{N*}$ can be obtained, respectively, as shown in Equations (12) and (13):

$$\Pi_m^{N*} = \frac{\left\{ \begin{array}{l} (-1+\delta)\left(4\delta\eta(1 - 8bh - 8b^2k + 8h\mu + 8bk\mu) + \gamma^2\left(-1+8b^2k + \delta - 8h\mu + 8b(h - k\mu)\right)\right) \\ +(\gamma^2 - 4\delta\eta)c_n^2 + 2(\gamma^2 - 4\delta\eta)c_n(-1+\delta - c_r) - 2\gamma^2(-1+\delta)c_r + (\gamma^2 - 4\eta)c_r^2 \end{array} \right\}}{8(-1+\delta)(-\gamma^2 + 4\delta\eta)} \tag{12}$$

$$\Pi_r^{N*} = -\frac{\left\{ \begin{array}{l} -((-1+\delta)(-\gamma^4(-1+\delta) + 8\gamma^2(-1+\delta)\delta\eta + 16\delta^2\eta^2)) + (\gamma^2 - 4\delta\eta)^2 c_n^2 + 2(\gamma^2 - 4\delta\eta)^2 \\ c_n(-1+\delta - c_r) - 2\gamma^2(-1+\delta)(\gamma^2 - 8\delta\eta)c_r + (\gamma^4 - 8\gamma^2\delta\eta + 16\delta\eta^2)c_r^2 \end{array} \right\}}{16(-1+\delta)(\gamma^2 - 4\delta\eta)^2} \tag{13}$$

At this time, the optimal total profit function of the entire CDW resource utilization supply chain is shown in Equation (14):

$$\Pi^{DN*} = -\frac{\left\{\left(\begin{array}{l} -\left((-1+\delta)\left(\begin{array}{c} -\gamma^4(-1+\delta) + 8\gamma^2 \\ (-1+\delta)\delta\eta + 16\delta^2\eta^2 \end{array}\right)\right) + (\gamma^2 - 4\delta\eta)^2 c_n^2 + 2(\gamma^2 - 4\delta\eta)^2 c_n(-1+\delta - c_r) - 2\gamma^2(-1+\delta)(\gamma^2 - 8\delta\eta)c_r + (\gamma^4 - 8\gamma^2\delta\eta + 16\delta\eta^2)c_r^2 + \\ 2(\gamma^2 - 4\delta\eta)\left((-1+\delta)\left(\begin{array}{c} 4\delta\eta(1 - 8bh - 8b^2k + 8h\mu + 8bk\mu) + \\ \gamma^2(-1 + 8b^2k + \delta - 8h\mu + 8b(h - k\mu)) \end{array}\right) + (\gamma^2 - 4\delta\eta)c_n^2 + 2(\gamma^2 - 4\delta\eta)c_n(-1+\delta - c_r) - 2\gamma^2(-1+\delta)c_r + (\gamma^2 - 4\eta)c_r^2\right) \end{array}\right)\right\}}{16(-1+\delta)(\gamma^2 - 4\delta\eta)^2} \tag{14}$$

4.1.2. Centralized Decision Model

In the centralized decision-making model (N model), building material manufacturers and retailers make joint decisions as a whole, and the goal of decision making is to maximize profits throughout the supply chain. At this point, the total profit function of the supply chain is shown in Equation (15):

$$\max_{(\mathcal{W}_n,\ \mathcal{W}_r)} \Pi^N = (\mathcal{W}_n - c_n)q_n + (\mathcal{W}_r - c_r)q_r + (\mu - b)G - \frac{1}{2}\eta g^2 \tag{15}$$

Among these variables, $p_n$, $p_r$, and $g$ are decision variables. Let $\frac{\partial \Pi^N}{\partial p_n} = 0$, $\frac{\partial \Pi^N}{\partial p_r} = 0$, $\frac{\partial \Pi^N}{\partial g} = 0$ to find the maximum values of $p_n$, $p_r$, and $g$, respectively.

$$p_n^* = \frac{\gamma^2 - \gamma^2\delta - 2\delta\eta + (\gamma^2 - 2\delta\eta)c_n + \gamma^2 c_r}{2(\gamma^2 - 2\delta\eta)} \tag{16}$$

$$p_r^* = \frac{-\delta^2\eta + (\gamma^2 - \delta\eta)c_r}{\gamma^2 - 2\delta\eta} \tag{17}$$

$$g^* = \frac{\gamma(-\delta + c_r)}{\gamma^2 - 2\delta\eta} \tag{18}$$

Bringing $p_n^*$, $p_r^*$, and $g^*$ into Equation (15), the optimal profit function of the entire supply chain can be obtained as shown in Equation (19):

$$\prod^{N*} = \frac{\left\{ \begin{array}{l} (-1+\delta)\left(2\delta\eta(1 - 4bh - 4b^2k + 4h\mu + 4bk\mu) + \gamma^2\left(-1+4b^2k + \delta - 4h\mu + 4b(h - k\mu)\right)\right) \\ +(\gamma^2 - 2\delta\eta)c_n^2 + 2(\gamma^2 - 2\delta\eta)c_n(-1 + \delta - c_r) - 2\gamma^2(-1+\delta)c_r + (\gamma^2 - 2\eta)c_r^2 \end{array} \right\}}{4(-1+\delta)(-\gamma^2 + 2\delta\eta)} \tag{19}$$

### 4.2. *With Government Subsidies*

#### 4.2.1. Decentralized Decision Model (VR Model)

Under the decentralized decision-making model (VR model), the closed-loop supply chain Stackelberg model composed of building material manufacturers and retailers is a master–slave game model, in which the manufacturer is the leader and the retailer is the follower. Manufacturers and retailers are independent of each other, and their decision-making goal is to maximize their own profits. In order to increase the enthusiasm of manufacturers to produce CDW remanufactured products, the government provides remanufacturing subsidies to manufacturers.

Profit function for building materials manufacturers:

$$\max_{(W_n,\ W_r)} \Pi_m^{VR} = (W_n - c_n)q_n + (W_r - c_r + v)q_r + (\mu - b)G - \frac{1}{2}\eta g^2 \tag{20}$$

Profit function for building materials retailer:

$$\max_{(p_n, p_r)} \Pi_r^{VR} = (p_n - W_n)q_n + (p_r - W_r)q_r \tag{21}$$

Since $\Pi_r^{VR}$ is a concave function about $p_n$ and $p_r$, we can obtain:

$$p_n^{VR*}(W_n, W_r, g) = \frac{1}{2}(1 + g\gamma + W_n) \tag{22}$$

$$p_r^{VR*}(W_n, W_r, g) = \frac{1}{2}(g\gamma + \delta + W_r) \tag{23}$$

We can substitute $p_n^{VR*}(W_n, W_r, g)$, $p_r^{VR*}(W_n, W_r, g)$ into the manufacturer's profit function $\Pi_m^{VR}$, and find the first-order and second-order derivatives of $W_n$, $W_r$, and $g$, respectively, so as to obtain the Hessian matrix of the manufacturer's profit $\Pi_m^{VR}$ with respect to $W_n$, $W_r$, and $g$.

$$H = \begin{bmatrix} -\frac{1}{1-\delta} & \frac{1}{1-\delta} & 0 \\ \frac{1}{1-\delta} & -\frac{1}{(1-\delta)\delta} & \frac{\gamma}{2\delta} \\ 0 & \frac{\gamma}{2\delta} & -\eta \end{bmatrix} = \frac{\gamma^2 - 4\delta\eta}{4(1-\delta)\delta^2}$$

As in Lemma 1, when the condition $\gamma^2 - 4\delta\eta < 0$ is satisfied, $\Pi_m^{VR}$ is a concave function about $\mathcal{W}_n$, $\mathcal{W}_r$, and $g$, so the optimal decision of the manufacturer can be obtained:

$$\mathcal{W}_n^{VR*} = \frac{\gamma^2 - v\gamma^2 - \gamma^2\delta - 4\delta\eta + (\gamma^2 - 4\delta\eta)c_n + \gamma^2 c_r}{2(\gamma^2 - 4\delta\eta)} \tag{24}$$

$$\mathcal{W}_r^{VR*} = \frac{-v\gamma^2 + 2v\delta\eta - 2\delta^2\eta + (\gamma^2 - 2\delta\eta)c_r}{\gamma^2 - 4\delta\eta} \tag{25}$$

$$g^{VR*} = -\frac{\gamma(v + \delta - c_r)}{\gamma^2 - 4\delta\eta} \tag{26}$$

Substituting $\mathcal{W}_n^{VR*}$, $\mathcal{W}_r^{VR*}$, and $g^{VR*}$ into $p_n^{VR*}(\mathcal{W}_n, \mathcal{W}_r, g)$, $p_r^{VR*}(\mathcal{W}_n, \mathcal{W}_r, g)$, the optimal decision of the retailer can be obtained as:

$$p_n^{VR*} = \frac{-3\gamma^2(-1 + v + \delta) - 12\delta\eta + (\gamma^2 - 4\delta\eta)c_n + 3\gamma^2 c_r}{4(\gamma^2 - 4\delta\eta)} \tag{27}$$

$$p_r^{VR*} = \frac{-v\gamma^2 + v\delta\eta - 3\delta^2\eta + (\gamma^2 - \delta\eta)c_r}{\gamma^2 - 4\delta\eta} \tag{28}$$

By substituting the optimal pricing into Equations (20) and (21), $\Pi_m^{VR*}$ and $\Pi_r^{VR*}$ can be obtained, respectively, as shown in Equations (29) and (30):

$$\Pi_m^{VR*} = \frac{\left\{\begin{array}{l} \gamma^2 - 8bh\gamma^2 - 8b^2k\gamma^2 - 2v\gamma^2 + v^2\gamma^2 - 2\gamma^2\delta + 8bh\gamma^2\delta + 8b^2k\gamma^2\delta + 2v\gamma^2\delta + \gamma^2\delta^2 - 4v^2\eta - 4\delta\eta + 32bh\delta\eta + 32b^2k\delta\eta \\ +4\delta^2\eta - 32bh\delta^2\eta - 32b^2k\delta^2\eta + 8h\gamma^2\mu + 8bk\gamma^2\mu - 8h\gamma^2\delta\mu - 8bk\gamma^2\delta\mu - 32h\delta\eta\mu - 32bk\delta\eta\mu + 32h\delta^2\eta\mu + \\ 32bk\delta^2\eta\mu + (\gamma^2 - 4\delta\eta)c_n^2 + 2(\gamma^2 - 4\delta\eta)c_n(-1 + v + \delta - c_r) - 2(\gamma^2(-1 + v + \delta) - 4v\eta)c_r + (\gamma^2 - 4\eta)c_r^2 \end{array}\right\}}{8(-1 + \delta)(-\gamma^2 + 4\delta\eta)} \tag{29}$$

$$\Pi_r^{VR*} = -\frac{\left\{\begin{array}{l} \gamma^4 - 2v\gamma^4 + v^2\gamma^4 - 2\gamma^4\delta + 2v\gamma^4\delta + \gamma^4\delta^2 - 8\gamma^2\delta\eta + 16v\gamma^2\delta\eta - 8v^2\gamma^2\delta\eta + 16\gamma^2\delta^2\eta - 16v\gamma^2\delta^2\eta \\ -8\gamma^2\delta^3\eta + 16v^2\delta\eta^2 + 16\delta^2\eta^2 - 16\delta^3\eta^2 + (\gamma^2 - 4\delta\eta)^2 c_n^2 + 2(\gamma^2 - 4\delta\eta)^2 c_n(-1 + v + \delta - c_r) \\ -2(\gamma^4(-1 + v + \delta) - 8\gamma^2\delta(-1 + v + \delta)\eta + 16v\delta\eta^2)c_r + (\gamma^4 - 8\gamma^2\delta\eta + 16\delta\eta^2)c_r^2 \end{array}\right\}}{16(-1 + \delta)(\gamma^2 - 4\delta\eta)^2} \tag{30}$$

At this time, the optimal total profit function of the entire CDW resource utilization supply chain is shown in Equation (31):

$$\Pi^{DVR*} = -\frac{\left\{\left(\left(\begin{array}{l} \gamma^4 - 2v\gamma^4 + v^2\gamma^4 - 2\gamma^4\delta + 2v\gamma^4\delta + \gamma^4\delta^2 - 8\gamma^2\delta\eta + 16v\gamma^2\delta\eta - 8v^2\gamma^2\delta\eta + 16\gamma^2\delta^2\eta - 16v\gamma^2\delta^2\eta - 8\gamma^2\delta^3\eta + 16v^2\delta\eta^2 + 16\delta^2\eta^2 - \\ 16\delta^3\eta^2 + (\gamma^2 - 4\delta\eta)^2 c_n^2 + 2(\gamma^2 - 4\delta\eta)^2 c_n(-1 + v + \delta - c_r) - 2\left(\begin{array}{c} \gamma^4(-1 + v + \delta) - 8\gamma^2\delta \\ (-1 + v + \delta)\eta + 16v\delta\eta^2 \end{array}\right)c_r + (\gamma^4 - 8\gamma^2\delta\eta + 16\delta\eta^2)c_r^2 + 2(\gamma^2 - 4\delta\eta) \\ \left(\begin{array}{c} \gamma^2 - 8bh\gamma^2 - 8b^2k\gamma^2 - 2v\gamma^2 + v^2\gamma^2 - 2\gamma^2\delta + 8bh\gamma^2\delta + 8b^2k\gamma^2\delta + 2v\gamma^2\delta + \gamma^2\delta^2 - 4v^2\eta - 4\delta\eta + 32bh\delta\eta + 32b^2k\delta\eta + \\ 4\delta^2\eta - 32bh\delta^2\eta - 32b^2k\delta^2\eta + 8h\gamma^2\mu + 8bk\gamma^2\mu - 8h\gamma^2\delta\mu - 8bk\gamma^2\delta\mu - 32h\delta\eta\mu - 32bk\delta\eta\mu + 32h\delta^2\eta\mu + 32bk\delta^2\eta\mu + \\ (\gamma^2 - 4\delta\eta)c_n^2 + 2(\gamma^2 - 4\delta\eta)c_n(-1 + v + \delta - c_r) - 2(\gamma^2(-1 + v + \delta) - 4v\eta)c_r + (\gamma^2 - 4\eta)c_r^2 \end{array}\right) \end{array}\right)\right)\right\}}{16(-1 + \delta)(\gamma^2 - 4\delta\eta)^2} \tag{31}$$

4.2.2. Centralized Decision Model

In the centralized decision-making model (VR model), the government subsidizes remanufacturing. Building materials manufacturers and retailers make joint decisions as a whole, with the goal of maximizing profits throughout the supply chain. At this point, the total profit function of the supply chain is shown in Equation (32):

$$max\Pi^{VR} = (p_n - c_n)q_n + (p_r - c_r + v)q_r + (\mu - b)G - \frac{1}{2}\eta g^2 \tag{32}$$

Among these variables, $p_n$, $p_r$, and $g$ are decision variables. Let $\frac{\partial \Pi^{VR}}{\partial p_n} = 0$, $\frac{\partial \Pi^{VR}}{\partial p_r} = 0$, $\frac{\partial \Pi^{VR}}{\partial g} = 0$ to find the maximum values of $p_n$, $p_r$, and $g$, respectively.

$$p_n^{**} = \frac{\gamma^2 - v\gamma^2 - \gamma^2\delta - 2\delta\eta + (\gamma^2 - 2\delta\eta)c_n + \gamma^2 c_r}{2(\gamma^2 - 2\delta\eta)} \tag{33}$$

$$p_r^{**} = \frac{-v\gamma^2 + v\delta\eta - \delta^2\eta + (\gamma^2 - \delta\eta)c_r}{\gamma^2 - 2\delta\eta} \tag{34}$$

$$g^{**} = -\frac{\gamma(v + \delta - c_r)}{\gamma^2 - 2\delta\eta} \tag{35}$$

Bringing $p_n^{**}$, $p_r^{**}$, and $g^{**}$ into Equation (32), the optimal profit function of the entire supply chain can be obtained as shown in Equation (36):

$$\Pi^{VR*} = \frac{\left\{\begin{array}{l} \gamma^2 - 4bh\gamma^2 - 4b^2k\gamma^2 - 2v\gamma^2 + v^2\gamma^2 - 2\gamma^2\delta + 4bh\gamma^2\delta + 4b^2k\gamma^2\delta + 2v\gamma^2\delta + \gamma^2\delta^2 - 2v^2\eta - 2\delta\eta + 8bh\delta\eta + 8b^2k\delta\eta \\ +2\delta^2\eta - 8bh\delta^2\eta - 8b^2k\delta^2\eta + 4h\gamma^2\mu + 4bk\gamma^2\mu - 4h\gamma^2\delta\mu - 4bk\gamma^2\delta\mu - 8h\delta\eta\mu - 8bk\delta\eta\mu + 8h\delta^2\eta\mu + \\ 8bk\delta^2\eta\mu + (\gamma^2 - 2\delta\eta)c_n^2 + 2(\gamma^2 - 2\delta\eta)c_n(-1 + v + \delta - c_r) - 2(\gamma^2(-1 + v + \delta) - 2v\eta)c_r + (\gamma^2 - 2\eta)c_r^2 \end{array}\right\}}{4(-1 + \delta)(-\gamma^2 + 2\delta\eta)} \tag{36}$$

## 5. Model Analysis

In this section, the following main conclusions were drawn by comparing the optimal solutions of the decentralized decision-making model under two different scenarios with and without government subsidies.

**Proposition 1.** *Wholesale prices for new and remanufactured products are related as follows:*

(1) $\mathcal{W}_n^{VR*} > \mathcal{W}_n^{N*}$.

(2) *When* $1 > \gamma > \sqrt{2\delta\eta}$, *that is,* $\mathcal{W}_r^{VR*} > \mathcal{W}_r^{N*}$;
*When* $0 < \gamma < \sqrt{2\delta\eta}$, *that is,* $\mathcal{W}_r^{VR*} < \mathcal{W}_r^{N*}$.

**Proof.** By comparing the solutions of the two decentralized models, i.e., with and without government subsidies, we obtain:

(1) $\mathcal{W}_n^{VR*} - \mathcal{W}_n^{N*} = \frac{\gamma^2 - v\gamma^2 - \gamma^2\delta - 4\delta\eta + (\gamma^2 - 4\delta\eta)c_n + \gamma^2 c_r}{2(\gamma^2 - 4\delta\eta)} - \frac{\gamma^2 - \gamma^2\delta - 4\delta\eta + (\gamma^2 - 4\delta\eta)c_n + \gamma^2 c_r}{2(\gamma^2 - 4\delta\eta)} = -\frac{v\gamma^2}{2\gamma^2 - 8\delta\eta} > 0$, *that is,* $\mathcal{W}_n^{VR*} > \mathcal{W}_n^{N*}$.

(2) $\mathcal{W}_r^{VR*} - \mathcal{W}_r^{N*} = \frac{-v\gamma^2 + 2v\delta\eta - 2\delta^2\eta + (\gamma^2 - 2\delta\eta)c_r}{\gamma^2 - 4\delta\eta} - \frac{-2\delta^2\eta + (\gamma^2 - 2\delta\eta)c_r}{\gamma^2 - 4\delta\eta} = -\frac{v(\gamma^2 - 2\delta\eta)}{\gamma^2 - 4\delta\eta}$. According to Lemma 1, $\gamma^2 - 4\delta\eta < 0$, $v > 0$; thus, when $1 > \gamma > \sqrt{2\delta\eta}$, $\mathcal{W}_r^{VR*} > \mathcal{W}_r^{N*}$; when $0 < \gamma < \sqrt{2\delta\eta}$, $\mathcal{W}_r^{VR*} < \mathcal{W}_r^{N*}$. □

Proposition 1 shows that the wholesale price of a new product with government subsidies is higher than the price without government subsidies; that is, government subsidies will cause manufacturers to increase the wholesale price of new products, and, the greater the government subsidies, the higher the wholesale price of new products. When consumers' green preferences are greater than $\sqrt{2\delta\eta}$, the wholesale price of remanufactured products with government subsidies is higher than that without government subsidies. Conversely, when consumers' green preferences are less than $\sqrt{2\delta\eta}$, the wholesale price of remanufactured products with government subsidies is lower than that without government subsidies. This suggests that, when consumers' green preferences are smaller, building material manufacturers sell more CDW remanufactured products by lowering wholesale prices in order to realize their own gain. When consumers' green preferences are greater, building material manufacturers will increase the wholesale price of remanufactured products in order to seek greater benefits.

**Proposition 2.** *The retail prices of new and remanufactured products are related as follows:*

(1) $p_n^{VR*} > p_n^{N*}$.

(2) *When* $1 > \gamma > \sqrt{\delta\eta}$, *that is,* $p_r^{VR*} > p_r^{N*}$;
*When* $0 < \gamma < \sqrt{\delta\eta}$, *that is,* $p_r^{VR*} < p_r^{N*}$.

**Proof.** By comparing the solutions of the two decentralized models, i.e., with and without government subsidies, we obtain:

(1) $p_n^{VR*} - p_n^{N*} = \frac{-3\gamma^2(-1+v+\delta)-12\delta\eta+(\gamma^2-4\delta\eta)c_n+3\gamma^2c_r}{4(\gamma^2-4\delta\eta)} - \frac{1}{4}\left(c_n + \frac{3(\gamma^2-\gamma^2\delta-4\delta\eta+\gamma^2c_r)}{\gamma^2-4\delta\eta}\right) = -\frac{3v\gamma^2}{4(\gamma^2-4\delta\eta)} > 0$, that is, $p_n^{VR*} > p_n^{N*}$.

(2) $p_r^{VR*} - p_r^{N*} = \frac{-v\gamma^2+v\delta\eta-3\delta^2\eta+(\gamma^2-\delta\eta)c_r}{\gamma^2-4\delta\eta} - \frac{-3\delta^2\eta+(\gamma^2-\delta\eta)c_r}{\gamma^2-4\delta\eta} = \frac{v(-\gamma^2+\delta\eta)}{\gamma^2-4\delta\eta}$. According to Lemma 1, $\gamma^2 - 4\delta\eta < 0$, $v > 0$; thus, when $1 > \gamma > \sqrt{\delta\eta}$, $p_r^{VR*} > p_r^{N*}$; when $0 < \gamma < \sqrt{\delta\eta}$, $p_r^{VR*} < p_r^{N*}$. □

Proposition 2 shows that the retail price of new products with government subsidies is higher than that without government subsidies; that is, government subsidies will lead retailers to increase the retail prices of new products, and, the greater the government subsidies, the higher the retail prices of new products. This shows that retailers resist the increase in wholesale prices of manufacturers by adjusting the retail price of new products. When consumers' green preferences are greater than $\sqrt{\delta\eta}$, the retail price of CDW remanufactured products with government subsidies is higher than that without government subsidies. Conversely, when consumers' green preferences are less than $\sqrt{\delta\eta}$, the retail price of CDW remanufactured products with government subsidies is lower than that without government subsidies. This suggests that, when consumers' green preferences are smaller, retailers sell more CDW remanufactured products by lowering retail prices in order to realize their own interests. When consumers' green preferences are greater, retailers will increase the retail price of CDW remanufactured products in order to seek greater benefits.

**Proposition 3.** *The relationship between the product green degree of new products and remanufactured products is:* $g^{VR*} > g^{N*}$.

**Proof.** By comparing the solutions of the two decentralized models, i.e., with and without government subsidies, we obtain:
$g^{VR*} - g^{N*} = -\frac{\gamma(v+\delta-c_r)}{\gamma^2-4\delta\eta} - \frac{\gamma(-\delta+c_r)}{\gamma^2-4\delta\eta} = -\frac{v\gamma}{\gamma^2-4\delta\eta} > 0$, that is, $g^{VR*} > g^{N*}$. □

Proposition 3 shows that the green degree of products is affected by government subsidies, and, with the increase in government subsidies, the green degree of products increases.

**Proposition 4.** *Using Model N, the effects of:* $c_n$ *and* $c_r$ *on the optimal pricing and product green degree of building material manufacturers and retailers are shown in Table 3.*

**Table 3.** The influence of the change in each parameter on the optimal decision under the decentralized decision-making model (N model).

| N | $p_n^{N*}$ | $p_r^{N*}$ | $w_n^{N*}$ | $w_r^{N*}$ | $g^{N*}$ |
|---|---|---|---|---|---|
| $c_n$ | + | / | + | / | / |
| $c_r$ | − | $0 < \gamma < \sqrt{\delta\eta}, +$<br>$\sqrt{\delta\eta} < \gamma < 1, -$ | − | $0 < \gamma < \sqrt{2\delta\eta}, +$<br>$\sqrt{2\delta\eta} < \gamma < 1, -$ | − |

Note: "+" means positive correlation, "−" means negative correlation, "/" means irrelevant.

**Proof.** From Table 3, it is easy to get the following verification:

(1) $\frac{\partial p_n^{N*}}{\partial c_n} = \frac{1}{4} > 0$. $\frac{\partial p_n^{N*}}{\partial c_r} = \frac{3\gamma^2}{4(\gamma^2 - 4\delta\eta)} < 0$, then, $p_n^{N*}$ is positively correlated with $c_n$ and negatively correlated with $c_r$

(2) $\frac{\partial p_r^{N*}}{\partial c_r} = \frac{\gamma^2 - \delta\eta}{\gamma^2 - 4\delta\eta}$. Evidently, $\frac{\gamma^2 - \delta\eta}{\gamma^2 - 4\delta\eta} > 0$ requires $\gamma^2 - \delta\eta < 0$, and $\frac{\gamma^2 - \delta\eta}{\gamma^2 - 4\delta\eta} < 0$ requires $\gamma^2 - \delta\eta > 0$.

(3) $\frac{\partial \mathcal{W}_n^{N*}}{\partial c_n} = \frac{1}{2} > 0$. $\frac{\partial \mathcal{W}_n^{N*}}{\partial c_r} = \frac{\gamma^2}{2(\gamma^2 - 4\delta\eta)} < 0$, then, $\mathcal{W}_n^{N*}$ is positively correlated with $c_n$ and negatively correlated with $c_r$.

(4) $\frac{\partial \mathcal{W}_r^{N*}}{\partial c_r} = \frac{\gamma^2 - 2\delta\eta}{\gamma^2 - 4\delta\eta}$. Evidently, $\frac{\gamma^2 - 2\delta\eta}{\gamma^2 - 4\delta\eta} > 0$ requires $\gamma^2 - 2\delta\eta < 0$, and $\frac{\gamma^2 - \delta\eta}{\gamma^2 - 4\delta\eta} < 0$ requires $\gamma^2 - 2\delta\eta > 0$.

(5) $\frac{\partial g^{N*}}{\partial c_r} = \frac{\gamma}{\gamma^2 - 4\delta\eta} < 0$, then, $g^{N*}$ is negatively correlated with $c_r$. □

Proposition 4 shows that the reduction in the manufacturing cost of new products reduces the wholesale and retail prices of new products, so that more consumers buy new products, and the sales of new products increase, but the sales of CDW remanufactured products decrease. This requires that remanufactured products compete with new products for market share by lowering their own selling prices. In addition, lower manufacturing costs for CDW remanufactured products lead to higher wholesale and retail prices of new products. If consumers' green preferences are low (0 < $0 < \gamma < \sqrt{2\delta\eta}/0 < \gamma < \sqrt{\delta\eta}$), in order to sell more CDW remanufactured products, the wholesale and retail prices of CDW remanufactured products are also reduced. If consumers have a high degree of green preference ($\sqrt{2\delta\eta} < \gamma < 1/\sqrt{\delta\eta} < \gamma < 1$), in order to maximize profits, the wholesale and retail prices of remanufactured products will increase. At the same time, the cost of remanufacturing is reduced, and manufacturers are more willing to improve the green degree of their products.

**Proposition 5.** *In Model VR, the effects of* $c_n$, $c_r$, and $v$ *on the optimal pricing and product green degree of building materials manufacturers and retailers are shown in Table 4.*

**Table 4.** The influence of the change in each parameter on the optimal decision under the decentralized decision-making model (VR model).

| *VR* | $p_n^{VR*}$ | $p_r^{VR*}$ | $\mathcal{W}_n^{VR*}$ | $\mathcal{W}_r^{VR*}$ | $g^{VR*}$ |
|---|---|---|---|---|---|
| $c_n$ | + | / | + | / | / |
| $c_r$ | − | $0 < \gamma < \sqrt{\delta\eta}$, +<br>$\sqrt{\delta\eta} < \gamma < 1$, − | − | $0 < \gamma < \sqrt{2\delta\eta}$, +<br>$\sqrt{2\delta\eta} < \gamma < 1$, − | − |
| $v$ | + | $\sqrt{\delta\eta} < \gamma < 1$, +<br>$0 < \gamma < \sqrt{\delta\eta}$, − | + | $\sqrt{2\delta\eta} < \gamma < 1$, +<br>$0 < \gamma < \sqrt{2\delta\eta}$, − | + |

Note:"+" means positive correlation, "−" means negative correlation, "/" means irrelevant.

**Proof.** The proof process is the same as that of Proposition 4. □

Proposition 5 shows that the effects of $c_n$ and $c_r$ on the optimal pricing and product green degree of building material manufacturers and retailers are the same as those in Proposition 4. Increased government subsidies for remanufactured products will lead to higher wholesale and retail prices of new products. The impact of government subsidies on the pricing of remanufactured products is also related to the range of consumer green preferences. When consumers' green preferences are low ($0 < \gamma < \sqrt{2\delta\eta}/0 < \gamma < \sqrt{\delta\eta}$), in order to sell more remanufactured products, the wholesale and retail prices of remanufactured products need to be reduced. If consumers have a high degree of green preference ($\sqrt{2\delta\eta} < \gamma < 1/\sqrt{\delta\eta} < \gamma < 1$), in order to maximize profits, the wholesale price and retail sales of remanufactured products are increased. At the same time, the increase in government subsidies will improve the green degree of the product.

## 6. Numerical Simulation and Discussion

This section presents the simulation and analysis of the game model using MATLAB2016b in order to provide a more intuitive illustration of the impact of consumer green preferences and government subsidies on the supply chain of construction waste reuse. The initial values of related parameters are shown in Table 5.

**Table 5.** Simulation parameter assignment.

| $\delta$ | $\eta$ | $h$ | $k$ | $b$ | $c_n$ | $c_r$ | $\mu$ |
|---|---|---|---|---|---|---|---|
| 0.7 | 0.6 | 0.4 | 0.5 | 0.2 | 0.6 | 0.3 | 0.25 |

The basis of these basic parameters was taken from a survey of China's CDW remanufacturing companies and the academic results of related papers [57,58]. Therefore, this paper found that, at the current development stage of China's CDW remanufacturing industry, the cost of remanufactured products is 45–55% of the cost of new products. Overall, we set the unit cost of building materials manufacturers to produce new products at 0.6, the unit cost of CDW remanufactured products at 0.3, $\delta = 0.4$, and $\eta = 0.5$.

Setting $v = 0.02$ and $\gamma \in [0, 1]$ allows us to clarify the impact of consumer green preferences on the profits of building materials manufacturers and retailers under decentralized decision making (as shown in Figure 2).

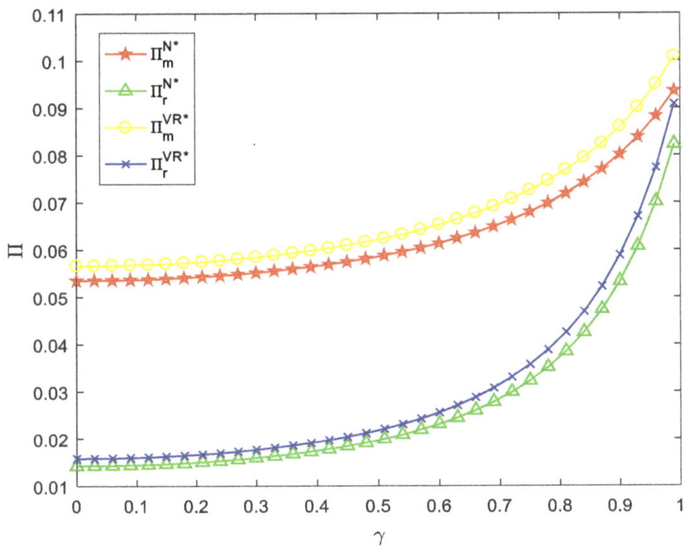


**Figure 2.** Profits of supply chain members under decentralized decision making.

Figure 2 shows that, under decentralized decision making, when consumers' green preferences increase, the profits of manufacturers and retailers in the CDW resource utilization supply chain also increase. This is because, with the increase in consumers' green preferences, the demand for green products increases, which promotes the increase in the sales revenue of manufacturers and retailers in the supply chain, and ultimately achieves an increase in the total profit of the supply chain. In addition, the change in the profits of building materials manufacturers and retailers increases; that is, the greater the green preferences of consumers, the more obvious the effect of improving profits. At the same time, the profit of the CDW resource utilization supply chain with government subsidies is obviously better than that without government subsidies. This is because government subsidies, to a certain extent, increase the sales of remanufactured products, and thus increase the profits of manufacturers and retailers in the supply chain. In addition, building materials manufacturers have greater profits than retailers, but retailers' profits have grown more than those of manufacturers.

Setting $v = 0.02$ and $\gamma \in [0, 1]$ enables us to clarify the impact of consumers' green preferences on the total profit of the construction waste resource utilization supply chain under centralized decision making (as shown in Figure 3).

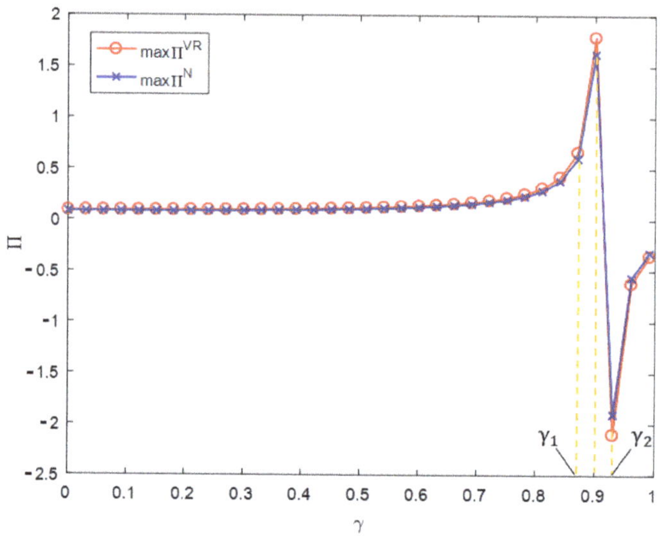


**Figure 3.** Impact on the total profit of the supply chain under centralized decision making.

Figure 3 shows that, under centralized decision making, when consumers' green preferences are $\gamma \in [0, \gamma_1]$, the total profit of the CDW resource utilization supply chain increases slightly but not significantly. When consumers green preferences are $\gamma \in [\gamma_1, 0.9]$, the total profit of the supply chain surges. When consumers' green preferences are 0.9, the profit reaches the maximum value. Immediately thereafter, the profit drops sharply between $\gamma \in [0.9, \gamma_2]$. Then, the profit starts to increase again between $\gamma \in [\gamma_2, 1]$. At the same time, when consumers' green preferences are between $[0, 0.9]$, the profits of the supply chain with government subsidies are slightly higher than those without government subsidies. However, when consumers' green preferences are between $[\gamma_2, 1]$, the profits of the supply chain without government subsidies are slightly higher than those with government subsidies. In general, with the increase in consumers' green preferences, the total profit of the supply chain first increases and then decreases, and the profit changes from a positive value to a negative value. Thus, the improvement in consumers' green preferences does not always improve the profits of the supply chain. At the same time, government subsidies do not necessarily increase the profits of the supply chain.

This result is slightly different from that of the study by Liu et al. [59]. They believe that, under the two models of decentralized and centralized decision making, the environmental protection level and profit of enterprises are positively correlated with consumers' environmental awareness. This study shows that the green degree of products is positively correlated with consumers' green preferences under both decision models. However, the changes in corporate profits are slightly different; that is, under decentralized decision making, corporate profits are proportional to consumers' green preferences; under centralized decision making, profits first increase and then decrease. This bias may be due to the fact that the manufacturers in this study produce both new and remanufactured products.

Figures 4 and 5 show the influence of consumers' green preferences on the total profit of the CDW resource utilization supply chain under decentralized and centralized decision making, respectively, without and with government subsidies. The total profits of the supply chain under decentralized decision making are $\Pi^{DN*}, \Pi^{DVR*}$; the total profits of the supply chain under centralized decision making are $\Pi^{N*}, \Pi^{VR*}$. Setting $\gamma \in [0,1]$, $v = 0.02$, enables us to compare the total profit of the CDW resource utilization supply chain under decentralized and centralized decision making without or with government subsidies.

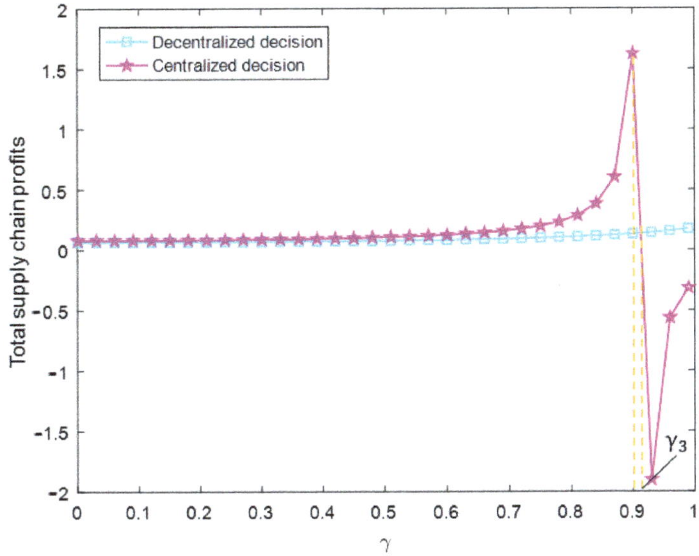


**Figure 4.** The impact of $\gamma$ on total CDW supply chain profit without government subsidies.

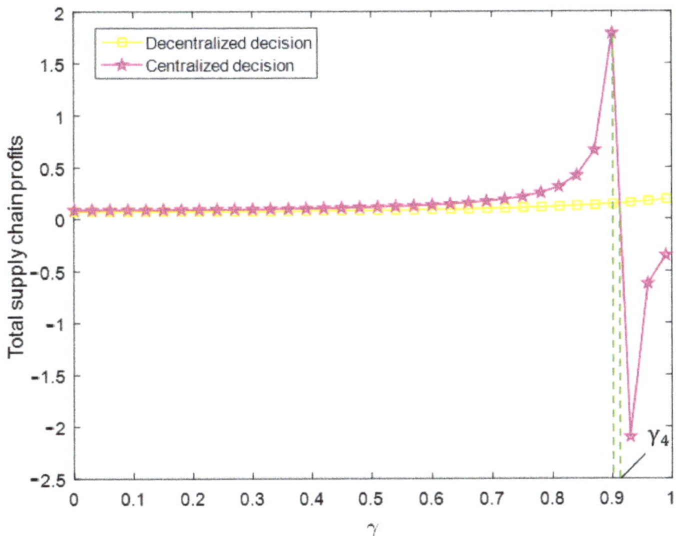


**Figure 5.** The impact of $\gamma$ on total CDW supply chain profit with government subsidies.

Figure 4 shows that, when consumers' green preferences are $\gamma \in [0, 0.9]$, the total profit of the construction waste resource utilization supply chain under centralized decision making is higher than that under decentralized decision making. When consumers' green preferences are 0.9, the total profit of the CDW resource utilization supply chain under centralized decision making reaches the maximum value, and then decreases sharply. When consumers' green preferences are $\gamma_3$, the total profit of the CDW resource utilization supply chain is equal under centralized and decentralized decision making. Then, when consumers' green preferences are $\gamma \in [\gamma_3, 1]$, the total profit of the CDW resource utilization supply chain under centralized decision making is lower than that under decentralized decision making. In general, with the increase in consumers' green preferences, the total profit of the CDW resource utilization supply chain under centralized decision making is first greater than, and then smaller than, that under decentralized decision making. Therefore, in the case of the absence of government subsidies, the centralized or decentralized decision-making model should be selected according to the degree of consumers' green preferences in order to maximize the total profit of the CDW resource utilization supply chain.

Figure 5 shows that, when consumers' green preferences are $\gamma \in [0, 0.9]$, the total profit of the CDW resource utilization supply chain under centralized decision making is higher than that under decentralized decision making. When consumers' green preferences are 0.9, the total profit of the CDW resource utilization supply chain under centralized decision-making reaches the maximum value, and then decreases sharply. When consumers' green preferences are $\gamma_4$, the total profits of the CDW resource utilization supply chain under centralized and decentralized decision making are equal. Then, when consumers' green preferences are $\gamma \in [\gamma_4, 1]$, the total profit of the CDW resource utilization supply chain under centralized decision making is lower than that under decentralized decision making. In general, with the increase in consumers' green preferences, the total profit of the CDW resource utilization supply chain under centralized decision making is first greater than, and then smaller than, that under decentralized decision making. Secondly, the comparison of Figures 4 and 5 shows that, regardless of whether there are government subsidies, the relationship between the total profits of the CDW resource utilization supply chain under two the different decision-making models will change with the increase in consumers' green preferences. Therefore, it is necessary to choose a centralized or decentralized decision-making model according to the degree of consumers' green preferences in order to maximize the total profit of the CDW resource utilization supply chain.

This result is different from that of Zhan et al. [60]. They believe that the overall profit of the supply chain under the centralized decision-making model is always higher than that under the decentralized decision-making model. However, this study shows that the relative size of the two is related to the size of consumers' green preferences. The reason for this bias may be that the previous authors only considered green products and not remanufactured products in the supply chain model. Therefore, the current study better expands the research on remanufactured products in the green supply chain.

## 7. Conclusions and Implications

### 7.1. Conclusions

In this study, a Stackelberg game model consisting of building materials manufacturers, retailers, and consumers was constructed and used to investigate the impact of consumers' green preferences and government subsidies on CDW resource utilization supply chain decisions. The following four supply chain models were developed: decentralized decision making without government subsidies, centralized decision making without government subsidies, decentralized decision making with government subsidies, and centralized decision making with government subsidies. The strategies of the members in these four models are discussed in this paper, and are analyzed using numerical simulations. The specific findings of the study can be summarized as follows:

First, government subsidies increase wholesale and retail prices of new products, but do not always lead to lower prices for remanufactured construction waste. Government subsidies are not maximized, depending on the range of consumers' green preferences. When consumer green preferences are low, government subsidies reduce building materials manufacturers' pricing of CDW remanufactured products. When consumers' green preferences are high, under government subsidies, building materials manufacturers will instead increase the price of CDW remanufactured products. Second, government subsidies have an impact on the green degree of products. With the increase in government subsidies, the green degree of products is improved. Moreover, with the increase in consumers' green preferences, the green degree of products also increases.

Second, under decentralized decision making, the optimal profit of the CDW resource utilization supply chain with government subsidies is always greater than that of the CDW resource utilization supply chain without government subsidies. However, under centralized decision making, the optimal profit of the supply chain with government subsidies is not always greater than the optimal profit of the supply chain without government subsidies, and is also related to the degree of consumers' green preferences. When consumers' green preferences are low, the supply chain profit is higher with government subsidies than without government subsidies, whereas, when consumers' green preferences are high, the supply chain profit is higher without government subsidies than with government subsidies.

Third, under decentralized decision making, as consumers' green preferences increase, the optimal profits of building materials manufacturers and retailers also increase gradually. In addition, the greater the green preference of consumers, the more obvious the effect of increasing profits. Furthermore, although the optimal profit of building materials manufacturers is greater than that of retailers, the growth rate of retailers is greater than that of building materials manufacturers.

Fourth, regardless of the implementation of government subsidies, when consumers' green preferences are small, the total profit of the CDW resource utilization supply chain under centralized decision making is higher than that under decentralized decision making. When consumers' green preferences are too high, the total profit of the CDW resource utilization supply chain under decentralized decision making is higher than that under centralized decision making.

### 7.2. Implications

This study simultaneously considered the two parameters of consumers' green preferences and government subsidies, thereby enriching the research on CDW resource utilization supply chain decision making. The research also provides effective management recommendations for enterprises and governments in the CDW resource utilization supply chain, namely:

(1) For building materials manufacturers and retailers, the degree of consumer green preferences in the CDW resource utilization market should be clarified, so as to choose whether to centralize or decentralize decision making. When choosing centralized decision making, both manufacturers and retailers need to formulate a coordinated plan for common benefits and costs, with the goal of maximizing the benefits of the entire CDW resource utilization supply chain. When choosing decentralized decision making, the goal of both parties should be to maximize their own interests. In conclusion, under different consumer green preferences, choosing different decision-making models is not only beneficial to improving the total profit of the CDW resource utilization supply chain, but also to improving the CDW reuse rate.

(2) For the government, in order to promote the efficient and sustainable development of the CDW resource utilization industry, relevant subsidy policies should be formulated to stimulate the total profit of the supply chain. When consumers' green preferences are low, subsidies for CDW remanufactured products are increased; when consumers' green preferences are high, subsidies for CDW remanufactured products can be appropriately reduced to reduce consumer losses. At the same time, the government can also subsidize consumers who buy CDW remanufactured products to increase the sales of CDW remanufactured products, thereby improving the utilization rate of CDW resources.

Although this study obtained some insights into the impact of consumers' green preferences and government subsidies on CDW resource utilization supply chain decision making, and made effective recommendations for CDW management, some limitations remain.

First, this study examined a CDW resource utilization supply chain that consisted of only one building material manufacturer and one retailer. In the future, researchers can consider multiple building materials manufacturers and retailers to make it more realistic; for example, both recycling-capable building materials manufacturers and non-recycling-capable building materials manufacturers exist in the supply chain model.

Second, the target of government subsidies in this study was relatively simple, and only the subsidized manufacturers of CDW remanufactured products were considered. In the future, researchers can also consider the situation in which the government subsidizes retailers and consumers, and compare the differences in CDW resource utilization supply chain profits under different government subsidy strategies, so as to provide better subsidy strategies for the government.

**Author Contributions:** Methodology, Validation, Formal analysis, Investigation, Resources, Data Curation, Writing—Original Draft, Writing—Review and Editing, Visualization, Y.H. (Yating Han) and H.Z.; Conceptualization, Methodology, Writing—Original Draft, Supervision, Project administration, X.L.; Writing—Review and Editing, Y.H. (Yicheng Huang). All authors have read and agreed to the published version of the manuscript.

**Funding:** This work was funded by Special Funds of the National Social Science Fund of China (grant number 18VSJ038), the Scientific Research Startup Foundation for Introducing Talents of Sichuan Agricultural University (grant number 2122996022), the Social Science Special Project of Sichuan Agricultural University Disciplinary Construction Dual Support Program (grant number 2021SYYB05).

**Institutional Review Board Statement:** Not applicable.

**Informed Consent Statement:** Not applicable.

**Data Availability Statement:** Not applicable.

**Conflicts of Interest:** The authors declare no conflict of interest.

## References

1. Banias, G.; Achillas, C.; Vlachokostas, C.; Moussiopoulos, N.; Papaioannou, I. A web-based Decision Support System for the optimal management of construction and demolition waste. *Waste Manag.* **2011**, *31*, 2497–2502. [CrossRef] [PubMed]
2. Chen, M.; Liu, W.; Tao, X. Evolution and assessment on China's urbanization 1960–2010: Under-urbanization or over-urbanization? *Habitat Int.* **2013**, *38*, 25–33. [CrossRef]
3. Duan, H.; Miller, T.R.; Liu, G.; Tam, V.W. Construction debris becomes growing concern of growing cities. *Waste Manag.* **2019**, *83*, 1–5. [CrossRef] [PubMed]
4. Erlandsson, M.; Levin, P. Environmental assessment of rebuilding and possible performance improvements effect on a national scale. *Build. Environ.* **2005**, *40*, 1459–1471. [CrossRef]
5. Jain, P.; Powell, J.; Tolaymat, T. *Methodology to Estimate the Quantity Composition, and Management of Construction and Demolition Debris in the United States*; US Environmental Protection Agency: Washington, DC, USA, 2015.
6. Ajayi, S.O.; Oyedele, L.O.; Akinade, O.O.; Bilal, M.; Owolabi, H.A.; Alaka, H.A.; Kadiri, K.O. Reducing waste to landfill: A need for cultural change in the UK construction industry. *J. Build. Eng.* **2016**, *5*, 185–193. [CrossRef]
7. Zheng, L.; Wu, H.; Zhang, H.; Duan, H.; Wang, J.; Jiang, W.; Dong, B.; Liu, G.; Zuo, J.; Song, Q. Characterizing the generation and flows of construction and demolition waste in China. *Constr. Build. Mater.* **2017**, *136*, 405–413. [CrossRef]
8. Esa, M.R.; Halog, A.; Rigamonti, L. Developing strategies for managing construction and demolition wastes in Malaysia based on the concept of circular economy. *J. Mater. Cycles Waste Manag.* **2017**, *19*, 1144–1154. [CrossRef]
9. Oliveira, T.C.; Dezen, B.G.; Possan, E. Use of concrete fine fraction waste as a replacement of Portland cement. *J. Clean. Prod.* **2020**, *273*, 123126. [CrossRef]
10. Widhiawati, I.A.R.; Astana, I.N.Y.; Indrayani, N.L.A. An analysis of building construction waste in Badung, Bali. In Proceedings of the MATEC Web of Conferences, Bali, Indonesia, 24–25 October 2018; EDP Sciences: Les Ulis, France, 2019; Volume 276, p. 02010.
11. Genc, T.S.; De Giovanni, P. Trade-in and save: A two-period closed-loop supply chain game with price and technology dependent returns. *Int. J. Prod. Econ.* **2017**, *183*, 514–527. [CrossRef]
12. Poon, C.S.; Chan, D. The use of recycled aggregate in concrete in Hong Kong. *Resour. Conserv. Recycl.* **2007**, *50*, 293–305. [CrossRef]

13. Ling, Y.Y.; Leo, K.C. Reusing timber formwork: Importance of workmen's efficiency and attitude. *Build. Environ.* **2000**, *35*, 135–143. [CrossRef]
14. Tam, V.W. On the effectiveness in implementing a waste-management-plan method in construction. *Waste Manag.* **2008**, *28*, 1072–1080. [CrossRef] [PubMed]
15. Abanda, F.H.; Tah, J.H.M.; Cheung, F.K.T. Mathematical modelling of embodied energy, greenhouse gases, waste, time–cost parameters of building projects: A review. *Build. Environ.* **2013**, *59*, 23–37. [CrossRef]
16. Wu, Z.; Ann, T.W.; Shen, L.; Liu, G. Quantifying construction and demolition waste: An analytical review. *Waste Manag.* **2014**, *34*, 1683–1692. [CrossRef] [PubMed]
17. Chitra, K. In search of the green consumers: A perceptual study. *J. Serv. Res.* **2007**, *7*, 173–191.
18. Chen, X.; Wang, X.; Zhou, M. Firms' green R&D cooperation behaviour in a supply chain: Technological spillover, power and coordination. *Int. J. Prod. Econ.* **2019**, *218*, 118–134.
19. Modak, N.M.; Ghosh, D.K.; Panda, S.; Sana, S.S. Managing green house gas emission cost and pricing policies in a two-echelon supply chain. *CIRP J. Manuf. Sci. Technol.* **2018**, *20*, 1–11. [CrossRef]
20. Qin, Q.; Jiang, M.; Xie, J.; He, Y. Game analysis of environmental cost allocation in green supply chain under fairness preference. *Energy Rep.* **2021**, *7*, 6014–6022. [CrossRef]
21. Zhao, J.; Sun, N. Government subsidies-based profits distribution pattern analysis in closed-loop supply chain using game theory. *Neural Comput. Appl.* **2020**, *32*, 1715–1724. [CrossRef]
22. Zhang, X.; Li, Q.; Qi, G. Decision-making of a dual-channel closed-loop supply chain in the context government policy: A dynamic game theory. *Discret. Dyn. Nat. Soc.* **2020**, *2020*, 2313698. [CrossRef]
23. Ilyas, S.; Hu, Z.; Wiwattanakornwong, K. Unleashing the role of top management and government support in green supply chain management and sustainable development goals. *Environ. Sci. Pollut. Res.* **2020**, *27*, 8210–8223. [CrossRef] [PubMed]
24. Louis, G.E. A historical context of municipal solid waste management in the United States. *Waste Manag. Res.* **2004**, *22*, 306–322. [CrossRef] [PubMed]
25. Li, L.; Jiang, T.; Chen, B.; Wen, J.; Yang, G. Integrated Utilization of Vanadium-Titanium Magnetite Tailings for Synthesis of Lightweight Foamed Ceramics: Effect of Chemical Composition on the Properties and Phase Evolution. *J. Sustain. Metall.* **2022**, *8*, 646–657. [CrossRef]
26. Brown, S. An interview with Wickham Skinner, emeritus professor at harvard business school. *Int. J. Oper. Prod. Manag.* **2013**, *33*, 104–110. [CrossRef]
27. Nguyen, H.P.; Mueller, A.; Nguyen, C.T. Development and characterization of lightweight aggregate recycled from construction and demolition waste mixed with other industrial by-products. *Constr. Build. Mater.* **2021**, *313*, 125472. [CrossRef]
28. Wu, H.; Zuo, J.; Zillante, G.; Wang, J.; Yuan, H. Status quo and future directions of construction and demolition waste research: A critical review. *J. Clean. Prod.* **2019**, *240*, 118163. [CrossRef]
29. Kabirifar, K.; Mojtahedi, M.; Wang, C.C. A systematic review of construction and demolition waste management in Australia: Current practices and challenges. *Recycling* **2021**, *6*, 34. [CrossRef]
30. Mohammed, M.; Shafiq, N.; Al-Mekhlafi, A.B.A.; Rashed, E.F.; Khalil, M.H.; Zawawi, N.A.; Muhammad, A.; Sadis, A.M. The Mediating Role of Policy-Related Factors in the Relationship between Practice of Waste Generation and Sustainable Construction Waste Minimisation: PLS-SEM. *Sustainability* **2022**, *14*, 656. [CrossRef]
31. Sun, H.; Wan, Y.; Zhang, L.; Zhou, Z. Evolutionary game of the green investment in a two-echelon supply chain under a government subsidy mechanism. *J. Clean. Prod.* **2019**, *235*, 1315–1326. [CrossRef]
32. Li, X.; Huang, R.; Dai, J.; Li, J.; Shen, Q. Research on the evolutionary game of construction and demolition waste (CDW) recycling units' green behavior, considering remanufacturing capability. *Int. J. Environ. Res. Public Health* **2021**, *18*, 9268. [CrossRef]
33. Hu, Q.; Su, P.; Xiong, F. Dynamic game strategies for construction and demolition waste recycling with learning and reference effects. *Environ. Impact Assess. Rev.* **2021**, *89*, 106592. [CrossRef]
34. Zheng, H.; Li, X.; Zhu, X.; Huang, Y.; Liu, Z.; Liu, Y.; Liu, J.; Li, X.; Li, Y.; Li, C. Impact of Recycler Information Sharing on Supply Chain Performance of Construction and Demolition Waste Resource Utilization. *Int. J. Environ. Res. Public Health* **2022**, *19*, 3878. [CrossRef] [PubMed]
35. Li, D.; Peng, Y.; Guo, C.; Tan, R. Pricing strategy of construction and demolition waste considering retailer fairness concerns under a governmental regulation environment. *Int. J. Environ. Res. Public Health* **2019**, *16*, 3896. [CrossRef] [PubMed]
36. Michael, R.T.; Becker, G.S. On the new theory of consumer behavior. *Swed. J. Econ.* **1973**, *75*, 378–396. [CrossRef]
37. Oduro, S.; Maccario, G.; De Nisco, A. Green innovation: A multidomain systematic review. *Eur. J. Innov. Manag.* **2021**, *25*, 567–591. [CrossRef]
38. He, J.; Lei, Y.; Fu, X. Do consumer's green preference and the reference price effect improve green innovation? A theoretical model using the food supply chain as a case. *Int. J. Environ. Res. Public Health* **2019**, *16*, 5007. [CrossRef]
39. Xu, Y.; Xie, H. Consumer environmental awareness and coordination in closed-loop supply chain. *Open J. Bus. Manag.* **2016**, *4*, 427. [CrossRef]
40. Abbey, J.D.; Meloy, M.G.; Guide, V.D.R., Jr.; Atalay, S. Remanufactured products in closed-loop supply chains for consumer goods. *Prod. Oper. Manag.* **2015**, *24*, 488–503. [CrossRef]
41. Alam-Tabriz, A.; Rajabani, N.; Farrokh, M. An integrated fuzzy DEMATEL-ANP-TOPSIS methodology for supplier selection problem. *Glob. J. Manag. Stud. Res.* **2014**, *1*, 85–99.

42. Shooshtarian, S.; Maqsood, T.; Khalfan, M.; Yang, R.J.; Wong, P. Landfill levy imposition on construction and demolition waste: Australian stakeholders' perceptions. *Sustainability* **2020**, *12*, 4496. [CrossRef]
43. Mak, T.M.; Iris, K.M.; Wang, L.; Hsu, S.C.; Tsang, D.C.; Li, C.N.; Yeung, T.L.; Zhang, R.; Poon, C.S. Extended theory of planned behaviour for promoting construction waste recycling in Hong Kong. *Waste Manag.* **2019**, *83*, 161–170. [CrossRef] [PubMed]
44. Yang, D.; Xiao, T. Pricing and green level decisions of a green supply chain with governmental interventions under fuzzy uncertainties. *J. Clean. Prod.* **2017**, *149*, 1174–1187. [CrossRef]
45. Chemama, J.; Cohen, M.C.; Lobel, R.; Perakis, G. Consumer subsidies with a strategic supplier: Commitment vs. flexibility. *Manag. Sci.* **2019**, *65*, 681–713. [CrossRef]
46. Li, Z.; Zhang, J.; Meng, Q.; Zheng, W.; Du, J. Influence of government subsidy on remanufacturing decision under different market models. *Math. Probl. Eng.* **2019**, *2019*, 9460315. [CrossRef]
47. He, P.; He, Y.; Xu, H. Channel structure and pricing in a dual-channel closed-loop supply chain with government subsidy. *Int. J. Prod. Econ.* **2019**, *213*, 108–123. [CrossRef]
48. Huang, Y.; Liang, Y. Exploring the strategies of online and offline recycling channels in closed-loop supply chain under government subsidy. *Environ. Sci. Pollut. Res.* **2022**, *29*, 21591–21602. [CrossRef] [PubMed]
49. Barman, A.; Das, R.; De, P.K.; Sana, S.S. Optimal pricing and greening strategy in a competitive green supply chain: Impact of government subsidy and tax policy. *Sustainability* **2021**, *13*, 9178. [CrossRef]
50. Yu, Y.; Han, X.; Hu, G. Optimal production for manufacturers considering consumer environmental awareness and green subsidies. *Int. J. Prod. Econ.* **2016**, *182*, 397–408. [CrossRef]
51. Ghosh, D.; Shah, J. A comparative analysis of greening policies across supply chain structures. *Int. J. Prod. Econ.* **2012**, *135*, 568–583. [CrossRef]
52. Swami, S.; Shah, J. Channel coordination in green supply chain management. *J. Oper. Res. Soc.* **2013**, *64*, 336–351. [CrossRef]
53. Li, B.; Zhu, M.; Jiang, Y.; Li, Z. Pricing policies of a competitive dual-channel green supply chain. *J. Clean. Prod.* **2016**, *112*, 2029–2042. [CrossRef]
54. Bakal, I.S.; Akcali, E. Effects of random yield in remanufacturing with price-sensitive supply and demand. *Prod. Oper. Manag.* **2006**, *15*, 407–420. [CrossRef]
55. Hong, I.H.; Lee, Y.T.; Chang, P.Y. Socially optimal and fund-balanced advanced recycling fees and subsidies in a competitive forward and reverse supply chain. *Resour. Conserv. Recycl.* **2014**, *82*, 75–85. [CrossRef]
56. Ferrer, G.; Swaminathan, J.M. Managing new and differentiated remanufactured products. *Eur. J. Oper. Res.* **2010**, *203*, 370–379. [CrossRef]
57. Chen, X.; Cao, J.; Kumar, S. Government regulation and enterprise decision in China remanufacturing industry: Evidence from evolutionary game theory. *Energy Ecol. Environ.* **2021**, *6*, 148–159. [CrossRef] [PubMed]
58. Long, X.; Ge, J.; Shu, T.; Liu, Y. Analysis for recycling and remanufacturing strategies in a supply chain considering consumers' heterogeneous WTP. *Resour. Conserv. Recycl.* **2019**, *148*, 80–90. [CrossRef]
59. Liu, Z.L.; Anderson, T.D.; Cruz, J.M. Consumer environmental awareness and competition in two-stage supply chains. *Eur. J. Oper. Res.* **2012**, *218*, 602–613. [CrossRef]
60. Zhan, S.; Shu, Z.; Jiang, H. Research on two-echelon green supply chain decision under government subsidy. *Am. J. Ind. Bus. Manag.* **2018**, *8*, 487–495. [CrossRef]

buildings

MDPI

Article

# How Does the Government Promote the Collaborative Innovation of Green Building Projects? An Evolutionary Game Perspective

**Xiaochun Zeng [1], Suicheng Li [1], Shi Yin [2,*] and Zeyu Xing [3,*]**

1 School of Economics and Management, Xi'an University of Technology, Xi'an 710048, China
2 College of Economics and Management, Hebei Agricultural University, Baoding 071001, China
3 School of Management, Zhejiang University of Technology, Hangzhou 310023, China
* Correspondence: shyshi0314@163.com (S.Y.); xingzeyusmile@zjut.edu.cn (Z.X.)

**Abstract:** In order to study how to promote the collaborative innovation of green building projects of construction enterprises, this paper establishes an evolutionary game model of suppliers' participation in collaborative innovation of green building projects of construction enterprises under government governance. Through a numerical simulation, our research analyzes the influence of government tax preference, government infrastructure construction, and environmental pollution punishment on the behavior of the government, suppliers, and construction enterprises. The empirical research shows that the government's tax incentives will continue to encourage suppliers to choose collaborative innovation. If the government's short-term tax incentives are small, construction enterprises will evolve in the direction of midway betrayal. When the government's long-term tax incentives are large, construction enterprises will evolve in the direction of the collaborative innovation of green building projects. Furthermore, the government's infrastructure support for suppliers to participate in collaborative innovation of green building projects of construction enterprises will encourage suppliers and construction enterprises to choose green building projects for collaborative innovation. With the continuous maturity of green building projects in the construction market, the government has evolved from positive governance to negative governance. The government will take the opportunity to give up infrastructure construction and turn to other supporting policies. Lastly, a low intensity of environmental pollution punishment makes it difficult to promote the construction enterprises to evolve in the direction of collaborative innovation. A moderate intensity of environmental pollution punishment can encourage construction enterprises to evolve in the direction of collaborative innovation, while high-intensity environmental pollution punishment can encourage construction enterprises to choose the direction of midway betrayal. Environmental pollution punishment has no significant impact on suppliers' selection of collaborative innovation of green building projects in the short term.

**Keywords:** green building projects; construction enterprises; collaborative innovation; evolutionary game

## 1. Introduction

As one of the pillar industries of China's economic development, the construction industry plays a very important role in promoting the rapid development of China's economy and society and the people's quality of life. However, the rapid expansion of China's construction industry has led to a large amount of energy consumption, as well as a large amount of construction waste and carbon emissions; accordingly, the green innovation of construction enterprises has attracted widespread attention to achieve sustainable development and reduce environmental pollution as much as possible [1,2]. In this way, the environmental pollution of buildings has been reduced as much as possible. Green innovation can not only reduce the impact of production activities on the environment, but also play a key role in creating a more sustainable economic and social value. Therefore, it is urgent to actively carry out green innovation activities of construction enterprises and develop corresponding green new products, so as to realize the rational use of resources, as well as sustainable development and growth, effectively improve the ecological environment, promote sustainable development of enterprises, and enhance the competitive advantage of green innovation [3,4].

Because the development of green technology innovation of Chinese construction enterprises is restricted by many factors, most Chinese construction enterprises choose to give up green technology innovation. As a result, the development of green technology innovation in China lags behind that of developed countries [5]. To reverse this situation, construction enterprises gradually carry out green innovation through collaborative cooperation. In the construction product innovation activities, the green building project innovation of construction enterprises is a complex innovation behavior, involving many subjects, including the government, other enterprises, and universities. It is necessary to cooperate with each other [5–7]. Construction enterprises are the main body of innovation, while the government is the corresponding promoter. With the support of government policies, construction enterprises cooperate closely with suppliers. The willingness of cooperation between construction enterprises and suppliers is increasing [8], whereby supplier participation in green project innovation of construction enterprises enables construction enterprises to solve the problem of insufficient knowledge, technology, and other related resources in the innovation process by consulting with suppliers to design products and jointly determine green raw material standards and design specifications, so as to carry out the collaborative innovation of green building projects among enterprises [9]. The participation of suppliers in green project innovation of construction enterprises can not only effectively reduce the innovation risk and cost but also accelerate the industrialization process of construction products [10]. Suppliers' participation in green project collaborative innovation has been widely recognized by construction enterprises. However, there are still many problems in the collaborative innovation system, such as a lack of collaborative innovation motivation, imperfect collaborative innovation infrastructure, and opportunistic behavior of suppliers (leakage of core knowledge, violation of resource allocation commitment, and violation of collaborative innovation norms).

The role of the government in the market is as a regulator, participant, and policymaker. The role of the government in the collaborative innovation system of green building projects of construction enterprises is to promote innovation or formulate government regulations [11]. It is necessary to formulate strict environmental laws and regulations to encourage innovation, support and encourage enterprises to invest in pollution-free technologies, upgrade corresponding equipment, and improve technologies [12,13], so as to stimulate enterprises to better carry out collaborative innovation of green building projects. As the top-level design and coordination elements, how to create a good policy environment, how to promote collaborative innovation of green building projects of construction enterprises, and how to enhance the enthusiasm of suppliers to participate in collaborative innovation of green building projects of construction enterprises have become urgent problems to be solved. This study analyzes the evolution process of the collaborative innovation of green building projects with respect to the equilibrium strategy of construction enterprises in different scenarios, as well as determines the government behavior factors and strategy selections that affect this system equilibrium.

The structure of this paper is as follows: Section 2 reviews the previous research and identifies the existing research gaps. Section 3 introduces the evolutionary game model of green building project collaborative innovation of construction enterprises, including its basic assumptions, model construction, and dynamic replication equation. Section 4 uses MATLAB to apply a numerical simulation to the construction enterprise green project coordinated green innovation game. Section 5 draws the corresponding conclusions and discusses the analysis through the evolutionary game simulation.

## 2. Literature Review

In the construction supply chain, the supply chain of integrated green building materials has become an important element for collaborative green innovation of construction enterprises [14]. It has been found that the partner choice of green technology innovation is an important factor affecting the green innovation among enterprises in the supply chain of integrated green building materials. Both internal and external green orientation of construction enterprises will promote green cooperation and innovation between enterprises and suppliers, as well as between enterprises and customers [15]. Suppliers participate in the collaborative innovation activities of green building projects, and enterprises fully acquire and utilize the professional knowledge and technology of suppliers. It is an enterprise that integrates external suppliers' knowledge, while expanding and updating its own knowledge base [16,17]. Colin, applying resource-based theory, knowledge-based theory, and capability theory, discussed the diversified sustainable development orientation of green entrants and the key factors affecting green innovation performance [9]. Other studies discussed the role of suppliers in successfully implementing green innovation capability in enterprise product development [18,19].

Olanipekun et al. believed that incentive policy is the key means to promote the development of green building market [20]. From the perspective of green building developers and consumers, Rehn and Ade introduced government subsidies, analyzed the dynamic evolution path of developers and consumers, and proposed incentive policies [21]. Zhao et al. established the corresponding evolutionary game model by studying the subsidies to enterprises in the green supply chain and giving appropriate subsidies to consumers when making green production decisions [22]. The study found that not only does direct financial support have an impact on the green technology innovation of small and medium-sized enterprises, but the construction of the regional environment is also very important for small and medium-sized enterprises [23]. Chang et al. pointed out that the government's incentive policies can promote the application of renewable energy in construction and infrastructure construction [24]. Environmental regulation has a positive impact on the green technology innovation of construction enterprises [25].

Considering a management system composed of the government and two competing enterprises which are producing green products, a game model was constructed to seek the optimal subsidy amount of the government [26]. The subsidies had a positive effect on green product innovation, whereas voluntary agreement had no effect on green product innovation, and tax laws and regulations on green product innovation produced a negative influence. This was mainly due to reduced financial input resources available for product innovation [27]. Daddi et al. (2010) proved through empirical analysis that the impact of environmental regulation on enterprise technological innovation is constantly changing with the adjustment of time and external factors, and there is no definite positive or negative impact in the long term [28]. As the external environment manager of the country, the government will be passively affected by the collaborative innovation behavior of enterprises, becoming a stakeholder of green collaborative innovation, and its environmental regulation means will indirectly affect the green innovation behavior of enterprises [29]. Cohen et al. (2017) believed that the government should provide incentive policies for builders and buyers, which would not bring too much cost to government departments [30]. Government grants and financial support are considered key to promoting the development of green building products, and government regulation, government procurement, and research and development are important components of the green building industry [31]. Green building infrastructure construction projects are an important governance mechanism for the rapid development of the green building material industry, and green innovation subsidies are a core governance mechanism for high-quality development. Pollution punishment and fraud compensation punishment for green building materials enterprises, green innovation subsidies for green building developers tax incentives, and purchase subsidies for consumers are conducive to the transformation of green production and consumption concepts. Infrastructure construction is conducive to promoting building through the development and purchase of green building materials [32].

Enterprises in the construction supply chain strive to adopt advanced science and technology to promote collaborative innovation. The collaborative innovation cooperation of green building projects promotes the research and development of new products, new technologies, new methods, etc. In the collaborative innovation process, the construction supply chain partners provide corresponding technology and resource support [1]. Enterprises can improve efficiency in the new knowledge developed by collaborative innovation activities, as well as launch innovative products faster with less research and development costs. Participating in collaborative innovation activities is one of the important behaviors [33] under the situation of open innovation, whereby more and more construction enterprises are crossing industrial fields, breaking regional restrictions, and building collaborative innovation networks that conform to their own development and system. Construction enterprises are project-driven enterprises, and all kinds of business activities of enterprises are carried out around engineering projects. There is coordination and interaction between construction enterprises and their stakeholders at every moment in the process from the initiation of engineering projects to the delivery to the construction party. The innovation of small project-based companies is usually closely related to their business activities and is driven by owners who use very scarce resources to make progress in normal business gaps [34]. Technological progress is the main means to drive the innovation of construction projects [35]. The motivation for a single enterprise to adopt innovation in construction projects is driven by the exchange and interaction between the relevant subjects involved in innovation [36]. By interacting with other enterprises in the supply chain of the construction industry, enterprises can find new opportunities and acquire new knowledge, which is also necessary knowledge to solve important problems and difficulties in construction projects [37]. By participating in collaborative innovation activities, knowledge integration and reorganization can be achieved through mutual exchange, and mutual cooperation can further increase collaborative innovation among participants in green building projects as an effective way to break through the obstacles of building development. It is difficult for a single enterprise to innovate independently and complete the project tasks. It is necessary to form a construction industry alliance with other enterprises.

This paper has several theoretical and practical contributions. Firstly, because the research on government behavior is still in the theoretical stage, not well combined with practice, and since there is less research on the dynamic change of government policy in government behavior, this paper further enriches the research on government regulation behavior and explores the dynamic change mechanism of government policy in government behavior. Second, although some scholars have studied the evolutionary behavior of green technology innovation of construction enterprises, there is no research on the evolutionary behavior of suppliers taking part in green project collaborative innovation of construction enterprises based on government behavior. This paper extends the micro-research on the impact of government behavior on suppliers' participation in collaborative innovation of green building projects of construction enterprises. Third, the evolutionary game theory can clearly describe the decision-making process and learning behavior of participants in the green project collaborative innovation of construction enterprises, reveal their dynamic evolution process, and analyze and predict the group behavior of the participants from the perspective of bounded rationality. Therefore, this paper uses evolutionary game theory to describe the decision-making process of green building project collaborative innovation by the government, supplier, and construction enterprise.

## 3. The Model

### 3.1. Analysis of the Interests of the Three Parties

Green project innovation cooperation among construction enterprises is mainly based on the actual needs of engineering cooperation projects [38]. In the collaborative innovation of green building projects of construction enterprises, construction enterprises and suppliers maintain active and effective communication. Both parties communicate product design information and product plans through corresponding channels, to realize product innovation. Through their own market capacity, suppliers will gradually guide enterprises to change toward the direction of green project innovation. Construction enterprises carry out green project innovation, produce an innovation effect, and gradually expand. The government attaches great importance to the collaborative innovation of green building projects of suppliers and construction enterprises. The government gives corresponding subsidies or preferential treatment to suppliers and construction enterprises' green project innovation, so as to gradually promote the diffusion of green innovative products. The government integrates environmental protection into the development of suppliers and construction enterprises. Because of the environmental pollution, the government has taken measures to punish the environmental pollution, gradually hindering the circulation of traditional nonenvironmental protection products on the market. Suppliers and construction enterprises have to change their product development means and carry out collaborative innovation of green projects. The government's governance can guide suppliers and construction enterprises to regulate their own behavior and obtain corresponding benefits.

### *3.2. Construction of Evolutionary Game Model for Collaborative Innovation of Green Building Projects in Construction Enterprises*

**Assumption 1.** *In the natural environment, without considering other constraints, while considering the overall system of the government, construction enterprises, and suppliers from the perspective of suppliers' participation in collaborative innovation of green building projects of construction enterprises, the participants are bounded rationally. In the process of collaborative innovation of green building projects, the participants have limited information, thus exhibiting incomplete symmetry [39–41].*

**Assumption 2.** *It is difficult to realize the collaborative innovation of green building projects solely relying on the market economy system, and the government needs to correct the market mechanism. The government's infrastructure construction can effectively offset the shortage of funds in the collaborative innovation of green building projects and build a green innovation platform. The government stimulates the development and production of green products of construction enterprises by improving the tax preferential policy system. When construction enterprises introduce new green technologies, it is conducive to the promotion and application of green technology innovation. However, if construction enterprises withdraw from the collaborative innovation of green building projects and fail to adopt green technologies, corresponding environmental pollution problems will emerge, and corresponding punishments will be applied. In order to better explore the role of the government, the government's role in the collaborative innovation of green building projects of construction enterprises is divided into tax incentives, government infrastructure construction, and environmental pollution punishment* $\alpha, \beta, \gamma$. *The cost of government consumption is denoted by* $\alpha S, \beta I, \gamma P$. *The collaborative innovation of green building projects of construction enterprises as a function of government governance can lead to innovation. The government's tax preferences for collaborative innovation of suppliers and construction enterprises are* $\alpha S_1, \alpha S_2$, *respectively. When the supplier and the construction enterprise are betrayed in the middle of the project, they are punished by the government for environmental pollution* $(\gamma P_1, \gamma P_2)$.

**Assumption 3.** *When suppliers participate in collaborative innovation of green building projects of construction enterprises, suppliers and construction enterprises are in a game for the collaborative innovation of green projects. When both suppliers and construction enterprises choose to quit halfway, the corresponding costs of suppliers and construction enterprises are* $C_{g1}$ *and* $C_{z1}$, *respectively. The profits obtained by both parties are* $R_g$ *and* $R_z$. *When suppliers and construction enterprises carry out collaborative innovation of green building projects according to certain rules, corresponding green technologies are adopted for green project development, while corresponding R&D and new equipment are introduced. The corresponding costs of suppliers and construction enterprises are* $C_{g2}$ *and* $C_{z2}$. *The profit is higher than the original* $\Delta R_1$ *and* $\Delta R_2$. *The government's revenue is* $R$. *The corresponding loss is* $L$. *In the collaborative innovation of green building projects between* *suppliers and construction enterprises, when the government takes the initiative to govern, the government supervises the collaborative innovation of green building projects between suppliers and construction enterprises to avoid the occurrence of intellectual property disputes and opportunism. At this time, the profits of suppliers increase ($\mu\Delta R_1'$). The income of construction enterprises also increases ($(1-\mu)\Delta R_1'$). When the government carries out passive governance, the government formulates corresponding policies to encourage the supplier and the construction enterprise to establish a good cooperative relationship, and the supplier's income decreases ($\mu\Delta R_2'$), while the profits of construction enterprises decrease ($(1-\mu)\Delta R_2'$).*

**Assumption 4.** *Due to the supplier dependence, there is mutual dependence between the supplier and the construction enterprise. However, once only one party carries out the green project collaborative innovation, the supplier and the construction enterprise will choose the strategy of midway betrayal for some reasons. Both sides agree that, when one party betrays halfway and chooses to quit, the other party's income from the other party is D. To achieve the goal of collaborative innovation in the green building project, if one party betrays the green building project, whereby the betrayer gives up the green building project and turns to the nongreen building project, this is punishable due to the corresponding environmental pollution.*

**Assumption 5.** *The participants in the game model include suppliers, construction enterprises, and the government. Each participant has two behavior choices, i.e., construction enterprises and suppliers choose collaborative innovation or midway betrayal, and the government chooses active or negative governance. Here, $x, y, z$ represent the selection probabilities of the corresponding strategies, and $x, y, z \in [0,1]$ is a function of time. In the process of learning and imitation, the three sides of the game adjust their own strategies through trial and error or choice, in order to find a better strategy, until a balance is reached.*

### 3.3. The Dynamic Replication Equation of Evolutionary Game Subjects

According to the above assumptions, the evolutionary game income payment matrix of suppliers ($x$), construction enterprises ($y$), and governments ($z$) is constructed in Tables 1–5.

**Table 1.** The tripartite game payment matrix of suppliers' participation in collaborative innovation of green building projects of construction enterprises ($x$).

| | Active Governance by Government (z) | | |
|---|---|---|---|
| | **Supplier Payments** | **Construction Enterprises Payments** | **Government Payments** |
| Collaborative innovation of construction enterprises ($y$) | $R_g + \Delta R_1 + \mu\Delta R_1' + \alpha S_1 - C_{g2}$ | $R_z + \Delta R_2 + (1-\mu)\Delta R_1' + \alpha S_2 - C_{z2}$ | $R - \alpha(S_1 + S_2) - \beta I$ |
| Betrayal of construction enterprises ($1-y$) | $R_g + \Delta R_1 + \alpha S_1 - C_{g2} - D$ | $R_Z - \gamma P_2 - C_{Z2} + D$ | $\gamma P_2 - \alpha S_1 - L$ |

**Table 2.** The tripartite game payment matrix of suppliers' participation in collaborative innovation of green building projects of construction enterprises ($x$).

| | Passive Governance by Government ($1-z$) | | |
|---|---|---|---|
| | **Supplier Payments** | **Construction Enterprises Payments** | **Government Payments** |
| Collaborative innovation of construction enterprises ($y$) | $R_g + \Delta R_1 - C_{g2}$ | $R_z + \Delta R_2 - C_{z2}$ | $R$ |
| Betrayal of construction enterprises ($1-y$) | $R_g + \Delta R_1 - C_{g2} - D$ | $R_Z - \gamma P_2 - C_{Z2} + D$ | $\gamma P_2 - L$ |

**Table 3.** The tripartite game payment matrix of supplier betrayal $(1-x)$.

| | Active Governance by Government $(z)$ | | |
|---|---|---|---|
| | **Supplier Payments** | **Construction Enterprises Payments** | **Government Payments** |
| Collaborative innovation of construction enterprises $(y)$ | $R_g - C_{g2} - \gamma P_1 - \alpha S_1 + D$ | $R_z + \Delta R_2 + \alpha S_2 - C_{z2} - D$ | $R - \alpha S_2 + \alpha S_1 + \gamma P_1$ |
| Betrayal of construction enterprises $(1-y)$ | $R_g - C_{g2} - \gamma P_1 - \alpha S_1$ | $R_Z - C_{Z2} - \gamma P_2 - \alpha S_2$ | $\gamma(P_1 + P_2) + \alpha(S_1 + S_2) - L$ |

**Table 4.** The tripartite game payment matrix of supplier betrayal $(1-x)$.

| | Passive Governance by Government $(1-z)$ | | |
|---|---|---|---|
| | **Supplier Payments** | **Construction Enterprises Payments** | **Government Payments** |
| Collaborative innovation of construction enterprises $(y)$ | $R_g - C_{g2} - \gamma P_1 + D$ | $R_z + \Delta R_2 - C_{z2} - D$ | $R + \gamma P_1$ |
| Betrayal of construction enterprises $1-y$ | $R_g - C_{g2} - \gamma P_1 - \mu\Delta R_2'$ | $R_Z - C_{Z2} - \gamma P_2 - (1-\mu)\Delta R_2'$ | $\gamma(P_1 + P_2) - L$ |

**Table 5.** Jacobian matrix.

| Equilibrium Point | Characteristic Value | Characteristic Value | Characteristic Value |
|---|---|---|---|
| $J_1(0,0,0)$ | $\Delta R_1 - D + \gamma P_1 + \mu\Delta R_2'$ | $\Delta R_2 - D + \gamma P_2 + (1-\mu)\Delta R_2'$ | 0 |
| $J_1(0,0,1)$ | $2\alpha S_1 + \Delta R_1 - D + \gamma P_1 + \mu\Delta R_2' - \mu\Delta R_2'$ | $(1-\mu)\Delta R_1' + \Delta R_z - D + \gamma P_2$ | 0 |
| $J_1(0,1,0)$ | $\Delta R_1 - D + \gamma P_1 + \mu\Delta R_2' - \mu\Delta R_2'$ | $-[(1-\mu)\Delta R_1' + \Delta R_z - D + \gamma P_2 + (1-\mu)\Delta R_2']$ | $R - \alpha S_2$ |
| $J_1(1,0,0)$ | $-(\Delta R_1 - D + \gamma P_1 + \mu\Delta R_2')$ | $[(1-\mu)\Delta R_1' + \Delta R_z - D + \gamma P_2 + (1-\mu)(1-x)\Delta R_2']$ | $-\alpha S_1$ |
| $J_1(0,1,1)$ | $\mu\Delta R_1' + 2\alpha S_1 + C_{g2} + \Delta R_1 - D + \gamma P_1$ | $-[(1-\mu)\Delta R_1' + \Delta R_z - D + \gamma P_2]$ | $-(R - \alpha S_2)$ |
| $J_1(1,0,1)$ | $-(2\alpha S_1 + \Delta R_1 - D + \gamma P_1)$ | $(1-\mu)\Delta R_1' + \Delta R_z - D - \alpha S_2 + \gamma P_2$ | $\alpha S_1$ |
| $J_1(1,1,0)$ | $-(\Delta R_1 - D + \gamma P_1)$ | $-[(1-\mu)\Delta R_1' + \Delta R_z - D + \gamma P_2]$ | $-\beta I + R - \alpha S_2$ |
| $J_1(1,1,1)$ | $-(\mu\Delta R_1' + 2\alpha S_1 + C_{g2} + \Delta R_1 - D + \gamma P_1)$ | $-[(1-\mu)\Delta R_1' + \Delta R_z - D - \alpha S_2 + \gamma P_2]$ | $\beta I - R + \alpha S_2$ |

Expected average revenue of suppliers:

$$E_1 = E_{11}x + E_{12}(1-x) \quad (1)$$

The benefits of collaborative innovation of green building projects selected by suppliers:

$$E_{11} = yz(R_g + \Delta R_1 + \mu\Delta R_1' + \alpha S_1 - C_{g2}) + z(1-y)(R_g + \Delta R_1 + \alpha S_1 - C_{g2} - D) + y(1-z)(R_g + \Delta R_1 - C_{g2}) + (1-z)(1-y)(R_g + \Delta R_1 - C_{g2} - D) \quad (2)$$

The benefits when suppliers choose green building projects to betray halfway:

$$E_{12} = zy(R_g - C_{g2} - \gamma P_1 - \alpha S_1 + D) + z(1-y)(R_g - C_{g2} - \gamma P_1 - \alpha S_1) + y(1-z)(R_g - C_{g2} - \gamma P_1 + D) + (1-y)(1-z)(R_g - C_{g2} - \gamma P_1 - \mu\Delta R_2') \quad (3)$$

The supplier's dynamic replication equation:

$$U(x) = x(1-x)(yz\mu\Delta R_1' + 2\alpha z S_1 + yzC_{g2} + \Delta R_1 - D + \gamma P_1 + \mu\Delta R_2' - z\mu\Delta R_2' - y\mu\Delta R_2' + zy\mu\Delta R_2') \quad (4)$$

Expected average income of construction enterprises:

$$E_2 = E_{21}y + E_{22}(1-y) \quad (5)$$

The benefits of collaborative innovation of green building projects selected by construction enterprises:

$$E_{21} = xz[R_z + \Delta R_2 + (1-\mu)\Delta R_1' + \alpha S_2 - C_{z2}] + x(1-z)(R_z + \Delta R_2 - C_{z2}) + z(1-x)(R_z + \Delta R_2 + \alpha S_2 - C_{z2} - D) + (1-x)(1-z)(R_z + \Delta R_2 - C_{z2} - D) \quad (6)$$

The benefits when construction enterprises choose to betray green building projects halfway:

$$E_{22} = xz(R_Z - \gamma P_2 - C_{Z2} + D) + x(1-z)(R_Z - \gamma P_2 - C_{Z2} + D) + z(1-x)(R_Z - C_{Z2} - \gamma P_2 - \alpha S_2) + (1-x)(1-z)[R_Z - C_{Z2} - \gamma P_2 - (1-\mu)\Delta R_2'] \quad (7)$$

The construction enterprise's dynamic replication equation:

$$U(y) = y(1-y)[(1-\mu)\Delta R_1' + \Delta R_z - D - \alpha xzS_2 + \gamma P_2 + (1-\mu)(1-x-z+xz)\Delta R_2'] \quad (8)$$

Expected average revenue of government:

$$E_3 = E_{31}z + E_{32}(1-z) \quad (9)$$

The benefits of government's active governance strategy:

$$E31 = xy[R - \alpha(S_1 + S_2) - \beta I] + x(1-y)(\gamma P_2 - \alpha S_1 - L) + (1-x)y(R - \alpha S_2 + \gamma P_1) + (1-x)(1-y)[\gamma(P_1 + P_2) - L] \quad (10)$$

The benefits of government's passive governance strategy:

$$E32 = xyR + x(1-y)(\gamma P_2 - L) + (1-x)y(R + \gamma P_1) + (1-x)(1-y)[\gamma(P_1 + P_2) - L] \quad (11)$$

The government's dynamic replication equation:

$$U(Z) = z(1-z)(\alpha xyS_1 - \beta Ixy - \alpha xS_1 + yR - \alpha yS_2) \quad (12)$$

### 3.4. Stability Analysis of the Three Parties Involved in the Game

Equilibrium refers to the realization of the maximum utility of each party in the game, i.e., the realization of each party's satisfaction with the game result, such that the actual utility and satisfaction degree of each party are different. In the evolution of a stable strategy, both sides adopt an optimal response strategy. Thus, an evolutionary game model of suppliers, construction enterprises, and the government can be established to analyze the stability of the three parties, with the following equilibrium points:

$$\begin{cases} U_1'(x) = 0 \\ U_2'(y) = 0 \\ U_3'(z) = 0 \end{cases} \quad (13)$$

According to Equation (13), in terms of income, $\{(x, y, z) \mid 0 \le x \le 1, 0 \le y \le 1, 0 \le z \le 1\}$, eight equilibrium points are obtained through calculation and analysis.

In order to simplify the formula, let

$$\begin{cases} T_1(y,z) = yz\mu\Delta R_1' + 2\alpha zS_1 + yzC_{g2} + \Delta R_1 - D + \gamma P_1 + \mu\Delta R_2' - z\mu\Delta R_2' - y\mu\Delta R_2' + zy\mu\Delta R_2' \\ T_2(x,z) = (1-\mu)\Delta R_1' + \Delta R_z - D - \alpha xzS_2 + \gamma P_2 + (1-\mu)(1-x-z+xz)\Delta R_2' \\ T_3(x,y) = \alpha xyS_1 - \beta Ixy - \alpha xS_1 + yR - \alpha yS_2 \end{cases} \quad (14)$$

In the above equilibrium point, the following equation is satisfied:

$$\begin{cases} T_1(y,z) = 0 \\ T_2(x,z) = 0 \\ T_3(x,y) = 0 \end{cases} \quad (15)$$

According to the evolutionary game theory, when $U_1'(x) < 0$, $U_2'(y) < 0$, and $U_3'(z) < 0$, Equation (13) is the stable strategy (ESS) of the tripartite game among suppliers, construction enterprises, and the government.

$$\begin{cases} U_1'(x) = (1-2x)T_1(y,z) \\ U_2'(y) = (1-2y)T_2(x,z) \\ U_3'(z) = (1-2z)T_3(x,y) \end{cases} \tag{16}$$

For suppliers, when $T_1(y,z) > 0$, $U_1'(0) > 0$, and $U_1'(1) < 0$, the supplier chooses the green building project collaborative innovation as a stable state. On the contrary, midway betrayal is a stable state. Because $x \in (0,1)$, $U_1'(x) > 0$, the evolutionary phase diagram of a supplier choosing to participate in green building project collaborative innovation depends on the curve shape of Equation (16).

For construction enterprises, if $T_2(x,z) > 0$, then $U_2'(0) > 0$, $U_2'(1) < 0$, indicating that construction enterprises choose green building project collaborative innovation as a stable state. On the contrary, construction enterprises choose to betray as a stable state. Because $x \in (0,1)$, $U_2'(y) > 0$, the evolution phase diagram of the stability of collaborative innovation of green building projects selected by construction enterprises depends on the curve form of Equation (16).

For the government, the dividing line determining a stable state is $T_3(x,y) = 0$. If $T_3(x,y) > 0$, then $U_3'(0) > 0$, and $U_3'(1) < 0$, revealing that the stable state of the government evolutionary game actively governs the suppliers' participation in the collaborative innovation development of green building projects of construction enterprises. On the contrary, the government's passive governance is a stable state. The evolution phase diagram of government stability depends on the curve form of Equation (16).

According to Lyapunov stability theory, the asymptotic stability of the system at the equilibrium point can be judged by analyzing the eigenvalues of the system Jacobian matrix. The Jacobian matrix of the system is obtained as follows:

$$J = \begin{vmatrix} \frac{du(x)}{dx} & \frac{du(x)}{dy} & \frac{du(x)}{dz} \\ \frac{du(y)}{dx} & \frac{du(y)}{dy} & \frac{du(y)}{dz} \\ \frac{du(z)}{dx} & \frac{du(z)}{dy} & \frac{du(z)}{dz} \end{vmatrix}$$

$$= \begin{vmatrix} U_1'(x) & x(1-x)(z\mu\Delta R_1' + zC_{g2} - \mu\Delta R_2' + y\mu\Delta R_2') & x(1-x)(y\mu\Delta R_1' + 2\alpha S_1 + yC_{g2} - \mu\Delta R_2' + y\mu\Delta R_2') \\ y(1-y)[-\alpha z S_2 + (1-\mu)(-1+x)\Delta R_2'] & U_2'(y) & y(1-y)[-\alpha x S_2 + (1-\mu)(-1+x)\Delta R_2'] \\ z(1-z)(\alpha y S_1 - \beta I y - \alpha S_1) & z(1-z)(\alpha x S_1 - \beta I x + R - \alpha S_2) & U_3'(z) \end{vmatrix} \tag{17}$$

In view of the fact that the asymptotic stable solution of the multigroup evolutionary game dynamic replication system must be a strict Nash equilibrium solution, the evolutionary system has eight pure strategy equilibrium points, which are $E_1(0,0,0)$, $E_2(0,0,1)$, $E_3(0,1,0)$, $E_4(1,0,0)$, $E_5(0,1,1)$, $E_6(1,0,1)$, $E_7(1,1,0)$, and $E_8(1,1,1)$. These eight can be substituted into the Jacobian matrix to solve the matrix eigenvalues. The results are shown in Table 5.

According to the evolutionary game theory, when $Det(J) > 0, Tr(J)_i < 0$, an evolutionary stable state will be formed, i.e., the equilibrium point will approach the local gradual stable state. At that time, it is in an unstable state, i.e., the equilibrium point will not approach the local gradual stable state. When $Det(J) > 0, Tr(J)_i = 0$ or is uncertain, the equilibrium point is the saddle point, i.e., the equilibrium point is stable in one direction and unstable in the other direction.

## 4. Numerical Simulations

According to the literature, in terms of green building project collaborative innovation between suppliers and construction enterprises, the data are limited, and the statistical data used are difficult to obtain. According to the basic paradigm of research on simulation parameters in the literature [42–44], this paper consulted experts who study simulations in supply chain-related fields and carried out a simulation combined with the actual research scope. By reading domestic and foreign studies and using the mature interview design method, the corresponding universities, architectural research institutes, construction industries, and their suppliers were selected as the research objects of the interview. The main information included the personal information of the interviewees and the survey topics related to this study. Among the interviewees, 81.02% were male and 18.98% were female; 31% of the interviewees had a doctor's degree, 20.21% had a master's degree, and 17.84% had a bachelor's degree or below; 21.56% had worked for 5 years or less, 32.64% had worked for more than 5 years but less than 10 years, and 45.80% had worked for 10 years or more. The main questions asked during the interview were as follows: What are the influencing factors of suppliers' participation in collaborative innovation of green production films of construction enterprises? What role does the government play in the collaborative innovation of green building projects between suppliers and construction enterprises? What is the impact of government tax incentives, government infrastructure construction, and environmental pollution punishment on the collaborative innovation of green building projects between suppliers and construction enterprises?

In the process of suppliers' participation in the collaborative innovation of green building projects of construction enterprises, the dynamic evolution behavior of the three parties in the evolutionary game model involving the government, suppliers, and construction enterprises can be more truly described. In combination with the dynamic replication equations of the three parties involved in the game, the evolutionary trajectory of can be established, which can be processed using the discrete method.

Using the timestep $\Delta t$, the following equations can be derived:

$$\frac{dx(t)}{dt} \approx \frac{x(t+\Delta t) - x(t)}{\Delta t} \tag{18}$$

$$\frac{dy(t)}{dt} \approx \frac{y(t+\Delta t) - y(t)}{\Delta t} \tag{19}$$

$$\frac{dz(t)}{dt} \approx \frac{z(t+\Delta t) - z(t)}{\Delta t} \tag{20}$$

According to Equations (18)–(20), the evolution trajectory of the three players participating in the game was simulated using MATLAB. In order to more truly reflect the evolution status of the players, the values of other parameters were determined according to the actual situation of green building project collaborative innovation between suppliers and construction enterprises, and the corresponding strategies of the players were set to 0.5. A numerical simulation was used to analyze the dynamic evolution of the three parties in the game.

### *4.1. The Influence of Tax Preference α on Suppliers'Participation in the Evolutionary Game of Collaborative Innovation of Green Building Projects in Construction Enterprises*

In order to make the simulation more objective, tax preferences were set as $\alpha = 0.2, 0.5$, or 0.8; the corresponding evolutionary game simulation results are shown in Figure 1.

The government's tax incentives will continue to encourage suppliers to choose collaborative innovation. If the government's short-term tax incentives are small, construction enterprises will evolve in the direction of midway betrayal. When the government's long-term tax incentives are large, construction enterprises will evolve in the direction of collaborative innovation of green building projects, with the construction enterprises having greater enthusiasm for collaborative innovation of green building projects. The short-term preferential tax policy has no obvious stimulating effect on the level of green building project collaborative innovation of construction enterprises, mainly because the construction enterprises are not willing to make a large short-term investment in innovation. Driven by the effect of long-term preferential tax policies of the government, construction enterprises will continue to evolve toward the direction of collaborative innovation of green building projects, because they enjoy the dividend of preferential tax policies. In the collaborative innovation of green construction projects, in order to stimulate suppliers and construction enterprises to better carry out collaborative innovation of green projects, the government can formulate corresponding incentive policies such as early tax incentives, thereby actively promoting suppliers and construction enterprises to evolve toward the collaborative innovation of green building projects. In the early stage of the policy, the government can implement a large number of tax incentives or even reduce the corresponding tax, such as the implementation of 3–5 years of preferential tax relief. With the continuous tightening of the preferential tax policies of the government and the construction enterprises, the corresponding level of innovation will gradually improve. When green building projects in the construction industry become stable, while the green building project technologies of suppliers and construction enterprises are in a mature stage, the government should cancel the corresponding preferential tax policies.

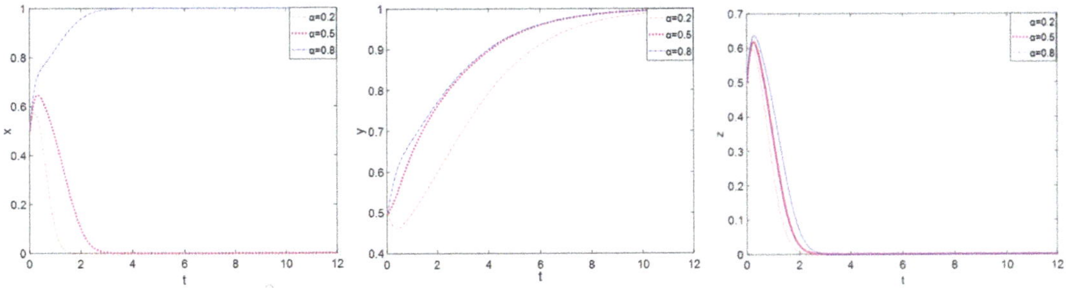

**Figure 1.** The influence of tax preference on the behavior of suppliers of collaborative innovation in green building projects of construction enterprises according to evolutionary game.

### *4.2. The Impact of Government Infrastructure Construction β on Suppliers' Participation in the Evolutionary Game of Collaborative Innovation of Green Building Projects in Construction Enterprises*

In order to make the simulation more objective, the government infrastructure construction was set as $\beta = 0.2, 0.5$, or $0.8$; the corresponding evolutionary game simulation results are shown in Figure 2.

The government's infrastructure support for suppliers to participate in collaborative innovation of green building projects of construction enterprises will encourage suppliers and construction enterprises to choose green building projects for collaborative innovation. With the continuous maturity of green building projects in the construction market, the government will take the opportunity to give up infrastructure construction and turn to other supporting policies. In order to support suppliers and construction enterprises to carry out collaborative innovation of green building projects, under different infrastructure construction efforts, the government can evolve from positive promotion to negative promotion. In order to stimulate the collaborative innovation and development of green building projects of suppliers and construction enterprises, the government can continuously increase the corresponding infrastructure construction for green building project innovation. For example, a green development foundation of the construction industry and a green collaborative innovation cooperation platform can be established. The government will continue to strengthen the construction of collaborative innovation infrastructure, so as to improve the resource supply in the process of collaborative innovation and enhance the performance of collaborative innovation. When the green building project innovation of suppliers and construction enterprises develops to a certain extent, the government should weaken or even reduce the corresponding infrastructure construction until completely withdrawing from the corresponding infrastructure construction.

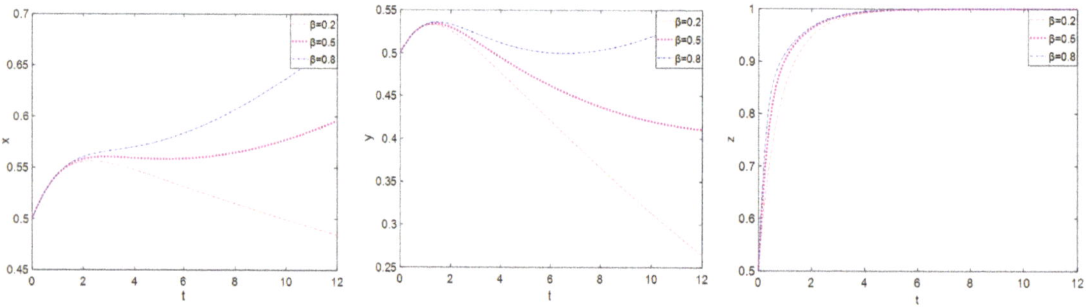


**Figure 2.** The impact of government infrastructure construction on the behavior of suppliers of collaborative innovation in green building projects of construction enterprises according to evolutionary game.

### 4.3. The Influence of Environmental Pollution Punishment on Supplier's Participation in Evolutionary Game Behavior of Green Building Project Collaborative Innovation of Construction Enterprises

In order to make the simulation more objective, the environmental pollution penalty was set to $\gamma = 0.2$, 0.5, or 0.8; the corresponding evolutionary game simulation results are shown in Figure 3.

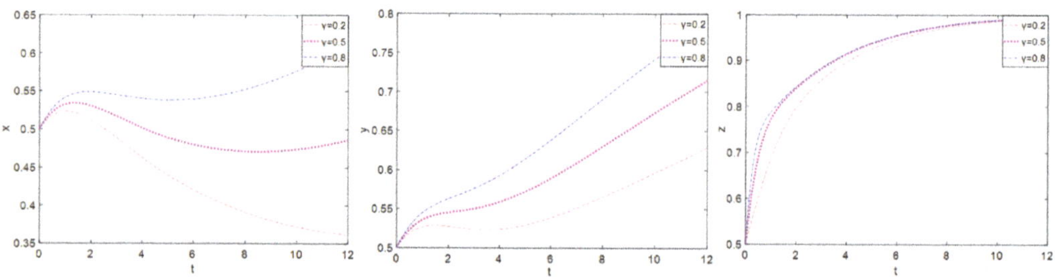


**Figure 3.** The impact of environmental pollution punishment on the behavior of suppliers of collaborative innovation in green building projects of construction enterprises according to evolutionary game.

A low-intensity environmental pollution punishment can hardly encourage construction enterprises to evolve toward collaborative innovation, whereas a medium-intensity environmental pollution punishment can encourage construction enterprises to evolve toward collaborative innovation, and a high-intensity environmental pollution punishment can encourage construction enterprises to choose the direction of midway betrayal. The environmental pollution punishment has no significant impact on suppliers' selection of green building project collaborative innovation in the short term. In the construction market, the means of environmental pollution punishment are through the collection of pollution discharge fees, environmental administrative penalties, financial expenditures for environmental protection, and the promulgation of environmental laws and regulations. The low-intensity environmental pollution punishment has no significant impact on the collaborative innovation of green building projects of construction enterprises. With the medium-intensity environmental pollution punishment, construction enterprises will continue to improve their R&D level and promote their green building project innovation. A high-intensity environmental pollution punishment prevents the construction enterprises from conducting green building project innovation. The high intensity is too much for the construction enterprises to bear, which affects the enthusiasm of the construction enterprises for green technology innovation. In the short term, environmental pollution punishment has no significant impact on suppliers' choice of green building project collaborative innovation.

## 5. Conclusions and Implications

Following early investigations based on limited rational game theory, this study constructed an evolutionary game model of suppliers' participation in green building project collaborative innovation of construction enterprises under government governance, before running a simulation using MATLAB according to the actual situation. In the development process of green building projects in the construction industry, the government plays the role of promoting collaborative innovation. Through government guidance, a mutual benefit can be established among the government, construction enterprises, and suppliers, thus achieving the innovation and cooperation goal of green building projects in the construction industry. The results show that the government's tax preference will continue to encourage suppliers to choose collaborative innovation. If the government's short-term tax preference is small, the construction enterprises will evolve toward midway betrayal. If the government's long-term tax preference is large, the construction enterprises will evolve toward collaborative innovation of green projects, and the construction enterprises have a high enthusiasm for collaborative innovation of green building projects. Furthermore, the government's support for suppliers' participation in the infrastructure construction of green building project collaborative innovation of construction enterprises will encourage suppliers and construction enterprises to choose green building project collaborative innovation. With the continuous maturity of green building projects in the construction market, the government can evolve from active governance to passive governance. The government can take the opportunity to give up infrastructure construction and turn to other supporting policies. Lastly, a low-intensity environmental pollution punishment can hardly encourage construction enterprises to evolve toward collaborative innovation, whereas a medium-intensity environmental pollution punishment can encourage construction enterprises to evolve toward collaborative innovation, and a high-intensity environmental pollution punishment can encourage construction enterprises to choose the direction of midway betrayal. Environmental pollution punishment has no significant impact on suppliers' selection of green building project collaborative innovation in the short term.

In order to improve the collaborative innovation research and development of green building projects in the construction market, countermeasures should be put forward. First, the strategic height of green building development should be further enhanced, preferential tax policies should be reasonably formulated, and the steady development of the green building market should be promoted. Second, the government can build an efficient sharing and transfer platform for green building technology, promote the rapid transformation of green building scientific and technological achievements, and maximize the benefits of the integration and allocation of green project resources in the construction industry. Third, a government governance mechanism that combines punishment with incentives should be implemented to thoroughly establish a green development strategy.

As the collaborative innovation of green building projects in construction enterprises is a complex behavior involving the interests of many subjects, several aspects will be studied in the future. First, the strategy selection mechanism of various game players under the influence of government subsidies, cost sharing, and other factors will be discussed, while deeply analyzing the driving factors of collaborative innovation of green building projects in construction enterprises. Second, the risk preference behavior of different subjects will be introduced, and the strategy choice of decision-makers under different risk preferences will be studied to deeply analyze the potential risk factors of collaborative innovation of green building projects in the construction market. Third, factors such as civic initiatives, driven by demands and energy prices, will be considered.

**Author Contributions:** Methodology, validation, and formal analysis, X.Z., S.L. and Z.X.; investigation and writing—original draft, S.Y. and X.Z.; visualization and resources, S.Y. and Z.X.; conceptualization, supervision, and project administration, S.Y. and Z.X.; writing—review and editing, S.L. and S.Y., S.Y. and Z.X. contributed equally to this work. All authors have read and agreed to the published version of the manuscript.

**Funding:** This research was funded by the Top Young Talents Scientific Research Project of Higher Education in Hebei Province (grant number BJ2021084), the Scientific Research Foundation for the Talents of Hebei Agricultural University (grant number YJ2020017), the Hebei Agricultural Industry System Project (HBCT2021230301), the 2021 First Class Discipline Climbing Engineering Project (GY21161320019), and the Zhejiang University of Technology Research Start-Up Project (2021132006129).

**Institutional Review Board Statement:** Not applicable.

**Informed Consent Statement:** Not applicable.

**Data Availability Statement:** Not applicable.

**Conflicts of Interest:** The authors declare no conflict of interest.

## References

1. Li, X.; Li, J.; Huang, Y.; He, J.; Liu, X.; Dai, J.; Shen, Q. Construction enterprises' adoption of green development behaviors: An agent-based modeling approach. *Humanit. Soc. Sci. Commun.* **2022**, *9*, 1–11. [CrossRef]
2. Yin, S.; Dong, T.; Li, B.; Gao, S. Developing a Conceptual Partner Selection Framework: Digital Green Innovation Management of Prefabricated Construction Enterprises for Sustainable Urban Development. *Buildings* **2022**, *12*, 721. [CrossRef]
3. Chang, C.H. The influence of corporate environmental ethics on competitive advantage: The mediation role of green innovation. *J. Bus. Ethics* **2011**, *104*, 361–370. [CrossRef]
4. Waqas, M.; Honggang, X.; Ahmad, N.; Khan, S.A.R.; Iqbal, M. Big data analytics as a roadmap towards green innovation, competitive advantage and environmental performance. *J. Clean. Prod.* **2021**, *323*, 128998. [CrossRef]
5. Yin, S.; Zhang, N.; Ullah, K.; Gao, S. Enhancing Digital Innovation for the Sustainable Transformation of Manufacturing Industry: A Pressure-State-Response System Framework to Perceptions of Digital Green Innovation and Its Performance for Green and Intelligent Manufacturing. *Systems* **2022**, *10*, 72. [CrossRef]
6. Jain, M.; Siva, V.; Hoppe, T.; Bressers, H. Assessing governance of low energy green building innovation in the building sector: Insights from Singapore and Delhi. *Energy Policy* **2020**, *145*, 111752. [CrossRef]
7. Yin, S.; Yu, Y. An adoption-implementation framework of digital green knowledge to improve the performance of digital green innovation practices for industry 5.0. *J. Clean. Prod.* **2022**, *363*, 132608. [CrossRef]
8. Yu, X.; Lan, Y.; Zhao, R. Strategic green technology innovation in a two-stage alliance: Vertical collaboration or co-development? *Omega* **2019**, *98*, 102116. [CrossRef]
9. Cheng, C.C.J. Sustainability orientation, green supplier involvement, and green innovation performance: Evidence from diversifying green entrants. *J. Bus. Ethics* **2020**, *161*, 393–414. [CrossRef]
10. Lee, V.H.; Ooi, K.B.; Chong, A.Y.L.; Lin, B.A. structural analysis of greening the supplier, environmental performance and competitive advantage. *Prod. Plan. Control.* **2015**, *26*, 116–130. [CrossRef]
11. Beerepoot, M.; Beerepoot, N. Government regulation as an impetus for innovation: Evidence from energy performance regulation in the Dutch residential building sector. *Energy Policy* **2007**, *35*, 4812–4825. [CrossRef]
12. Eiadat, Y.; Kelly, A.; Roche, F.; Roche, F.; Eyadat, H. Green and competitive? An empirical test of the mediating role of environmental innovation strategy. *J. World Bus.* **2008**, *43*, 131–145.
13. Qi, G.Y.; Shen, L.Y.; Zeng, S.X.; Jorge, O.J. The drivers for contractors' green innovation: An industry perspective. *J. Clean. Prod.* **2010**, *18*, 1358–1365. [CrossRef]
14. Yevu, S.K.; Ann, T.W.; Darko, A. Digitalization of construction supply chain and procurement in the built environment: Emerging technologies and opportunities for sustainable processes. *J. Clean. Prod.* **2021**, *322*, 129093. [CrossRef]
15. Cellini, R.; Lambertini, L.; Mantovani, A. Persuasive advertising under Bertrand competition: A differential game. *Oper. Res. Lett.* **2008**, *36*, 381–384. [CrossRef]
16. Zhou, K.Z.; Li, C.B. How knowledge affects radical innovation: Knowledge base, market knowledge acquisition, and internal knowledge sharing. *Strateg. Manag. J.* **2012**, *33*, 1090–1102. [CrossRef]
17. Huo, B.; Haq, M.Z.U.; Gu, M. The impact of information sharing on supply chain learning and flexibility performance. *Int. J. Prod. Res.* **2021**, *59*, 1411–1434. [CrossRef]
18. Andersén, J. A relational natural-resource-based view on product innovation: The influence of green product innovation and green suppliers on differentiation advantage in small manufacturing firms. *Technovation* **2021**, *104*, 102254. [CrossRef]
19. AL Shaar, I.M.; Khattab, S.; Alkaied, R.; Al-Abbadi, L. Supply chain integration and green innovation, the role of environmental uncertainty: Evidence from Jordan. *Uncertain Supply Chain Manag.* **2022**, *10*, 657–666. [CrossRef]

20. Olubunmi, O.A.; Xia, P.B.; Skitmore, M. Green building incentives: A review. *Renew. Sustain. Energy Rev.* **2016**, *59*, 1611–1621. [CrossRef]
21. Rehm, M.; Ade, R. Construction costs comparison between 'green' and conventional office buildings. *Build. Res. Inf.* **2013**, *41*, 198–208. [CrossRef]
22. Zhao, S.; Zhu, Q.; Cui, L. A decision-making model for remanufacturers: Considering both consumers' environmental preference and the government subsidy policy. *Resour. Conserv. Recycl.* **2018**, *128*, 176–186. [CrossRef]
23. Kaufmann, A.; Tödtling, F. How effective is innovation support for SMEs? An analysis of the region of Upper Austria. *Technovation* **2002**, *22*, 147–159. [CrossRef]
24. Chang, R.; Soebarto, V.; Zhao, Z.; Zillante, G. Facilitating the transition to sustainable construction: China's policies. *J. Clean. Prod.* **2016**, *131*, 534–544. [CrossRef]
25. Li, X.; He, J.; Huang, Y.; Li, J.; Liu, X.; Dai, J. Predicting the factors influencing construction enterprises' adoption of green development behaviors using artificial neural network. *Humanit. Soc. Sci. Commun.* **2022**, *9*, 1–12. [CrossRef]
26. Yang, R.; Tang, W.; Zhang, J. Technology improvement strategy for green products under competition: The role of government subsidy. *Eur. J. Oper. Res.* **2021**, *289*, 553–568. [CrossRef]
27. Stucki, T.; Woerter, M.; Arvanitis, S.; Peneder, M.; Rammer, C. How different policy instruments affect green product innovation: A differentiated perspective. *Energy Policy* **2018**, *114*, 245–261. [CrossRef]
28. Daddi, T.; Testa, F.; Iraldo, F. A cluster-based approach as an effective way to implement the Environmental Compliance Assistance Programme: Evidence from some good practices. *Local Environ.* **2010**, *15*, 73–82. [CrossRef]
29. Hoen, K.M.R.; Tan, T.; Fransoo, J.C. Effect of Carbon Emission Regulations on Transport Mode Selection under Demand. *Flex. Serv. Manuf. J.* **2014**, *26*, 170–195. [CrossRef]
30. Cohen, C.; Pearlmutter, D.; Schwartz, M. A game theory-based assessment of the implementation of green building in Israel. *Build. Environ.* **2017**, *125*, 122–128. [CrossRef]
31. Xing, Z.; Cao, X. Promoting Strategy of Chinese Green Building Industry: An Evolutionary Analysis Based on the Social Network Theory. *IEEE Access* **2019**, *7*, 67213–67221. [CrossRef]
32. Yin, S.; Li, B.; Xing, Z. The governance mechanism of the building material industry (BMI) intransformation to green BMI: The perspective of green building. *Sci. Total Environ.* **2019**, *677*, 19–33. [CrossRef] [PubMed]
33. Lawson, B.; Krause, D.; Potter, A. Improving supplier new product development performance: The role of supplier development. *J. Prod. Innov. Manag.* **2015**, *32*, 777–792. [CrossRef]
34. Barrett, P.; Sexton, M. Innovation in small, project-based construction firms. *Br. J. Manag.* **2006**, *17*, 331–346. [CrossRef]
35. Tatum, C.B. Potential mechanisms for construction innovation. *J. Constr. Eng. Manag.* **1986**, *112*, 178–191. [CrossRef]
36. Dulaimi, M.F.; Ling, F.Y.Y.; Bajracharya, A. Organizational motivation and inter-organizational interaction in construction innovation in Singapore. *Constr. Manag. Econ.* **2003**, *21*, 307–318. [CrossRef]
37. Cruz-Gonzalez, J.; Lopez-Saez, P.; Navas-López, J.E. Absorbing knowledge from supply-chain, industry and science: The distinct moderating role of formal liaison devices on new product development and novelty. *Ind. Mark. Manag.* **2015**, *47*, 75–85. [CrossRef]
38. Zan, A.; Yao, Y.; Chen, H. University–Industry Collaborative Innovation Evolutionary Game and Simulation Research: The Agent Coupling and Group Size View. *IEEE Trans. Eng. Manag.* **2019**, *68*, 1406–1417. [CrossRef]
39. Ma, L.; Zhang, L. Evolutionary game analysis of construction waste recycling management in China. *Resour. Conserv. Recycl.* **2020**, *161*, 104863. [CrossRef]
40. Sun, H.; Wan, Y.; Zhang, L.; Zhou, Z. Evolutionary game of the green investment in a two-echelon supply chain under a government subsidy mechanism. *J. Clean. Prod.* **2019**, *235*, 1315–1326. [CrossRef]
41. Liu, Y.; Zuo, J.; Pan, M.; Ge, Q.; Chang, R.; Feng, X.; Fu, Y.; Dong, N. The incentive mechanism and decision-making behavior in the green building supply market: A tripartite evolutionary game analysis. *Build. Environ.* **2022**, *214*, 108903. [CrossRef]
42. Wu, W.; Sheng, L.; Tang, F.; Zhang, A.; Liu, J. A system dynamics model of green innovation and policy simulation with an application in Chinese manufacturing industry. *Sustain. Prod. Consum.* **2021**, *28*, 987–1005. [CrossRef]
43. Huang, Z.; Liao, G.; Li, Z. Loaning scale and government subsidy for promoting green innovation. *Technol. Forecast. Soc. Chang.* **2019**, *144*, 148–156. [CrossRef]
44. Lai, Z.; Lou, G.; Zhang, T.; Fan, T. Financing and coordination strategies for a manufacturer with limited operating and green innovation capital: Bank credit financing versus supplier green investment. *Ann. Oper. Res.* **2021**, 1–35. [CrossRef]

buildings

MDPI

Article

# Green Home Buying Intention of Malaysian Millennials: An Extension of Theory of Planned Behaviour

**Mohammad Masukujjaman [1], Cheng-Kun Wang [2], Syed Shah Alam [1,*], Chieh-Yu Lin [3,*], Yi-Hui Ho [3] and Abu Bakkar Siddik [4]**

1 Graduate School of Business, Universiti Kebangsaan Malaysia, UKM, Bangi 43600, Selangor, Malaysia
2 College of Management, Chang Jung Christian University, Tainan 711301, Taiwan
3 Department of International Business, Chang Jung Christian University, Tainan 711301, Taiwan
4 School of Management, University of Science and Technology of China (USTC), Hefei 230026, China
* Correspondence: shahalam@ukm.edu.my (S.S.A.); jylin@mail.cjcu.edu.tw (C.-Y.L.)

**Abstract:** The empirical study aimed to examine the factors affecting green home buying intention in Malaysia. The study adopted the popular theory of individual decision-making, the theory of planned behaviour, but extended it with a few new variables to get better predictability. Based on the cross-sectional survey method and quota sampling technique, the instrument-like questionnaire was distributed to the potential buyers who had the intention to buy a home. Data were collected from 741 potential buyers who are young Malaysian students. Collected data were analysed using partial least squares (PLS) modelling (SmartPLS 3.0) in order to measure the strength of the independent variable effects on the dependent variable. The results indicate that attitude, perceived behavioural control subjective norms, perceived value, reasonable pricing and green building knowledge had a positive and significant effect on green home buying intention, with attitude showing the highest effect; the only exception is an environmental concern. This study also validated the theory of planned behaviour (TPB), which provides important insight for the firms involved with the housing development industry for their future business strategy.

**Keywords:** theory of planned behaviour; green home buying intention; Millennials; Malaysia

## 1. Introduction

Environmental degradation is an issue driven by rapid economic growth. Due to natural disasters, climate change, waste generation, destruction of the ecosystem and air pollution, the world's environment has changed dramatically, which affects people's economic status and their livelihoods. The Malaysian economy is the largest in Southeast Asia; rapid industrial development is resulting in environmental degradation through over-utilization and over-consumption of natural resources. In such a situation, consumers are aware of their environment. To reduce the impact on the environment, consumers can make a decision to choose the right product or move to sustainable consumption [1]. Thus, academicians, government institutions and private regulatory bodies have taken serious consideration of the environmental problem and its adverse impact on human health issues [2,3]. Because the housing sector accounts for approximately 70 percent of all structures and because it is anticipated that annual residential energy consumption would increase by 1.1% between the years 2008 and 2035, it is of utmost importance to promote environmentally friendly housing on a worldwide scale [4,5].

There has been a rise in both interest and demand for eco-friendly dwellings during the past two decades [6]. In response to the global call, Malaysia has embraced the idea of green home design. Although the prospect of development in this sector is high, the growth is still in the early stage in Malaysia [7]. Housing developers feel that the demand for green homes in Malaysia is very minimal. There are very few high-end green buildings in Malaysia [8]. Researchers such as Young et al. [9] stated that although consumers express their concern towards the environment, it does not transform into green buying behaviour. Other researchers [10,11] have argued that even though consumers express a favourable attitude to green products but it does not mean the consumers buy green products. It is not obvious, as of yet, which factors have a substantial impact on the customer's intention to make a purchase, nor what the connection is between these factors and the consumer's desire to buy green home (GH) products [12]. Given the need to get people to choose more environmentally friendly housing, more research is needed to find out how potential homebuyers feel about sustainability certifications for homes and what makes them want or not want to buy a home with a sustainability certification [12]. These studies have indicated research gaps on the green behaviour of consumers in buying homes.

Currently, most studies on the psychological factors contributing to pro-environmental behaviours have concentrated on relatively low-investment behaviours such as like reducing overall consumption, increasing recycling, or selecting 'green' alternatives in the context of organic food, energy-efficient appliances, and green mobile phones. sustainable lodgings, green banking or finance, green supply chain, hybrid vehicles, and power companies (e.g., [3,11,13–20]). Many studies [4,12,21–27] have investigated the impact of environmental considerations on consumers' choices regarding one of the largest lifetime purchases they will ever make (the purchase of a dwelling). Some academics have investigated the factors that influence people's decisions to buy eco-friendly homes [23,28,29]. There have been previous attempts to anticipate what features and qualities will appeal to buyers and influence their decisions about environmentally friendly homes [24]. However, research on green home-buying in Malaysia is rare. The only study found, given the limitation of the researchers was conducted by Tan et al. [30] on green residential buying intention. The study is limited as they did not use any guiding theory; there are some constructs which, however, proved insignificant against the global norms. Thus, it is important to retest the study in other areas in Malaysia and justify the findings again.

The previous studies on green building (GB) consumption focused on general consumers as the primary population of interest [29,31,32]. To the best of our knowledge, very little study was undertaken on the unique home-buying customer group, with young people's perspectives and values being especially ignored. Zhang et al. [23] conducted their home-buying intention study on the young generation, while Kota et al. [21] explored Generation Z's perceptions of green homes in the USA perspectives. Youth in Malaysia's cohort makes up 43 percent of the total population and they are the main target market for the green movement, and (still under research) specifically attitudes toward environmental issues [33]. On the other hand, young people have a strong desire for housing that meets their needs. Thus, it is likely that they will become possible buyers in the housing industry. To date, however, there has been no study conducted on the topic of GH purchase intent among young consumers in Malaysia.

The theory of planned behaviour, the values–beliefs–norms theory, the technology acceptance model, and the norm activation model are just some of the theories and models that are frequently used to investigate the perceptions, preferences, motivations, and purchase intentions of end-users of sustainable housing [23,25,28,34,35]. In order to analyse the data, this study bases its theoretical decisions on the idea of planned behaviour [36]. Sustainable residential housing often uses TPB [7,37,38]. Green consumer behaviour is complicated and affected by many things, especially when it comes to big-ticket items like houses and cars [12]. Moreover, judging a young mind is harder because there are more things to think about. So, TPB may not be enough to fully explain the use of GH purchase intention as a behaviour that helps the environment [39,40]. Although researchers have confirmed that TPB is one of the important models in behavioural research, many other researchers have included additional constructs such as price [41,42], environmental concern [1,3], perceived value [43,44], and environmental knowledge [1,3]. Moreover, Ajzen [36] suggested that TPB can be extended if a further important variable is identified. This argument allows researchers to add other important constructs to extend the TPB model. Therefore, price, environmental concern, and perceived value are included in this research as additional constructs. Previous research has not adequately defined "knowledge" in the context of GHs or their function in determining purchase intention. Existing studies used environmental knowledge [23,25] as a predictor of green home-buying intention. None of the studies adopted green building knowledge covering the level of knowledge a buyer holds about the factual, conceptual and technical aspects of green homes instead of overall environmental knowledge. Moreover, the dimension of reasonable pricing had not been included in any research to predict green home buying intention so far. Thus, new research is required which will explain GH buying decisions to combine an expanded view of TPB with other components. Therefore, the following research questions are to be answered:

RQ1: What are the predictors of green home buying intention for Millennial consumers?
RQ2: To what extent does the extended TPB validate green home buying in the Malaysian context?

To address the gaps and research questions above, the present study aims to investigate the factors responsible for the prediction of green home buying intentions of Malaysian Millennials empirically. The main novelty of this paper is as follows. Firstly, it provides an extension of the TPB model in the case of eco-homes buying intention, providing both the cognitive and context-specific factors. It introduces a new cognitive construct "green building knowledge" with a new scale customized to the green building context. It also integrates a few other constructs such as environmental concern, perceived value and reasonable price. Secondly, most of the previous studies focused on green home purchasing failed to consider young consumers. The research is conducted using the response of Millennials which will help the academic with a new understanding of a particular segment of the consumer in diffusing green homes. Understanding what people want to buy can help fill in this research gap. Moreover, for green homes to work, it would be important for policymakers to come up with good policy strategies. Thirdly, the study is contextualized in Malaysia where this particular type of study is rare. Hence, the study's results can offer policymakers and other groups in developing countries some useful examples and lessons learned. Particularly, this will help us to figure out how different types and levels of policies affect the prices, values, and level of awareness of green housing on the market. It will give us a new way to look at how government policies promote green housing and make sure it is of good quality.

The remaining sections of this work are structured as follows. In Section 2, related hypotheses and a research framework are presented. Section 3 then introduces the research methodologies. The results and data analysis are presented in Section 4. In Section 5, we explore the results and explain their practical relevance and consequences. The final portion offers the study's conclusion and limitations.

## 2. Literature Review and Hypotheses Development

### 2.1. Green Homes

Green building (GB), which saves energy, water, land, and materials while protecting the environment, is an effective way to cut carbon emissions, promote sustainable development, and improve quality of life [45,46]. Darko et al. [6] and Zheng et al. [23] show that "green buildings" (GBs) are buildings that take into account and try to minimize their effects on the natural environment and human health while also boosting their positive effects over a building's entire life cycle. There are a lot of different GB standards all over the world, and most of them agree that GBs should try to reduce negative effects on the environment, society, and economy [23]. As a branch of GB, green home (GH) meets all of the general requirements of GBs and puts a strong focus on the health, comfort, and safety of residents in the context of a place to live. It is an architectural idea for meeting modern development needs. It does not require a certain type of housing, and it does not make a difference between regions [47]. It is called having a smaller "ecological footprint" because it uses less energy at the beginning and throughout its life, makes less waste or treats some of it, and improves the quality of the indoor environment to make sure people are healthy and productive at work [48]. Table 1 shows a summary of some empirical research on green home buying intention.

**Table 1.** Empirical research on the intention of buying green homes.

| Sources | Research Method/Sample Size/Country | Analysis Tools | Research Model | Significant Factors |
|---|---|---|---|---|
| [4] | Empirical/ Questionnaire survey/Construction participants/180/China | Logistic regression Model-SPSS | TPB | Environmental awareness, green home comfort, neighbors'/friends' assessment and government incentive |
| [12] | Empirical/Online survey/residents/330/Australia | SEM (AMOS) | TPB | Perceived behavioural control, green consumer identity, subjective norms, and attitudes |
| [21] | Empirical/Student Interview/304/ USA | Multinomial logistic regression | Dual-Inheritance Theory and Normative Motivation | Intrinsic, instrumental, non-normative motivation, barriers |
| [22] | Empirical/Questionnaire survey/208/ China | SEM (AMOS) | TPB | Attitude, perceived behavioural control and subjective norms |
| [23] | Empirical/ Online survey/Young consumers/241/China | SEM (AMOS) | Extended TPB | Subjective norms, attitude, subjective knowledge, environmental concern, governmental initiatives |
| [24] | Empirical/Online survey/Urban residents/728/China | PLS-SEM | No existing model used | Perceived value, perceived benefits, environmental concern, perceived risk |
| [25] | Empirical/Face-to-face interview/Resident of Dhaka/319/Bangladesh | SEM (AMOS) | TPB | Environmental knowledge, environmental concern, subjective norms, perceived behavioural control, attitude |
| [30] | Empirical/Questionnaire survey/304/ Malaysia | PLS-SEM | No existing model used | Attitude, moral obligation, environmental concern, perceived value, perceived self-identity, financial risk |

Note: PLS = Partial Least Squares; AMOS = Analysis of Moment Structures, TPB = Theory of Planned Behaviour, SEM = Structural Equation Modeling, SPSS = Statistical Package for Social Sciences.

Green building reduces the negative impact of human activities and protects the natural environment. Green building provides safety, comfort, and healthy living to humans. According to Tan [8], one of the main aims for green and sustainable homes is to improve healthy living. Tan [7] stated that houses become 'green' if housing developer uses recycled roof systems, certified energy-efficient appliances, recyclable timber products, recyclable kitchen cabinets, less energy-consuming fluorescent lamps and light-emitting diode lighting systems. Nowadays green products are introduced due to the increasing concern about local and global pollution, overflowing wastes, global warming and diminishing natural reserves [49]. Green consumers are those who are more interested in and aware of environmental problems. In essence, it is important to create awareness and a shared sense of responsibility to the environment, so that everybody enjoys an environmentally sustainable lifestyle. Thus, in this research, we have tried to identify the factors those influence green and sustainable home-buying intention.

The research framework used in this study is shown in Figure 1. There are 13 hypotheses, drawn from eight constructs, namely, attitude towards buying a green and sustainable home, attitude, perceived behavioural control, subjective norm, green building knowledge, environmental concern, perceived value, reasonable price and buying intention for a green home.

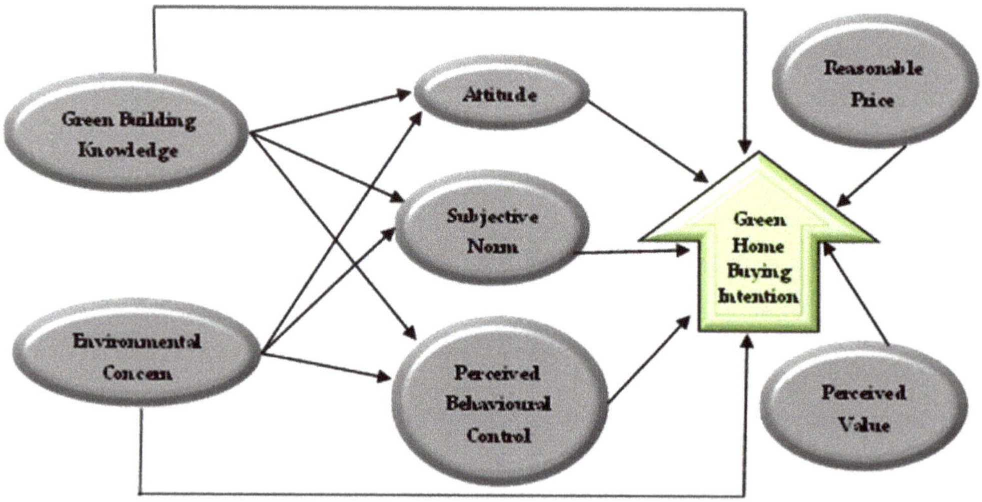


**Figure 1.** Proposed research framework.

### 2.2. *The Component of the Extended TPB Model*

Ajzen came up with the TPB model in 1991. The theory of planned behaviour (TPB) [36,50] is an extension of the theory of reasoned action (TRA) [51]. Researchers have used it for the past 20 years and found that it can predict a wide range of intentions and actions. Regardless of strong support by the research for the TPB model, there remains a proportion of unaccounted variance. Adding additional variables with strong theoretical justification would be able to explain unique variance in intention [36].

Young consumers are different from other groups in a number of ways, particularly in Malaysia. Most of them come from families with only one child. Their parents usually help them buy a home, which can greatly improve their ability to make purchases [52]. Furthermore, getting a college degree means that they may care more about the environment. To make sustainable incentive strategies [53], it is important to find out what young consumers' plans are for buying green homes since they are the future consumers and the future of society [54]. These young people are also the ones who may come up with new ideas for the sustainability market, such as green hotel stays [14]. There may be some other specific predictors of intention to achieve a green and sustainable home context. Thus, in this study, some other variables such as perceived value, environmental concern, environmental knowledge and reasonable price are included and extended to the TPB model, as they may affect behavioural intention.

### 2.3. *Attitude*

Attitude is the appraisal of engaging in a certain behaviour involving the object of the attitude, such as purchasing a product [55]. Attitude refers to the extent to which a person has a positive or negative view or assessment of the behaviour in question. In general, the more favourable an individual's attitude, the stronger their intention to engage in the conduct under discussion. Based on the expectancy-value model, an individual's attitude toward action is defined by the complete set of accessible behavioural beliefs tying the conduct to a variety of outcomes and other characteristics. In light of this, Ajzen [56] asserts that attitude might be regarded as a significant factor in predicting and explaining human behaviour.

Recently, it has been demonstrated that attitude includes perceived consequences, which are related to conduct [57]. Wu et al. [22] argued that mindset is the most significant predictor of the propensity to purchase a green home. Chen and Tung [58] believed that consumers' behavioural intentions would be favourable if their perceived attitude was positive. According to the studies cited, the psychological mood is transmitted through consumers' ratings. Other research, such as the study by Han and Yoon [59], which focused on ecologically friendly beverage packaging, indicated a favourable link between attitude and behavioural intent. In the context of the green home, it has been observed that consumers' attitudes have a substantial impact on their intentions to behave in certain ways [4,35]. Likewise, Judge et al. [12] and Zahan et al. [25] explored the same in their empirical study on green home-buying intention.

Thus, the following hypothesis needs to be substantiated:

**H1.** *Attitude has a strong and positive association with intention when it comes to the buying of green homes.*

### 2.4. Subjective Norm (SN)

In the TPB, the idea of "subjective norm" is seen as the second "conceptually independent" factor that affects a person's decision to do or not do something. It refers to "the perceived social influence to do or not do the behaviour" ([36], p. 188). People think of a subjective norm as a social factor made up of normative beliefs [60]. The predictor social factor called "subjective norm" is the perceived social influence to conform to expectations about the behaviour and how this behaviour should affect the person's decision to do it or not [61].

Several studies [62,63] in the field of consumer behaviour and marketing have found that subjective norm is the most important factor in predicting how people will act. In these studies, a positive connection between SN and intention was found. Wu et al. [22] also found that there is a link between SN and the intention to buy a green home. SN is a necessary part of the intention to buy a greenhouse [35], but Zahan et al. [25] found no connection between the two. Zhang et al. [4] and Judge et al. [12] also found a strong link between subjective norms and the intention to buy green housing in China. In this study, a consumer's perception of social pressure is a subjective norm that affects their decision to buy a green and sustainable home. That means the higher the external influence or peer recommendations in favour of environmental protection, the higher the consumers' intention to buy green homes. So, for this situation, we suggest:

**H2.** *There is a significant and positive relationship between subjective norms and intentions to purchase green and sustainable homes.*

### 2.5. Perceived Behavioural Control (PBC)

PBC is the third and most essential factor influencing intentions (PBC). Ajzen [36] defines PBC as the extent to which an individual feels capable of engaging in the behaviour. It includes two components: how much control a person has over the behaviour and how confident they are in their ability to perform or not perform the activity. It is determined by the individual's perceptions of the influence of both situational and internal elements on the performance of the behaviour. In this research, perceived behavioural control denotes how much control the respondents have even if the purchase appears as slightly expensive or people discourage buying or there are issues in terms of affordability and operating skills for a planned purchase of green buildings.

Studies have demonstrated a favourable and significant association between PBC and recycling [64], waste separation behaviour [65], and the intention to consume green items [1,66]. Wu et al. [22] discovered a favourable connection between PBC and the intention to purchase a green dwelling. Judge et al. [12] and Zahan et al. [25] similarly discovered a substantial relationship between PBC and the intention to purchase a green home; however, Zhang et al. [4] and Tan et al. [30] discovered an insignificant relationship. In light of the preceding, we can conclude that a person's likelihood of purchasing halal food increases in proportion to the degree to which he or she believes he or she has control over the decision. In this study, PBC is the ability to purchase sustainable and environmentally friendly dwellings. Consequently, the hypothesis is:

**H3.** *PBC and intention to buy green homes are positively linked.*

### 2.6. Environmental Concern (EC)

Crosby et al. define environmental concern as "a strong attitude toward protecting the environment" [67]. Environmental concern is one of the most important cognitive constructs for predicting green products [68]. Research by Yadav and Pathak [69], suggests that a positive outlook on environmentally friendly practices might have a positive impact on the world around us. In an empirical assessment of Indian clients, Jaiswal and Kant [68] revealed that EC was significantly related to attitudes about green products In addition to this, a number of prior studies incorporated this cognitive component into the TPB paradigm. Combining environmental concern (EC) with an expanded TPB framework, Chen and Tung found that EC had a positive effect on the attitude, subjective norms (SN), and perceived behavioural control (PBC) of guests at eco-friendly hotels [58]. Similar results were also found in the study by Paul et al. [1], which found that EC was favourably associated with sustainable products and TPB characteristics.

Concern for the environment is the subject of numerous research studies on product selection, such as green purchase behaviour [70], electric vehicle usage intention [71], and organic food [72]. Karatu and Mat [73] and Suki [74] assert that if a person is environmentally conscious, they will have favourable sentiments toward environmentally friendly products. The research of Davis [75] identified a modest or negligible link between EC and behaviour. Consequently, based on the findings of Davis's study, the current study aims to investigate further the relationship between EC and buyers' propensity to purchase green homes.

It was discovered that environmental concern strongly influences consumers' attitudes towards green products and their purchasing intentions [1]. Environmentally conscious consumers with favourable opinions toward GH showed high level GH purchasing intentions [4]. Tan et al. [30] showed that environmental concern was significantly related to green home buying intention. Zahan et al. [25] explored whether environmental concern affects significantly attitude and PBC in the case of green buying intention, finding no relationship with behavioural intention in the context of Bangladesh. Hence, consumers' elevated concern for environmental degradation and its impact on mankind foster consumers' favourable attitude, endorsement, and self-control in decision-making and GH buying intentions. The following hypotheses are proposed:

**H4.** *EC has a favourable relationship with attitude.*

**H5.** *EC is correlated positively with SN.*

**H6.** *The relationship between EC and PBC is positive.*

**H7.** *EC is positively associated with the intention to purchase green homes.*

### 2.7. Green Building Knowledge (GBK)

In consumer research, it is acknowledged that knowledge influences all phases of the decision-making process [16]. In particular, knowledge is an important and relevant construct that influences how consumers collect and organize data, how much information is employed in decision-making, and how consumers assess products and services. Green building knowledge in the present study indicates the know-how of green building in terms of three domains of knowledge such as factual, conceptual and procedural knowledge [76]. The factual information about the eco-home design relates to building codes, wind turbines, recycled content materials, dual-flush toilets, rainwater harvesting, water-saving landscape features/plumbing fixtures, furniture made out of salvaged materials, etc., whereas conceptual knowledge indicates how to maintain indoor air quality, size and window locations, ecological impacts, etc. [76]. The procedural knowledge of green building features abilities such as picking eco-friendly furniture, and knowledge of monitoring cooling systems or solar panels or using water-saving building fixtures [76].

Researchers have argued that environmental knowledge has a significant impact on environmental issues [77]. Scholars [11] indicated that consumers' green product buying intention is influenced by environmental knowledge. Other researchers highlighted that consumers who have environmental knowledge will have a more positive attitude towards green products, further influencing their purchase intention [11]. Consumer choice is influenced by subjective awareness which in turn inspired them to act on the knowledge they have [78]. The researcher argued that people know something when others expect them to know about environmental issues [79]. Scholars [4,25] remarked that information also increases the idea that one is in control of a certain situation, hence enhancing perceived behavioural control.

**H8.** *GBK is positively linked to attitude.*

**H9.** *GBK is significantly linked to SN.*

**H10.** *GBK is significantly linked to PBC.*

**H11.** *GBK is significantly linked to green home purchase intention.*

### 2.8. Perceived Value (PV)

Perceived value (PV) is a key factor in deciding whether or not to buy something [80]. In the green home context, perceived value refers to additional environmental benefits that are ensured or environmental functionality supporting new values or environmental performance meeting the consumer expectation for using green homes. Numerous prior studies have demonstrated that perceived value is a distinguishing factor that affects purchasing intentions [81–83]. Chen and Chang [83] revealed that in the setting of an eco-friendly purchase, green perceived value positively influences green purchasing intentions. Moreover, Chen et al. [84] emphasized that perceived value is the primary determinant of purchase intent for hydrogen-electric vehicles. The research by Zhao and Chen [24] found that PV is significantly related to green home buying intention and Tan et al. [30] observed the same result from the perspective of Malaysia. Liu et al. [85] suggested that there is a link between the ecological value of a person's attitude toward staying in a green hotel and their sense of control over their behaviour, which makes them want to stay in a green hotel. Salehzadeh and Pool [86], Ahmed and Zhang [87], and Wei et al. [88] found that there is a strong link between green perceived value and the intention to buy green online. In the Value-Attitude-Behaviour Model, the authors suggested that perceived value influences purchasing intent via attitude [89]. When consumers find proper environmental value addition in buying green homes compared to traditional homes, they intend to buy them. Thus, the following hypothesis developed is:

**H12.** *PV has a positive relationship with green home buying intention.*

### 2.9. Reasonable Price (RP)

Price is an essential factor before the purchase of any product. It is more relevant for sustainable homes as consumers frequently remark that the price of eco-friendly products is higher than that of conventional products [90]. We know that using green fixtures, furniture, designs or maintaining self-sustaining energy and water system may entail extra costs. In this context, a reasonable price is not indicating a cheaper price but points out the price against the benefits (value) sought. It stresses whether the price of green homes is reasonable or not to the consumers on these grounds.

According to the Massachusetts Department of Environmental Protection [91], customers are extremely price-sensitive despite their general environmental concerns. Customers are reluctant to pay greater prices as a result [92]. Simmons [93] stated that ecologically concerned consumers may be willing to cut consumption levels of traditional products. Studies have demonstrated that pricing has a substantial effect on green purchasing intentions [94]. Lee et al. [95] commented that if a producer offers a reasonable price for a green product, buyers with a positive attitude toward environmental protection will have a strong intent to acquire green items. Although the price of sustainable homes is considerable, it may be claimed that buyers who care about the environment will purchase them if they can relate price to added value. This resulted in the subsequent hypothesis:

**H13.** *There is a correlation between RPand consumer intent to purchase green homes.*

## 3. Research Design

As stated previously, the primary aim of this study is to discover the factors affecting Millennials' intention to purchase green homes in Malaysia. To meet the aims of the study, an empirical survey was conducted. In the sections that follow, several methodological concerns pertaining to sample and technique, variable measurement, and assessing the reliability and validity of the factors are discussed.

### 3.1. Sample and Procedure

This study's data was collected using the primary data gathering method of a consumer survey delivered to highly educated individuals enrolled in at least one public university. The rationale for choosing university students is that the complexity of the green context is difficult for minors and illiterate persons to understand [96]. Other researchers also highlighted that less educated people find it difficult to understand the topic compared to those with higher education [97]. Thus, the ideal sample for this study consisted of adults aged over 18 years as we focused on Millennial people. In this study, a quota sampling technique was used and respondents chosen for this study were at least studying at the university level.

Before conducting the final survey, five academics from a Malaysian institution were interviewed in depth in order to pick the most important questions to include in the questionnaire. 800 respondents were contacted, but only 741 responses were deemed suitable for data analysis. The majority of rejected questionnaires were incomplete, with respondents failing to answer questions about the factors of interest. The majority of responders were female on average (59.92 percent). Chinese make up the second-highest proportion of contributors (after Malay, at 55.87 percent) with 34.95 percent.

### 3.2. Measures and Common Method Bias

All items used to measure variables in the present investigation were derived from previously validated instruments except the green building knowledge (Appendix A). The constructs of attitude, perceived behavioural control, and the behavioural intention was measured using scales developed by Alam and Sayuti [98] and Alam et al. [99]. The scales were updated to accommodate the buying of a green and sustainable home. Vermier and Verbeke's [100] subjective norm was chosen for our study. The concept of perceived value was taken from Rizwan et al. [101]. Items for a reasonable price were incorporated and changed from Masukujjaman et al. [102], Cole [76] and Maichum et al. [2]. The questionnaire included items measuring subjective norms (two), perceived behavioural control (three), attitude (five), perceived value (five), reasonable price (four), environmental concern (four), green building knowledge (three), and purchasing intention (three items). All questions employed a six-point Likert scale: 1 for "strongly disagree", 2 for "disagree", 3 for "slightly disagree", 4 for "somewhat agree", 5 for "agree", and 6 for "strongly agree".

Based on the suggested guidelines by [103] in this research, common method bias was tested by utilizing exploratory factor analysis. Kaiser–Meyer–Olkin (KMO) standard was adopted to evaluate the sampling adequacy for factor analysis. The analysis findings demonstrated that all values in the matrix's diagonal were above 0.5, while the KMO value was at 0.811. Moreover, the Kaiser–Guttman standard and the screen test were applied to distinguish the number of existing variables. The evaluation demonstrated that eight factors have more principal eigenvalues than one, which accounted for 68.331% of the variance. In contrast, the primary factor represented 27.2% of the variance in the factors. As per the factor analysis, more than one single factor showed up, and most of the variance was not represented by one general factor.

### *3.3. Data Analysis Methods*

This research employed Partial Least Squares (PLS) path modelling, also known as Partial Least Squares Structural Equation Modeling (PLS-SEM), to evaluate the hypothetical model. According to Ringle et al. [104], the PLS path modelling approach to investigating causal linkages in route models is a broad model implicitly assessed by multiple indicators. Both principal components factor analysis and PLS utilize a component-based methodology (Compeau et al. [105]). PLS has been utilized in marketing, strategic management, and management system research. Aibinu and Al-Lawati [106] used PLS-SEM for modeling the willingness of construction organizations to participate in e-bidding. Afzal et al. [107] conducted an empirical study on the factors of organizational flexibility in the construction industry. PLS analysis has also been used in a variety of other research on entrepreneurship [108]. Other HRM practices research [109] identified PLS as a significant analysis technique. However, the green home purchasing intention study is still in its infancy.

SmartPLS3.0 statistical software was utilized to compute the PLS route model discussed in this study, as suggested by Ringle et al. [110]. PLS model criteria ((i) evaluation of construct validity and (ii) evaluation of the path model) were calculated using a two-states analysis technique.

## 4. Analysis of Outcomes

### *4.1. Assessment of Construct Validity and Discriminant Validity*

To evaluate internal consistency, construct validity was evaluated. Using composite reliability scores (CR), Cronbach's alpha, and the average variance extracted (AVE) test, the construct validity of measured constructs was determined. The results of convergent and discriminant validity assessments are displayed in Table 2. Fornell and Larcker [111] propose that every structure's AVE should be greater than 0.5. All average variance extracted (AVE) values were more than 0.5, indicating that all constructs fulfilled the acceptable criteria for convergent validity [111,112]. In addition, the values of AVE in the diagonal for all constructs were greater than the squared correlation with other constructs off-diagonal, indicating that all constructs matched the criteria for discriminant validity [111].

**Table 2.** Correlation of latent variables and square roots of AVE.

| Variable | ATT | EC | GBK | BI | PBC | PV | RP | SN | CR | AVE | Alpha |
|---|---|---|---|---|---|---|---|---|---|---|---|
| ATT | **0.842** | | | | | | | | 0.924 | 0.709 | 0.898 |
| EC | 0.686 | **0.908** | | | | | | | 0.934 | 0.825 | 0.894 |
| GBK | 0.658 | 0.590 | **0.845** | | | | | | 0.881 | 0.715 | 0.796 |
| BI | 0.745 | 0.621 | 0.775 | **0.853** | | | | | 0.914 | 0.729 | 0.875 |
| PBC | 0.628 | 0.512 | 0.704 | 0.701 | **0.890** | | | | 0.919 | 0.792 | 0.868 |
| PV | 0.745 | 0.658 | 0.756 | 0.842 | 0.654 | **0.836** | | | 0.921 | 0.700 | 0.892 |
| RP | 0.663 | 0.596 | 0.655 | 0.685 | 0.652 | 0.681 | **0.866** | | 0.923 | 0.750 | 0.888 |
| SN | 0.320 | 0.322 | 0.314 | 0.411 | 0.367 | 0.331 | 0.304 | **0.943** | 0.942 | 0.890 | 0.877 |

Notes: In the table bold elements show the square root of AVE). ATT = Attitude, BI = Behavioural Intention, EC = Environmental concern, GBK = Green Building knowledge, PBC = Perceived Behavioural Control, PV = Perceived Value, SN = Subjective Norm, RP = Reasonable Price.

### 4.2. Reliability

In internal consistency measurement, reliability is the most frequently employed psychometric measurement for evaluating survey instruments and scales [113]. Composite reliability ratings (CR) and Cronbach's alpha are the most prevalent methods for assessing reliability. CR value can range from 0 to 1. According to Akter et al. [114], the composite dependability value must be greater than 0.7 for a model to be considered reliable. Alternatively, Nunnally stated that a minimum alpha of 0.6 suffices for the early phases of the study, and if the value exceeds 0.7, the data is regarded as highly acceptable [115]. The above discussion can be summed up by noting that the cutoff values for CR and Chronbach alpha are 0.7 and 0.7, respectively, for adequate construct validity. Table 1 demonstrates that the CR and Chronbach alpha values exceeded the corresponding cutoff values of 0.70 and 0.70. With evidence of sufficient reliability and convergent validity, the construct is therefore deemed good.

### 4.3. Normality and Multicollinearity

First, the assumptions of multivariate normal distribution, error independence, and variance equality were examined. This study has a relatively high sample size (393 respondents); therefore, the Central Limit Theorem could be implemented and there is no doubt regarding the normality of the data. In this work, a single method was employed to assess the existence of multicollinearity among independent variables. This method required the determination of the Variance Inflation Factor (VIF) [116]. As indicated by the main effect regression models with variance inflation factors (VIF range: 1.356–3.241), multicollinearity was not a concern with this data set, as it is much below 10. In this study, the measures chosen for evaluating independent variables do not reach levels indicative of multicollinearity.

### 4.4. Coefficient of Determination

Researchers such as Santosa et al. [117] recommended that the model must be evaluated by determining the coefficient of determination ($R^2$) of the dependent variable in order to test its explanatory power. Falk and Miller [118] argue that the R-squared of an endogenous variable should be 0.10. Other researchers have also proposed that the R-squared of an endogenous variable might be considered large when its value is 0.26, moderate when it is 0.13, and weak when it is 0.02 [119]. The value of $R^2$ of all endogenous values was determined to be in excess of the conditions in the PLS analysis proposed by Falk and Miller [118], as shown in Table 3. The model's $R^2$ is greater than the threshold value; hence, the model's performance is quite excellent.

**Table 3.** Coefficient of determination ($R^2$).

| Endogenous Variables | $R^2$ |
|---|---|
| Behavioural Intention | 0.454 |
| Attitude | 0.294 |
| Perceived Behavioural Control | 0.302 |
| Subjective Norm | 0.281 |

### 4.5. Path Analysis

In this study, the TPB model was assessed independently before the whole model was tested. The key conclusions of this study concern the applicability of TPB in a developing nation like Malaysia. Results indicate that antecedents comprising attitude, subjective standards, and PBC have a substantial influence on the intention to purchase a green and sustainable home, as shown in Table 4. Therefore, the results of this study are congruent with those of earlier research [120,121].

**Table 4.** Results of the TPB model.

| Hypotheses | Path Coefficients (β) | t-Value | *p*-Value |
|---|---|---|---|
| Attitude → Intention | 0.451 | 15.616 ** | 0.001 |
| Subjective Norm → Intention | 0.090 | 3.702 ** | 0.001 |
| Perceived Behavioural Control → Intention | 0.385 | 13.102 ** | 0.001 |

Note: ** Significant at 0.001.

Figure 2 shows the path analysis results of the structural model. The path coefficients of the PLS structural model validate the correlations between constructs that are assumed theoretically [122]; individual path coefficients quantify the strength of the causal connection between two constructs. They can be understood as standardized beta coefficients of ordinary least square regressions. Table 5 summarizes the hypothesis testing results.

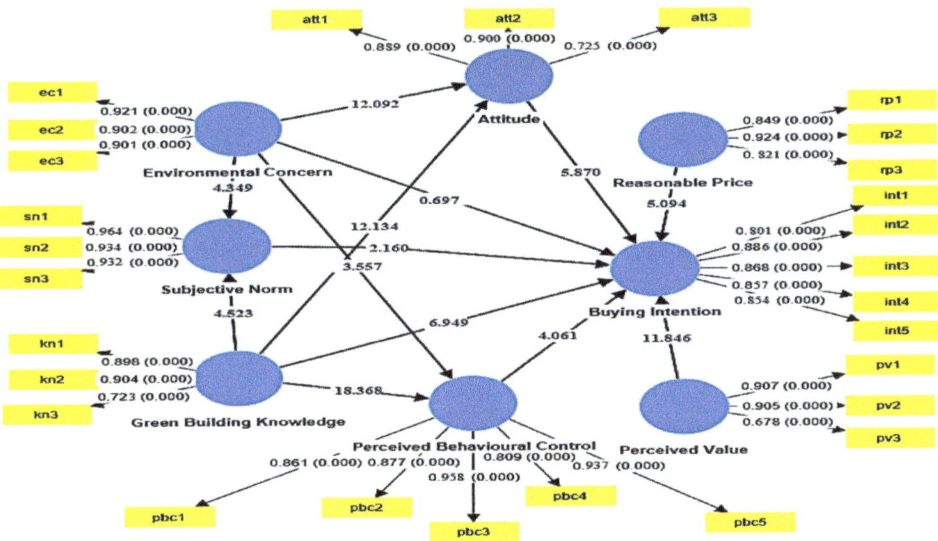


**Figure 2.** Structural model.

**Table 5.** Results of hypothesis testing.

| Hypotheses | t-Value | *p*-Value | Comment |
|---|---|---|---|
| H1: ATT → BI | 5.870 ** | 0.001 | Significant |
| H2: SN → BI | 2.160 * | 0.031 | Significant |
| H3: PBC → BI | 4.061 ** | 0.001 | Significant |
| H4: EC → ATT | 12.092 ** | 0.001 | Significant |
| H5: EC → SN | 4.349 ** | 0.001 | Significant |
| H6: EC → PBC | 3.557 ** | 0.010 | Significant |
| H7: EC → BI | 0.697 | 0.485 | Not Significant |
| H8: GBK→ ATT | 12.134 ** | 0.001 | Significant |
| H9: GBK→ SN | 4.523 ** | 0.001 | Significant |
| H10: GBK → PBC | 18.368 ** | 0.001 | Significant |
| H11: GBK → BI | 6.949 ** | 0.001 | Significant |
| H12: PV →BI | 11.846 ** | 0.001 | Significant |
| H13: RP → BI | 5.094 ** | 0.001 | Significant |

* Significant at 0.05, ** Significant at 0.001.

## 5. Discussion

When green building knowledge was added to the TPB model, it explained 46% of the differences in intentions. Attitudes, PBC, PV, concern for the environment, and a reasonable price were all good predictors of buying intentions. It was found that perceived value was the greatest predictor of whether or not someone would buy sustainable housing. Purchase intentions were also significantly predicted by environmental concern, subjective norms and green building knowledge. This supports the model which we found to be a good fit with the proposed extensions.

### *5.1. Predictors of Green Home Buying Intention*

The results of the PLS-SEM show that attitude is a good predictor of behavioural intention (t-value 5.870, significant at $p < 0.001$) (H1). This finding is in line with the study of Han and Yoon [59], who found the same thing in a different setting. Other studies [4,12,22,25,30] also agree with the results of the present study but contend that attitude is the most important predictor of the intention to buy a green home. In line with Chan and Hsu's [63] study, this analysis revealed that SN was related to intention in a way that was both positive and significant (H2). The result of the PLS analysis (t-value 2.160, significant $p$-value < 0.031) shows that subjective norms are the most important factor in deciding how someone will behave. Other studies in different fields such as marketing, covering variables such as online buying intention [123], halal food purchasing intention [98], organic food purchasing intention [62], and green product purchasing intention [63], have also confirmed the results of this study. A positive link was found between SN and intention in these studies. This is what some other researchers [4,12,22] have found to be true about people's plans to buy green homes. However, the current result goes against what certain studies [25,35] found, which was that there was no link between them.

Consistent with the literature, our study indicated that there is a positive correlation between PBC and the intent to purchase a green home (t-value = 4.061, $p < 0.001$) (H3). The findings in the present study are also confirmed by studies in different areas such as recycling [64], waste separation behaviour [65], and green product consumption intention [1,66]. Studies [12,22,25] on their green residence buying intention accentuated the same results. However, Zhang et al. [4] and Tan et al. [30] found no relationships. In accordance with the findings of studies [86,89], the current study discovered that perceived value positively and significantly (t-value = 11.846, $p = 0.001$) affects the intention to purchase a green home (H12). In addition, the conclusion is corroborated by studies [24,30], which indicate that PV is one of the most influential elements of purchasing intent.

The findings also show that EC is not a significant predictor of the intention to purchase a green home (t-value 0.697, $p < 0.485$) (H7). This was unexpected because previous research has repeatedly demonstrated that the environment has a large and favourable effect on purchasing intent [25,74]. The PLS-SEM results also reveal that knowledge about green homes has a positive relationship with green home buying intentions (H11), and that this relationship is significant (t-value = 6.949, $p$-value = 0.001). This is in accordance with the previous results of Zhang et al. [4]; and Zahan et al. [25]. Hence, proper knowledge of green building leads to a consumer's positive buying intention. The intention to buy a green home is also linked to a reasonable price (H13). The results of the PLS path coefficient analysis (t-value = 5.094, $p$-value = 0.001) mean that there is a positive and significant link between a reasonable price and the intention to buy a green and sustainable home. This research backs up the earlier findings that a reasonable price has a positive and significant effect on the likelihood of buying [124,125]. The price of a green home now shows how consumers feel about buying a green and sustainable home. Manufacturers who make green and sustainable homes at a fair price may be able to sell more of them.

### 5.2. Environmental Concern and Its Impact on TPB Constructs

According to the present study's PLS path coefficient findings, environmental concern has a positive and statistically significant link with attitude, SN, and PBC (H4, 5 and 6). The association between EC and attitude is significant: t-value 12.092 with significant *p*-value = 0.001, t-value 4.349 with significant *p*-value = 0.001, and t-value 3.557 with significant *p*-value = 0.01. The findings are consistent with earlier research [4,25,30]. This indicates that the higher the environmental concern in the consumer's mind, the higher the tendency to have a favourable opinion, peer influence and behavioural control of buying GH.

### 5.3. Green Building Knowledge and Its Impact on TPB Constructs

The greater the green building knowledge, the more positive the attitude (H8). The results of the PLS analysis (t-value = 12.134 and *p*-value = 0.001) indicate a positive and statistically significant association between knowledge and attitude. This study confirms the earlier findings that green building-related knowledge has a favourable and significant influence on attitudes toward green home purchasing [58]. The research results show that environmental knowledge has a significant influence on perceived behavioural control and subjective norms (t-value = 18.368, *p*-value = 0.001, and t-value = 4.523, *p*-value = 0.001, respectively) (H9 and 10). This result endorses the previous result of Zhang et al. [4]; and Zahan et al. [25]. It implies that the greater the knowledge about green building, the greater the individual's self-control over their decisions in this regard and the more likely the external influence on the decision making.

## 6. Implication of the Study

The findings of this study have various implications that may help in developing comprehensive strategies for green and sustainable home-buying intentions. This could help increase consumers' buying intention for green homes and ultimately positively impact the environment within Malaysia

The study contributes significantly to theoretical development. First, the study extended the popular and contemporary theory of planned individual behaviour. It successfully integrates new constructs (cognitive) in the model such as green building knowledge. The green building knowledge construct includes three domains of knowledge, namely, factual, conceptual and procedural (skill) knowledge, which has been missing in the green home buying literature and empirical testing. The scales generated here may be replicated by future researchers. We explored environmental concern as a cognitive variable that influences residents' perception of GHs. In eco-innovation adoption studies, environmental concern is one of the most studied personality factors. We discovered that these two psychological characteristics have distinct effects on inhabitants' views and intentions to acquire GHs. This research provides insight into the effects of psychological factors on the intention to purchase GH. The study also adds context-specific factors, i.e., values and reasonable prices. Second, studies on green home buying intention from Malaysian Millennials' perspectives are rare. Targeting this consumer group means that the present results will certainly improve the literature depth in this arena. Third, in terms of new results, we did not find any link between expressed environmental concern and buying intention of green homes for Millennials. This supplies a challenge for new studies to counter or validate such findings in the future, leading to a deepening of the literature in this area.

The outcome of this study has important managerial implications. Our study found a significant relationship between green building knowledge and purchase intention. It is necessary to popularize the knowledge pertaining to eco-homes, their designs, and their purchase because the green home is a newer form of housing. This knowledge includes the features of the green home as well as issues demanding attention in the selection and purchase. The knowlege can increase positive attitudes of the individual, and their ability to exert behavioural control. Thus, it is important for green home businesses to use a subject-specific environmental message to communicate with potential buyers through different media. The concerned marketing team of this sector should emphasize this with vigour. They can arrange displays in various public places, in Television Commercials (TVC), industry symposiums and elsewhere. Likewise, the methods described above should be put into practice by embracing more focused ways of approaching different types of customers. For example, marketing to women might focus on the alteration of their attitudes, while marketing to men might focus on the improvement of their perceived behavioural control. This study also suggests targeting customers based on their generations. Social media boosting based on the Millennial target group could also bring good results. Fourth, perceived value, attitude, and subjective norms were found to be significant factors in choosing green homes. Enhancing the reputation of eco-friendly homes, such as receiving high praise from consumers' family and friends, receiving extensive and supportive remarks from the media, and receiving moderate guidance from policies, can improve consumers' attitudes toward buying behaviour and subjective norms, and thus impact buyers to buy green residences, which is important for increasing the buying intention of customers for eco-friendly residences. Government, entrepreneurs, the private sector and marketers should develop programmes that will show the value for money when buying a green home. The complementary actions may also include quality assurance. The maintenance of the building code of conduct in green home construction is not only necessary but also requires publicity to improve the positive perceptions linked to branding. Again, the target-based promotional matters should relate to quality assurance.

In addition, the present study has policy consequences. Reasonable pricing is found to be significant in our result. This emphasises the need to ensure a fair market pricing policy, through the housing authority in Malaysia. The consumer associations that work for the consumers' rights should also be more active, including in the case of any mishaps that may have occurred. We would like to bring to the attention of policymakers the fact that instrumental motivations, which in the context of this study include the provision of tax incentives, did not appear to be valued nearly as high as other types of motivation (intrinsic and non-normative), by the Millennial potential homebuyers. As a result, policymakers should reconsider how they encourage the development of a built environment in light of the generation of consumers who appear to be influenced more by internal and non-normative factors. Lastly, our data reveal that cognitive processes play crucial roles in determining how residents view green buildings. Concern for the environment positively affects people's attitude, subjective norms, and perceived behavioural control. Thus, it is vital for governments and companies to start public relations efforts to underline the gravity of the current environmental challenges, with a focus on the pollution caused by the construction and operation of conventional housing.

## 7. Conclusions, Limitations and Future Research Directions

The objective of the study was to diagnose the factors affecting green and sustainable home-buying intention in Malaysia with the proposed extension of TPB. Attitude, perceived behavioural control and subjective norm had a high positive and significant effect on green and sustainable home buying intention with attitude showing the highest effect, followed by PBC and finally SN. Compared to subjective norms, the variable "attitude" shows a stronger influence, so marketers might try to attract Malaysian consumers towards green homes using advertising commercials, as this could help in affecting consumers' green home buying intention. Environmental concern had no evident significant influence on green home buying intention, but may in fact be widespread among Millennials. It showed an indirect relationship via the three TPB. Environmental knowledge and perceived value are the two constructs that had a strong relationship with green and sustainable home-buying intentions.

Even though the study accomplished its major aims, the results should be viewed with caution because they are based on a single data collection process (cross-sectional survey). Thus, respondents may gain knowledge as time progresses, which may affect how their reported thinking is put to use. This paves the way for even more in-depth studies, such as longitudinal studies, which may produce more precise data that more accurately represents the development of eco-friendly behaviours over time. Moreover, the model measured behavioural intention instead of actual behaviour. The study could have been stepped up to a willingness to pay so that the reader could understand the barrier of adoption in more detail. Hence, there is scope for advanced study in the future. The intention-behaviour gap in making a green decision in the housing sector may also be better addressed in future research if the moderating or mediating variables are more fully accounted for. Variables such as facilitating conditions and perceived risk should be tested in future research. In conclusion, the current research was only conducted with participants from the Millennial generation who live in Malaysia; as a result, any attempts to extrapolate the findings to other cultures and environments should be performed with extreme caution. Future research could come up with a multi-group analysis or comparative study by generations within Malaysia or the region.

**Author Contributions:** Conceptualization, M.M., C.-K.W., S.S.A. and C.-Y.L.; methodology, S.S.A., C.-Y.L. and Y.-H.H.; software, M.M. and A.B.S.; validation, S.S.A., C.-K.W., Y.-H.H. and A.B.S.; resources, M.M. and Y.-H.H.; writing—original draft preparation, M.M., C.-K.W., S.S.A., C.-Y.L. and Y.-H.H.; writing—review and editing, C.-Y.L. and Y.-H.H. All authors have read and agreed to the published version of the manuscript.

**Funding:** This study received no external funding.

**Institutional Review Board Statement:** Not applicable.

**Informed Consent Statement:** Informed consent was obtained from all subjects interviewed involved in this study.

**Data Availability Statement:** The data that support the findings of this study are available from the corresponding authors upon reasonable request.

**Conflicts of Interest:** The authors declare no conflict of interest.

## Appendix A Survey Questionnaire

| | Strongly Disagree | | | | | Strongly Agree |
|---|---|---|---|---|---|---|
| 1. Attitude | | ⟵ | | ⟶ | | |
| Green homes will help to save the environment because it uses environmentally friendly materials and processes | 1 | 2 | 3 | 4 | 5 | 6 |
| Buying a green home that has Green Building Index (GBI) ratings or similar international ratings is favourable | 1 | 2 | 3 | 4 | 5 | 6 |
| Living in a green home would be good for me because these homes improve our quality of living | 1 | 2 | 3 | 4 | 5 | 6 |
| 2. Subjective Norm | | | | | | |
| My family think that I should purchase a green home rather than a normal home | 1 | 2 | 3 | 4 | 5 | 6 |
| My close friends think that I should buy a green home | 1 | 2 | 3 | 4 | 5 | 6 |
| People who influence my behaviour think that I should purchase a green home. | 1 | 2 | 3 | 4 | 5 | 6 |

| | Strongly Disagree | | | | | Strongly Agree |
|---|---|---|---|---|---|---|
| 3. Perceived behavioural control | | | | | | |
| I am confident that I would purchase a green home even if it is slightly more expensive | 1 | 2 | 3 | 4 | 5 | 6 |
| I am confident that I would purchase a green home even if the other person advises me the opposite | 1 | 2 | 3 | 4 | 5 | 6 |
| I am sure I would be able to make difference to use PV solar energy | 1 | 2 | 3 | 4 | 5 | 6 |
| Buying a green home is entirely within my control | 1 | 2 | 3 | 4 | 5 | 6 |
| I have the resources, and ability to purchase green homes. | 1 | 2 | 3 | 4 | 5 | 6 |
| 4. Reasonable Price | | | | | | |
| I would buy a green home if the price is reasonable. | 1 | 2 | 3 | 4 | 5 | 6 |
| The price of green homes is normally higher than that of conventional homes. | 1 | 2 | 3 | 4 | 5 | 6 |
| It is easy to justify the price and benefits of green homes. | 1 | 2 | 3 | 4 | 5 | 6 |
| 5. Environmental Concerns | | | | | | |
| I prefer to check the Green Building Index (GBI) or other green ratings and certifications on the homes before purchase. | 1 | 2 | 3 | 4 | 5 | 6 |
| I want to have a deeper insight into the inputs, processes and impacts of the green home on the environment before the purchase. | 1 | 2 | 3 | 4 | 5 | 6 |
| I would prefer to gain substantial information on green homes before purchasing. | 1 | 2 | 3 | 4 | 5 | 6 |
| 6. Green Building Knowledge | | | | | | |
| I know factual information about eco-home design (e.g., building codes, wind turbines, recycled content materials, and dual-flush toilets etc.). | 1 | 2 | 3 | 4 | 5 | 6 |
| I have the conceptual knowledge of green building design (e.g., how to maintain indoor air quality, size and window locations, ecological impacts etc.) | 1 | 2 | 3 | 4 | 5 | 6 |
| I have procedural knowledge of green building operations (e.g., picking eco-friendly furniture, monitoring cooling systems or solar panels or using water-saving building fixtures). | 1 | 2 | 3 | 4 | 5 | 6 |
| 7. Perceived Value | | | | | | |
| Green home's environmental functions provide very good value for me | 1 | 2 | 3 | 4 | 5 | 6 |
| Green home's environmental performance meets my expectations | 1 | 2 | 3 | 4 | 5 | 6 |

| | Strongly Disagree | | | | | Strongly Agree |
|---|---|---|---|---|---|---|
| Buying a green home has more environmental benefits than other conventional homes | 1 | 2 | 3 | 4 | 5 | 6 |
| 8. Purchase Intention | | | | | | |
| I am planning to buy a green home in future | 1 | 2 | 3 | 4 | 5 | 6 |
| I will try to purchase a green and sustainable home in future | 1 | 2 | 3 | 4 | 5 | 6 |
| I will make an effort to purchase a green and sustainable home in future | 1 | 2 | 3 | 4 | 5 | 6 |
| I intend to purchase green homes next time because of their positive environmental contribution | 1 | 2 | 3 | 4 | 5 | 6 |
| I will consider switching to a green home for ecological reasons | 1 | 2 | 3 | 4 | 5 | 6 |

## References


1. Paul, J.; Modi, A.; Patel, J. Predicting Green Product Consumption Using Theory of Planned Behavior and Reasoned Action. *J. Retail. Consum. Serv.* **2016**, *29*, 123–134. [CrossRef]
2. Maichum, K.; Parichatnon, S.; Peng, K.-C. Application of the Extended Theory of Planned Behavior Model to Investigate Purchase Intention of Green Products among Thai Consumers. *Sustainability* **2016**, *8*, 1077. [CrossRef]
3. Yadav, R.; Pathak, G.S. Young Consumers' Intention towards Buying Green Products in a Developing Nation: Extending the Theory of Planned Behavior. *J. Clean. Prod.* **2016**, *135*, 732–739. [CrossRef]
4. Zhang, L.; Chen, L.; Wu, Z.; Xue, H.; Dong, W. Key Factors Affecting Informed Consumers' Willingness to Pay for Green Housing: A Case Study of Jinan, China. *Sustainability* **2018**, *10*, 1711. [CrossRef]
5. Briefing, U.S. International Energy Outlook 2013. *US Energy Inf. Adm.* **2013**, *506*, 507.
6. Darko, A.; Zhang, C.; Chan, A.P.C. Drivers for Green Building: A Review of Empirical Studies. *Habitat Int.* **2017**, *60*, 34–49. [CrossRef]
7. Tan, T.H. Use of Structural Equation Modeling to Predict the Intention to Purchase Green and Sustainable Homes in Malaysia. *Asian Soc. Sci.* **2013**, *9*, 181. [CrossRef]
8. Tan, P.-N.; Chawla, S.; Ho, C.K.; Bailey, J. *Advances in Knowledge Discovery and Data Mining, Part II, Proceedings of the 16th Pacific-Asia Conference, PAKDD 2012, Kuala Lumpur, Malaysia, 29 May–1 June 2012*; Springer: Berlin/Heidelberg, Germany, 2012; Volume 7302, ISBN 3642302203.
9. Young, W.; Hwang, K.; McDonald, S.; Oates, C.J. Sustainable Consumption: Green Consumer Behaviour When Purchasing Products. *Sustain. Dev.* **2010**, *18*, 20–31. [CrossRef]
10. Wiederhold, M.; Martinez, L.F. Ethical Consumer Behaviour in Germany: The Attitude-behaviour Gap in the Green Apparel Industry. *Int. J. Consum. Stud.* **2018**, *42*, 419–429. [CrossRef]
11. Li, Y.; Siddik, A.B.; Masukujjaman, M.; Wei, X. Bridging Green Gaps: The Buying Intention of Energy Efficient Home Appliances and Moderation of Green Self-Identity. *Appl. Sci.* **2021**, *11*, 9878. [CrossRef]
12. Judge, M.; Warren-Myers, G.; Paladino, A. Using the Theory of Planned Behaviour to Predict Intentions to Purchase Sustainable Housing. *J. Clean. Prod.* **2019**, *215*, 259–267. [CrossRef]
13. Carfora, V.; Caso, D.; Sparks, P.; Conner, M. Moderating Effects of Pro-Environmental Self-Identity on pro-Environmental Intentions and Behaviour: A Multi-Behaviour Study. *J. Environ. Psychol.* **2017**, *53*, 92–99. [CrossRef]
14. Verma, V.K.; Chandra, B. An Application of Theory of Planned Behavior to Predict Young Indian Consumers' Green Hotel Visit Intention. *J. Clean. Prod.* **2018**, *172*, 1152–1162. [CrossRef]
15. Yazdanpanah, M.; Forouzani, M. Application of the Theory of Planned Behaviour to Predict Iranian Students' Intention to Purchase Organic Food. *J. Clean. Prod.* **2015**, *107*, 342–352. [CrossRef]
16. Alshebami, A.S. Evaluating the Relevance of Green Banking Practices on Saudi Banks' Green Image: The Mediating Effect of Employees' Green Behaviour. *J. Bank. Regul.* **2021**, *22*, 275–286. [CrossRef]
17. Novitasari, M.; Alshebami, A.S.; Sudrajat, M.A. The Role of Green Supply Chain Management in Predicting Indonesian Firms' Performance: Competitive Advantage and Board Size Influence. *Indones. J. Sustain. Account. Manag.* **2021**, *5*, 137–149. [CrossRef]
18. Zheng, G.-W.; Siddik, A.B.; Masukujjaman, M.; Alam, S.S.; Akter, A. Perceived Environmental Responsibilities and Green Buying Behavior: The Mediating Effect of Attitude. *Sustainability* **2020**, *13*, 35. [CrossRef]
19. Chen, J.; Siddik, A.B.; Zheng, G.-W.; Masukujjaman, M.; Bekhzod, S. The Effect of Green Banking Practices on Banks' Environmental Performance and Green Financing: An Empirical Study. *Energies* **2022**, *15*, 1292. [CrossRef]

20. Zheng, G.-W.; Siddik, A.B.; Masukujjaman, M.; Fatema, N. Factors Affecting the Sustainability Performance of Financial Institutions in Bangladesh: The Role of Green Finance. *Sustainability* **2021**, *13*, 10165. [CrossRef]
21. Kota, B.R.; Debs, L.; Davis, T. Exploring Generation Z's Perceptions of Green Homes. *Sustainability* **2022**, *14*, 10148. [CrossRef]
22. Wu, C.; Liang, S.; Wu, W.; Hong, Y. Practicing Green Residence Business Model Based on TPB Perspective. *Sustainability* **2021**, *13*, 7379. [CrossRef]
23. Zhang, L.; Chen, L.; Wu, Z.; Zhang, S.; Song, H. Investigating Young Consumers' Purchasing Intention of Green Housing in China. *Sustainability* **2018**, *10*, 1044. [CrossRef]
24. Zhao, S.; Chen, L. Exploring Residents' Purchase Intention of Green Housings in China: An Extended Perspective of Perceived Value. *Int. J. Environ. Res. Public Health* **2021**, *18*, 4074. [CrossRef]
25. Zahan, I.; Chuanmin, S.; Fayyaz, M.; Hafeez, M. Green Purchase Behavior towards Green Housing: An Investigation of Bangladeshi Consumers. *Environ. Sci. Pollut. Res.* **2020**, *27*, 38745–38757. [CrossRef]
26. Cao, X.; Zhao, T.; Xing, Z. How Do Government Policies Promote Green Housing Diffusion in China? A Complex Network Game Context. *Int. J. Environ. Res. Public Health* **2022**, *19*, 2238. [CrossRef] [PubMed]
27. Rosner, Y.; Amitay, Z.; Perlman, A. Consumer's Attitude, Socio-Demographic Variables and Willingness to Purchase Green Housing in Israel. *Environ. Dev. Sustain.* **2022**, *24*, 5295–5316. [CrossRef]
28. He, C.; Yu, S.; Han, Q.; de Vries, B. How to Attract Customers to Buy Green Housing? Their Heterogeneous Willingness to Pay for Different Attributes. *J. Clean. Prod.* **2019**, *230*, 709–719. [CrossRef]
29. Zalejska-Jonsson, A. Stated WTP and Rational WTP: Willingness to Pay for Green Apartments in Sweden. *Sustain. Cities Soc.* **2014**, *13*, 46–56. [CrossRef]
30. Tan, W.-L.; Goh, Y.-N. The Role of Psychological Factors in Influencing Consumer Purchase Intention towards Green Residential Building. *Int. J. Hous. Mark. Anal.* **2018**, *11*, 788–807. [CrossRef]
31. Yau, Y. Eco-labels and Willingness-to-pay: A Hong Kong Study. *Smart Sustain. Built Environ.* **2012**, *1*, 277–290. [CrossRef]
32. Hu, H.; Geertman, S.; Hooimeijer, P. The Willingness to Pay for Green Apartments: The Case of Nanjing, China. *Urban Stud.* **2014**, *51*, 3459–3478. [CrossRef]
33. Ahmad, A.L.; Rahim, S.A.; Pawanteh, L.; Ahmad, F. The Understanding of Environmental Citizenship among Malaysian Youths: A Study on Perception and Participation. *Asian Soc. Sci.* **2012**, *8*, 85. [CrossRef]
34. Jia, J.; Wu, H.-Q.; Nie, H.-G.; Fan, Y. Modeling the Willingness to Pay for Energy Efficient Residence in Urban Residential Sector in China. *Energy Policy* **2019**, *135*, 111003. [CrossRef]
35. Liu, Y.; Hong, Z.; Zhu, J.; Yan, J.; Qi, J.; Liu, P. Promoting Green Residential Buildings: Residents' Environmental Attitude, Subjective Knowledge, and Social Trust Matter. *Energy Policy* **2018**, *112*, 152–161. [CrossRef]
36. Ajzen, I. The Theory of Planned Behavior. *Organ. Behav. Hum. Decis. Process.* **1991**, *50*, 179–211. [CrossRef]
37. Hauge, Å.L.; Thomsen, J.; Löfström, E. How to Get Residents/Owners in Housing Cooperatives to Agree on Sustainable Renovation. *Energy Effic.* **2013**, *6*, 315–328. [CrossRef]
38. Lee, E.; Allen, A.; Kim, B. Interior Design Practitioner Motivations for Specifying Sustainable Materials: Applying the Theory of Planned Behavior to Residential Design. *J. Inter. Des.* **2013**, *38*, 1–16. [CrossRef]
39. Zhang, X.; Geng, G.; Sun, P. Determinants and Implications of Citizens' Environmental Complaint in China: Integrating Theory of Planned Behavior and Norm Activation Model. *J. Clean. Prod.* **2017**, *166*, 148–156. [CrossRef]
40. Han, H.; Hyun, S.S. Drivers of Customer Decision to Visit an Environmentally Responsible Museum: Merging the Theory of Planned Behavior and Norm Activation Theory. *J. Travel Tour. Mark.* **2017**, *34*, 1155–1168. [CrossRef]
41. Alam, S.S.; Nik Hashim, N.H.; Rashid, M.; Omar, N.A.; Ahsan, N.; Ismail, M.D. Small-Scale Households Renewable Energy Usage Intention: Theoretical Development and Empirical Settings. *Renew. Energy* **2014**, *68*, 255–263. [CrossRef]
42. Eze, U.C.; Ndubisi, N.O. Green Buyer Behavior: Evidence from Asia Consumers. *J. Asian Afr. Stud.* **2013**, *48*, 413–426. [CrossRef]
43. Karimy, M.; Zareban, I.; Araban, M.; Montazeri, A. An Extended Theory of Planned Behavior (TPB) Used to Predict Smoking Behavior among a Sample of Iranian Medical Students. *Int. J. High Risk Behav. Addict.* **2015**, *4*, e24715. [CrossRef] [PubMed]
44. Karatu, V.M.H.; Mat, N.K.N. The Mediating Effects of Green Trust and Perceived Behavioral Control on the Direct Determinants of Intention to Purchase Green Products in Nigeria. *Mediterr. J. Soc. Sci.* **2015**, *6*, 256. [CrossRef]
45. Chang, R.D.; Soebarto, V.; Zhao, Z.Y.; Zillante, G. Facilitating the Transition to Sustainable Construction: China's Policies. *J. Clean. Prod.* **2016**, *131*, 534–544. [CrossRef]
46. Ofek, S.; Akron, S.; Portnov, B.A. Stimulating Green Construction by Influencing the Decision-Making of Main Players. *Sustain. Cities Soc.* **2018**, *40*, 165–173. [CrossRef]
47. Li, Q.; Long, R.; Chen, H. Differences and Influencing Factors for Chinese Urban Resident Willingness to Pay for Green Housings: Evidence from Five First-Tier Cities in China. *Appl. Energy* **2018**, *229*, 299–313. [CrossRef]
48. Gabay, H.; Meir, I.A.; Schwartz, M.; Werzberger, E. Cost-Benefit Analysis of Green Buildings: An Israeli Office Buildings Case Study. *Energy Build.* **2014**, *76*, 558–564. [CrossRef]
49. Srivastava, S.K. Green Supply-chain Management: A State-of-the-art Literature Review. *Int. J. Manag. Rev.* **2007**, *9*, 53–80. [CrossRef]
50. Ajzen, I. From Intentions to Actions: A Theory of Planned Behavior BT. In *Action Control: From Cognition to Behavior*; Kuhl, J., Beckmann, J., Eds.; Springer: Berlin/Heidelberg, Germany, 1985; pp. 11–39. ISBN 978-3-642-69746-3.
51. Ajzen, I.; Fishbein, M. *Understanding Attitudes and Predicting Social Behaviour*; Prentice-Hall: Englewood Cliffs, NJ, USA, 1980.

52. Liere, K.D.V.; Dunlap, R.E. The Social Bases of Environmental Concern: A Review of Hypotheses, Explanations and Empirical Evidence. *Public Opin. Q.* **1980**, *44*, 181–197. [CrossRef]
53. Taufique, K.M.R.; Vaithianathan, S. A Fresh Look at Understanding Green Consumer Behavior among Young Urban Indian Consumers through the Lens of Theory of Planned Behavior. *J. Clean. Prod.* **2018**, *183*, 46–55. [CrossRef]
54. Hume, M. Compassion without Action: Examining the Young Consumers Consumption and Attitude to Sustainable Consumption. *J. World Bus.* **2010**, *45*, 385–394. [CrossRef]
55. Blackwell, C.K.; Lauricella, A.R.; Wartella, E.; Robb, M.; Schomburg, R. Adoption and Use of Technology in Early Education: The Interplay of Extrinsic Barriers and Teacher Attitudes. *Comput. Educ.* **2013**, *69*, 310–319. [CrossRef]
56. Ajzen, I. *Attitudes, Personality, and Behavior*, 6th ed.; McGraw-Hill Education: London, UK, 1988.
57. Ramayah, T.; Rahbar, E. Greening the Environment through Recycling: An Empirical Study. *Manag. Environ. Qual. Int. J.* **2013**, *24*, 782–801. [CrossRef]
58. Chen, M.F.; Tung, P.J. Developing an Extended Theory of Planned Behavior Model to Predict Consumers' Intention to Visit Green Hotels. *Int. J. Hosp. Manag.* **2014**, *36*, 221–230. [CrossRef]
59. Han, H.; Yoon, H.J. Hotel Customers' Environmentally Responsible Behavioral Intention: Impact of Key Constructs on Decision in Green Consumerism. *Int. J. Hosp. Manag.* **2015**, *45*, 22–33. [CrossRef]
60. Ajzen, I.; Driver, B.L. Application of the Theory of Planned Behavior to Leisure Choice. *J. Leis. Res.* **1992**, *24*, 207–224. [CrossRef]
61. Bagozzi, R.P.; Wong, N.; Abe, S.; Bergami, M. Cultural and Situational Contingencies and the Theory of Reasoned Action: Application to Fast Food Restaurant Consumption. *J. Consum. Psychol.* **2000**, *9*, 97–106. [CrossRef]
62. Irianto, H. Consumers' Attitude and Intention towards Organic Food Purchase: An Extension of Theory of Planned Behavior in Gender Perspective. *Int. J. Manag. Econ. Soc. Sci.* **2015**, *4*, 17–31.
63. Chan, A.; Hsu, C.H.C.; Baum, T. The Impact of Tour Service Performance on Tourist Satisfaction and Behavioral Intentions: A Study of Chinese Tourists in Hong Kong. *J. Travel Tour. Mark.* **2015**, *32*, 18–33. [CrossRef]
64. Ramayah, T.; Lee, J.W.C.; Lim, S. Sustaining the Environment through Recycling: An Empirical Study. *J. Environ. Manag.* **2012**, *102*, 141–147. [CrossRef]
65. Zhang, D.; Huang, G.; Yin, X.; Gong, Q. Residents' Waste Separation Behaviors at the Source: Using SEM with the Theory of Planned Behavior in Guangzhou, China. *Int. J. Environ. Res. Public Health* **2015**, *12*, 9475–9491. [CrossRef]
66. Hoppe, A.; Vieira, L.M.; de Barcellos, M.D. Consumer Behaviour towards Organic Food in Porto Alegre: An Application of the Theory of Planned Behaviour. *Rev. Econ. Sociol. Rural* **2013**, *51*, 69–90. [CrossRef]
67. Crosby, L.A.; Gill, J.D.; Taylor, J.R. Consumer/Voter Behavior in the Passage of the Michigan Container Law. *J. Mark.* **1981**, *45*, 19–32. [CrossRef]
68. Jaiswal, D.; Kant, R. Green Purchasing Behaviour: A Conceptual Framework and Empirical Investigation of Indian Consumers. *J. Retail. Consum. Serv.* **2018**, *41*, 60–69. [CrossRef]
69. Yadav, R.; Pathak, G.S. Determinants of Consumers' Green Purchase Behavior in a Developing Nation: Applying and Extending the Theory of Planned Behavior. *Ecol. Econ.* **2017**, *134*, 114–122. [CrossRef]
70. Mark, N.; Law, M. Encouraging Green Purchase Behaviours of Hong Kong Consumers. *Asian J. Bus. Res. ISSN* **2015**, *5*, 2015.
71. Sang, Y.-N.; Bekhet, H.A. Modelling Electric Vehicle Usage Intentions: An Empirical Study in Malaysia. *J. Clean. Prod.* **2015**, *92*, 75–83. [CrossRef]
72. Basha, M.B.; Mason, C.; Shamsudin, M.F.; Hussain, H.I.; Salem, M.A. Consumers Attitude towards Organic Food. *Procedia Econ. Financ.* **2015**, *31*, 444–452. [CrossRef]
73. Karatu, V.M.H.; Nik Mat, N.K. Determinants of Green Purchase Intention in Nigeria: The Mediating Role of Green Perceived Value. In Proceedings of the International Conference on Accounting Studies (ICAS) 2015, Johor, Malaysia, 17–20 August 2015.
74. Suki, N.M. Consumer Environmental Concern and Green Product Purchase in Malaysia: Structural Effects of Consumption Values. *J. Clean. Prod.* **2016**, *132*, 204–214. [CrossRef]
75. Davis, J.J. The Effects of Message Framing on Response to Environmental Communications. *J. Mass Commun. Q.* **1995**, *72*, 285–299. [CrossRef]
76. Cole, L.B. Green Building Literacy: A Framework for Advancing Green Building Education. *Int. J. STEM Educ.* **2019**, *6*, 1–13. [CrossRef]
77. Zheng, G.W.; Akter, N.; Siddik, A.B.; Masukujjaman, M. Organic Foods Purchase Behavior among Generation y of Bangladesh: The Moderation Effect of Trust and Price Consciousness. *Foods* **2021**, *10*, 2278. [CrossRef] [PubMed]
78. Moorman, C.; Diehl, K.; Brinberg, D.; Kidwell, B. Subjective Knowledge, Search Locations, and Consumer Choice. *J. Consum. Res.* **2004**, *31*, 673–680. [CrossRef]
79. Yang, Z.J.; Kahlor, L. What, Me Worry? The Role of Affect in Information Seeking and Avoidance. *Sci. Commun.* **2013**, *35*, 189–212. [CrossRef]
80. Wu, H.-C. An Empirical Study of the Effects of Service Quality, Perceived Value, Corporate Image, and Customer Satisfaction on Behavioral Intentions in the Taiwan Quick Service Restaurant Industry. *J. Qual. Assur. Hosp. Tour.* **2013**, *14*, 364–390. [CrossRef]
81. Hsu, C.-L.; Lin, J.C.-C. What Drives Purchase Intention for Paid Mobile Apps?—An Expectation Confirmation Model with Perceived Value. *Electron. Commer. Res. Appl.* **2015**, *14*, 46–57. [CrossRef]
82. Ponte, E.B.; Carvajal-Trujillo, E.; Escobar-Rodríguez, T. Influence of Trust and Perceived Value on the Intention to Purchase Travel Online: Integrating the Effects of Assurance on Trust Antecedents. *Tour. Manag.* **2015**, *47*, 286–302. [CrossRef]

83. Chen, Y.S.; Chang, C.H. Enhance Green Purchase Intentions: The Roles of Green Perceived Value, Green Perceived Risk, and Green Trust. *Manag. Decis.* **2012**, *50*, 502–520. [CrossRef]
84. Chen, J.; Lobo, A. Organic Food Products in China: Determinants of Consumers' Purchase Intentions. *Int. Rev. Retail. Distrib. Consum. Res.* **2012**, *22*, 293–314. [CrossRef]
85. Liu, T.; Jin, S.; Zhu, H.; Qi, X. Construction of Revised TPB Model of Customer Green Be-Havior: Environmental Protection Purpose and Ecological Values Perspectives. In Proceedings of the IOP Conference Series: Earth and Environmental Science, Barcelona, Spain, 11–13 March 2018; IOP Publishing: Bristol, UK, 2018; Volume 167, p. 12021.
86. Salehzadeh, R.; Pool, J.K. Brand Attitude and Perceived Value and Purchase Intention toward Global Luxury Brands. *J. Int. Consum. Mark.* **2017**, *29*, 74–82. [CrossRef]
87. Ahmad, W.; Zhang, Q. Green Purchase Intention: Effects of Electronic Service Quality and Customer Green Psychology. *J. Clean. Prod.* **2020**, *267*, 122053. [CrossRef]
88. Wei, J.; Zhao, X.; Liu, Y.; Yang, X. Measuring Purchase Intention towards Green Power Certificate in a Developing Nation: Applying and Extending the Theory of Planned Behavior. *Resour. Conserv. Recycl.* **2021**, *168*, 105363. [CrossRef]
89. Zhao, R.; Siddik, A.B.; Masukujjaman, M.; Hamayun, M.; Ibrahim, A.M. Bi-Dimensional Values and Attitudes Toward Online Fast Food-Buying Intention During the COVID-19 Pandemic: An Application of VAB Model. *Front. Nutr.* **2022**, *9*, 2111. [CrossRef]
90. Bonini, S.; Oppenheim, J. Cultivating the Green Consumer. *Stanford Soc. Innov. Rev.* **2008**, *6*, 56–61.
91. Massachusetts Department of Environmental Protection. Barrier/Motivation Inventory No. 3. 2003. Available online: http://www.state.ma.us (accessed on 5 October 2022).
92. Ottman, J.A.; Stafford, E.R.; Hartman, C.L. Avoiding Green Marketing Myopia: Ways to Improve Consumer Appeal for Environmentally Preferable Products. *Environ. Sci. Policy Sustain. Dev.* **2006**, *48*, 22–36. [CrossRef]
93. Simmons, R.; Koenig, S. Probabilistic Robot Navigation in Partially Observable Environments. In Proceedings of the 14th International Joint Conference on Artificial Intelligence, Montreal, QC, Canada, 20–25 August 1995; Volume 2, pp. 1080–1087.
94. Shi, Y.; Siddik, A.B.; Masukujjaman, M.; Zheng, G.; Hamayun, M.; Ibrahim, A.M. The Antecedents of Willingness to Adopt and Pay for the IoT in the Agricultural Industry: An Application of the UTAUT 2 Theory. *Sustainability* **2022**, *14*, 6640. [CrossRef]
95. Lee, S.; Illia, A.; Lawson-Body, A. Perceived Price Fairness of Dynamic Pricing. *Ind. Manag. Data Syst.* **2011**, *111*, 531–550. [CrossRef]
96. Chan, V. Determining Students' Language Needs in a Tertiary Setting. *Engl. Teach. Forum* **2001**, *39*, 16–27. Available online: https://americanenglish.state.gov/files/ae/resource_files/01-39-3-d_0.pdf (accessed on 5 October 2022).
97. Hedlund, T. The Impact of Values, Environmental Concern, and Willingness to Accept Economic Sacrifices to Protect the Environment on Tourists' Intentions to Buy Ecologically Sustainable Tourism Alternatives. *Tour. Hosp. Res.* **2011**, *11*, 278–288. [CrossRef]
98. Shah Alam, S.; Mohamed Sayuti, N. Applying the Theory of Planned Behavior (TPB) in Food Purchasing. *Int. J. Commer. Manag.* **2011**, *21*, 8–20. [CrossRef]
99. Alam, S.S.; Janor, H.; Zanariah, C.; Ahsan, M.N. Is Religiosity an Important Factor in Influencing the Intention to Undertake Islamic Home Financing in Klang Valley. *World Appl. Sci. J.* **2012**, *19*, 1030–1041.
100. Vermeir, I.; Verbeke, W. Sustainable Food Consumption among Young Adults in Belgium: Theory of Planned Behaviour and the Role of Confidence and Values. *Ecol. Econ.* **2008**, *64*, 542–553. [CrossRef]
101. Tian, H.; Siddik, A.B.; Masukujjaman, M. Factors Affecting the Repurchase Intention of Organic Tea among Millennial Consumers: An Empirical Study. *Behav. Sci.* **2022**, *12*, 50. [CrossRef] [PubMed]
102. Masukujjaman, M.; Alam, S.S.; Siwar, C.; Halim, S.A. Purchase Intention of Renewable Energy Technology in Rural Areas in Bangladesh: Empirical Evidence. *Renew. Energy* **2021**, *170*, 639–651. [CrossRef]
103. Harman, H.H. *Modern Factor Analysis*; University of Chicago Press: Chicago, IL, USA, 1976; ISBN 0226316521.
104. Ringle, C.M.; Sarstedt, M.; Mooi, E.A. Response-Based Segmentation Using Finite Mixture Partial Least Squares. In Proceedings of the Data Mining; Springer: Berlin/Heidelberg, Germany, 2010; pp. 19–49.
105. Compeau, D.; Higgins, C.A.; Huff, S. Social Cognitive Theory and Individual Reactions to Computing Technology: A Longitudinal Study. *MIS Q.* **1999**, *23*, 145–158. [CrossRef]
106. Aibinu, A.A.; Al-Lawati, A.M. Using PLS-SEM Technique to Model Construction Organizations' Willingness to Participate in e-Bidding. *Autom. Constr.* **2010**, *19*, 714–724. [CrossRef]
107. Afzal, F.; Lim, B.; Prasad, D. An Investigation of Corporate Approaches to Sustainability in the Construction Industry. *Procedia Eng.* **2017**, *180*, 202–210. [CrossRef]
108. Ferreira, J.J.; Raposo, M.L.; Rodrigues, R.G.; Dinis, A.; Paço, A.d. A Model of Entrepreneurial Intention: An Application of the Psychological and Behavioral Approaches. *J. Small Bus. Enterp. Dev.* **2012**, *19*, 424–440. [CrossRef]
109. Triguero-Sánchez, R.; Peña-Vinces, J.C.; Sánchez-Apellániz, M. Hierarchical Distance as a Moderator of HRM Practices on Organizational Performance. *Int. J. Manpow.* **2013**, *34*, 794–812. [CrossRef]
110. Ringle, C.M.; Wende, S.; Will, S. *SmartPLS 2.0 (M3) Beta*; University of Hamburg: Hamburg, Germany, 2005.
111. Fornell, C.; Larcker, D.F. Structural Equation Models with Unobservable Variables and Measurement Error. *J. Mark. Res.* **1981**, *18*, 39–50. [CrossRef]
112. Barclay, D.; Higgings, C.; Thompson, R. The Partial Least Squares (PLS) Approach to Casual Modeling: Personal Computer Adoption and Use as an Illustration. *Technol. Stud.* **1995**, *2*, 285–309.

113. Zhang, J.; Weaver, S.E.; Hamill, A.S. Risks and Reliability of Using Herbicides at Below-Labeled Rates. *Weed Technol.* **2000**, *14*, 106–115. [CrossRef]
114. Akter, S.; D'ambra, J.; Ray, P. An Evaluation of PLS Based Complex Models: The Roles of Power Analysis, Predictive Relevance and GoF Index. 2011. Available online: https://ro.uow.edu.au/commpapers/3126 (accessed on 5 October 2022).
115. Wong, P.S.P.; Cheung, S.O. Structural Equation Model of Trust and Partnering Success. *J. Manag. Eng.* **2005**, *21*, 70–80. [CrossRef]
116. Kleinbaum, D.G.; Kupper, L.L.; Nizam, A.; Rosenberg, E.S. *Applied Regression Analysis and Other Multivariable Methods*; Cengage Learning: Boston, MA, USA, 2013; ISBN 128596375X.
117. Santosa, P.I.; Wei, K.K.; Chan, H.C. User Involvement and User Satisfaction with Information-Seeking Activity. *Eur. J. Inf. Syst.* **2005**, *14*, 361–370. [CrossRef]
118. Falk, R.F.; Miller, N.B. *A Primer for Soft Modeling*; University of Akron Press: Akron, OH, USA, 1992; ISBN 0962262846.
119. Wetzel, H.A.; Hammerschmidt, M.; Zablah, A.R. Gratitude versus Entitlement: A Dual Process Model of the Profitability Implications of Customer Prioritization. *J. Mark.* **2014**, *78*, 1–19. [CrossRef]
120. Saleem, F.; Gopinath, C. Antecedents of Environmental Conscious Purchase Behaviors. *Middle East J. Sci. Res.* **2013**, *14*, 979–986.
121. Teo, W.L.; Abd Manaf, A.; Choong, P.L.F. Information Technology Governance: Applying the Theory of Planned Behaviour. *J. Organ. Manag. Stud.* **2013**, *2013*. [CrossRef]
122. Adham, S.; Hussain, A.; Minier-Matar, J.; Janson, A.; Sharma, R. Membrane Applications and Opportunities for Water Management in the Oil & Gas Industry. *Desalination* **2018**, *440*, 2–17.
123. Shah Alam, S.; Masukujjaman, M.; Sayeed, M.S.; Omar, N.A.; Ayob, A.H.; Wan Hussain, W.M.H. Modeling Consumers' Usage Intention of Augmented Reality in Online Buying Context: Empirical Setting with Measurement Development. *J. Glob. Mark.* **2022**. [CrossRef]
124. Ansar, N. Impact of Green Marketing on Consumer Purchase Intention. *Mediterr. J. Soc. Sci.* **2013**, *4*, 650. [CrossRef]
125. Zhen, J.S.S.; Mansori, S. Young Female Motivations for Purchase of Organic Food in Malaysia. *Int. J. Contemp. Bus. Stud.* **2012**, *3*, 61–72.

buildings

MDPI

*Article*

# Green Evaluation for Building Interior Decoration Based on BIM-BN Technology

**Wenhan Fan [1,2,†], Baofeng Yan [3], Quanxi Bao [3], Yueqin Zhao [1] and Jianliang Zhou [1,*,†]**

1 School of Mechanics and Civil Engineering, China University of Mining and Technology, Xuzhou 221116, China
2 School of Engineering and Art Design, Shanxi Vocational and Technical College of Finance and Trade, Taiyuan 030031, China
3 The Third Construction Co. Ltd. of China Construction Eighth Engineering Division, Xuzhou 221100, China
* Correspondence: zhoujianliang@cumt.edu.cn; Tel.: +86-1589-521-2115
† These authors contributed equally to this work.

**Abstract:** The popularity of green building and BIM technology has increased globally, with strong government support in China. However, the integration of green requirements into interior decoration poses practical difficulties. Despite the few studies on the combination of green evaluation and BIM in building decoration, current methods largely rely on expert scores after completion. This research proposes a green evaluation index system for building interior decoration, examining the inter-index relationships and contribute to the final green degree of the project. Additionally, a green evaluation method based on BIM technology and the Bayesian network is explored, aimed at evaluating the green degree of design schemes, providing feedback, and supporting the realization of green interior decoration. With a focus on green evaluation, the current methods that rely solely on expert scores after completion will be improved. The results will provide technical support for the realization of green decoration and offer a reference for the improvement of green evaluation methods in the future.

**Keywords:** building interior decoration; green evaluation; BIM; bayesian network; BIM-BN technology

## 1. Introduction and Literature Review

Green building development has become a worldwide popular topic in recent years [1]. Due to the different factors such as the level of economic development, geographical environment, resource availability, and so on, there is no single, widely accepted definition for green building [2,3]. The concept of green building in China was developed from the "energy-saving and land-saving residential buildings" launched by the central government in 2004 [4]. According to the Chinese national standard "Evaluation Standard for Green Building," formulated in 2006, green buildings can be defined as buildings that maximize resource conservation (energy, land, water, materials), protect the environment, and reduce pollution throughout the life cycle, so as to provide people with healthy, suitable and efficient use of space and live in harmony with nature [5].

Society is continuously developing, and the living qualities of people are constantly improving. The requirements of the inhabitants for a drawing room are not confined to shelter from the elements. However, what is more important, is a good place for the study and work of mankind, and a quiet and comfortable resting place, which is singularly needed [6]. In China, the total energy consumption is huge, while the resource endowment is relatively poor, with high external dependence [7]. Green building (GB) certification has been conducted as a major policy for reducing the energy consumption and greenhouse gas (GHG) emissions of buildings in developed countries [8]. Therefore, the decoration design of green buildings emerged in response to the proper time and conditions. Currently, there are several green building evaluation standards for interior decoration. However, research on green evaluation of building interior decoration with BIM Technology is rare [9]. In addition, the existing green evaluation index system rarely considers the internal relationship between the indexes and the contribution mechanism to the green degree. Therefore, exploring green evaluation methods based on BIM technology is of great significance to the development of green evaluation in China.

It is complicated work to determine the weight of the index system, which greatly affects the accuracy of evaluation results [10]. Comparative analysis [11,12], whole life cycle theory [13–16], as well as environmental, economic, and social evaluation [17], are mainly used in the study of green building evaluation index system in China; Methods to give the weight of green building indicators mainly include the analytic hierarchy process [18–21], expert investigation method [22–24] and simple correlation function [25]; several commonly used evaluation models are grey clustering evaluation method [26–31], fuzzy comprehensive evaluation method [32,33], principal component analysis method [25,34,35], BP neural network [20,36,37].

As the main theme of the informatization development of China's construction industry in the next decade, a series of applications based on BIM has become an irreversible trend. Many scholars around the world have also conducted relevant studies on the application depth of BIM technology in green engineering evaluation, including green evaluation application of BIM in the whole life cycle of buildings [38], using BIM models to evaluate indoor energy-saving optimization design of green buildings [39], the application of BIM in architectural lighting design [40], there are also comparisons between the BIM model in conventional design and the requirements of green building regulations [41]. In addition, by studying the defects of BIM in green evaluation [9], it is found that most of the green BIM research focuses on the design and construction stage [42]. At the design stage, the integration method of BIM and LEED system [43,44] and the Green Building Assessment Tool (GBAT) [45] can be used to extract necessary data from the BIM model and calculate green rating in order to provide feedback for further evaluation.

With the continuous acceleration of economic development, environmental pollution is becoming more and more serious. In addition,, the contradiction between economic development and environmental protection continues to escalate. Thus, it is extremely urgent to vigorously promote green decoration. Considering that the application of BIM technology in the field of decoration is of great significance to the industrialization and sustainable development of Chinese housing. The application and trend of BIM technology in the field of decoration are deeply discussed by using the huge potential of software parametric modeling and integrated database. The main research direction of the next step is to apply BIM technology to assist the green evaluation of building interior decoration and to optimize the interior design scheme in order to truly achieve green decoration and create an environmentally friendly and comfortable indoor environment.

Bayesian networks were originally proposed in 1981 by R. Howard and J. Matheson. Early Bayesian networks were primarily used in expert systems to express uncertain expert knowledge. Since the 1990s, great progress has been made in the study of Bayesian network learning methods. Domestic scholars have also conducted relevant studies on Bayesian networks. The research area mainly focuses on security risks, including the application of Bayesian networks. In terms of analyzing the mechanism of risk events caused by risk factors [46], human reliability assessment on building construction work at the height [47], and dam risk analysis, a new method for dam risk analysis is provided [48]. Since the Bayesian network is a kind of uncertain knowledge expression and reasoning technology, it can determine the relationship between different influencing factors and then provide decision support [49]. Thus, the Bayesian network can be applied for risk management, fault diagnosis, and data mining research. As a very active field, it presents great advantages in analyzing the influence mechanism between things and opens up new ideas for studying the internal relationship between the green evaluation indicators of the decoration project and the final greenness contribution mechanism.

However, the existing green evaluation methods are relatively reliant on score-giving by experts after completion. In addition, it is unclear of the internal relationship between the evaluation indicators. There are many necessary data in BIM models for calculating the green rating and providing feedback for further evaluation. With BIM data support, designers can optimize the building's green design in the very early stages and produce a better solution. By the Bayesian networks support, it can simulate the contribution mechanism of each indicator to the green level. This paper aims to use BIM-BN technology to construct an indicator system specifically for the evaluation of green degree by the case study of building interior decoration, analyze the internal relationship between indicators and its contribution mechanism to the final green level, feed the evaluation results back into the decoration designer to provide technical support for the realization of green decoration. The research results not only help to improve the pertinence of green evaluation of decoration projects, provide a reference for improving green evaluation methods, but also promote the realization of the goals of saving resources, protecting the environment, and improving economic benefits.

## 2. Research Methodology

### *2.1. Rebuild the Green Evaluation Index System of Building Interior Decoration Based on BIM*

The Ministry of Construction of the People's Republic of China first proposed the concept of The Four Saving, One Environmental Protection in the "Green Construction Guideline", released in 2007. The Four Saving, One Environmental Protection means energy saving, land saving, water saving, material saving, and environmental protection. According to the definition and evaluation principle of "The Four Saving, One Environmental Protection" for green buildings, the Assessment standard for green interior decoration (T/CBDA 2-2016) was officially implemented in China on 1 December 2016 [50]. However, T/CBDA 2-2016 only gives the approximate weight of the first-level evaluation indicators (Mainly including: environmental protection, resource conservation, process management, etc.). The second-level indicators are only listed in the form of articles in which corresponding scores are given, and the setting of indicator levels is not clear enough. In addition, some articles are complicated and repetitive, which is not conducive to the development of green decoration evaluation and the promotion of green decoration projects.

In T/CBDA 2-2016, the evaluation components of green decoration include energy conservation and energy utilization, water conservation and water resource utilization, material conservation and material resource utilization, land conservation and space efficient utilization, indoor environmental quality, green construction management, and operation management. Each type of component involves evaluation indexes of control items and scoring items. The evaluation results of control items are either met or not, and the evaluation results of scoring items are scored as numbers. The overall evaluation process is shown in Figure 1. Firstly, experts give the corresponding score for each indicator in T/CBDA 2-2016 according to the submitted project materials and documents. Then the total score will be calculated for all evaluation components based on the experts scoring. After that, the total score of green evaluation of building interior decoration will be recalculated by adding the weights of all evaluation components. Finally, the evaluation grade is determined according to the recalculated total score.

Through an in-depth study of the green evaluation provisions of the green building interior decoration evaluation standard, valuable BIM indicators were selected through the BIM model, BIM schedule, and Ecotect secondary analysis. For example, the basic data of material saving, water saving, energy saving, and indoor air quality could be extracted from the BIM element properties and BIM schedule report. The other basic environmental data, such as indoor thermal, light, acoustic, and space use, could be output by the Ecotech secondary analysis. Figure 2 shows the green evaluation indicator system of building interior decoration based on BIM technology. The weights of each indicator are based on the corresponding weights in the China national standard Assessment standard for green interior decoration (T/CBDA 2-2016).

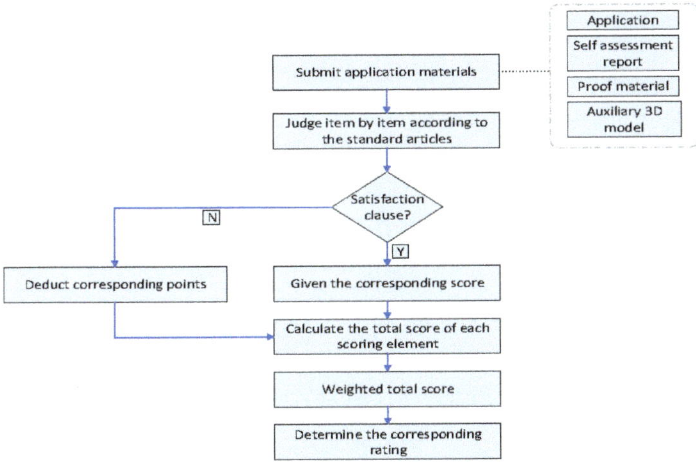


**Figure 1.** Scoring method for building interior decoration based on green evaluation standards.

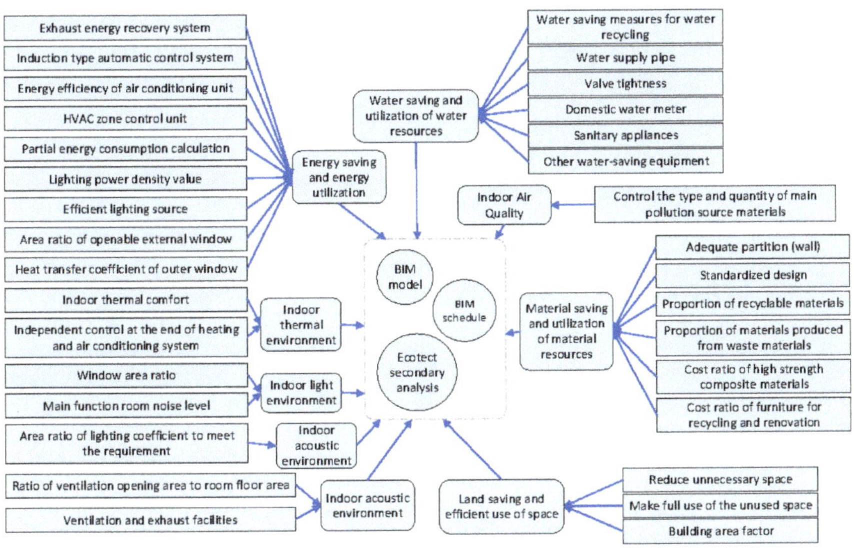


**Figure 2.** Green evaluation index system of building interior decoration based on BIM.

## *2.2. Green Evaluation Method of Building Interior Decoration Based on BIM-BN Technology*

Bayesian network (BN) is a combination of graph theory and probability theory, which can be used to describe the dependence between variables. At present, it can be divided into three categories: discrete Bayesian network, continuous Bayesian network, and hybrid Bayesian network. In general, learning Bayesian networks follows a 'five-step' principle: defining node variables, determining node states, BN structure learning, BN parameter learning, and BN inference optimization.

The green degree can be defined as the level of "The Four Saving, One Environmental Protection" of building interior decoration. According to the evaluation grade division principle in green building interior decoration evaluation standard, the ultimate green degree of building interior decoration is divided into four grades from state0 to state3, which the four states corresponding to target nodes in the green evaluation model of building interior decoration based on BIM-BN technology are constructed as the following Figure 3.

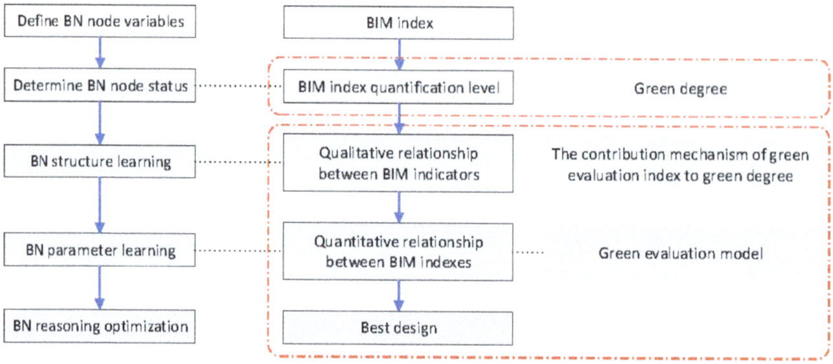


**Figure 3.** Evaluation method for building interior decoration based on BIM-BN Technology.

In Figure 3, the BIM index of the interior building decoration green evaluation indicator is defined as the BN node variable. After the quantitative grade of the BIM index is determined as the BN node state, the qualitative relationship between BIM indexes can be determined by BN structure learning, and the quantitative relationship between BIM indexes also can be determined by using BN parameter learning. Therefore, the green evaluation model of building interior decoration based on BIM-BN is constructed. This model-based combination of BIM and BN can then be used to explore the contribution mechanism of green evaluation indicators to the green degree of the project.

### *2.3. Green Evaluation Process of Building Interior Decoration Based on BIM-BN Technology*

For the green goal to be achieved, the combination of BN and BIM can provide evaluation feedback to improve and optimize the design. The BIM can provide parameter data and information for the green evaluation of interior decoration. The BN can provide a breakthrough point for the application of results-oriented theory in the green evaluation of building interior decoration. Its principles of backward reasoning and forward verification can be used to realize not only the green design optimization of interior decoration but also help to reduce the energy consumption of interior decoration.

As the following Figure 4, the green degree evaluation process of interior decoration is divided into two stages: sub-item evaluation and overall evaluation. The former is the basis, and the latter is the purpose. First, all information of the BIM model, BIM schedule, and Ecotect secondary analysis are used to evaluate the green evaluation indicators of the interior decoration project based on BIM. Then the final green degree of the interior decoration is evaluated by the BN model. For the green goal to be achieved, the reasoning function of the BN model is used to improve the design scheme through evaluation feedback and then obtain the optimal design scheme to ensure the realization of green interior decoration and reduce decoration energy consumption.

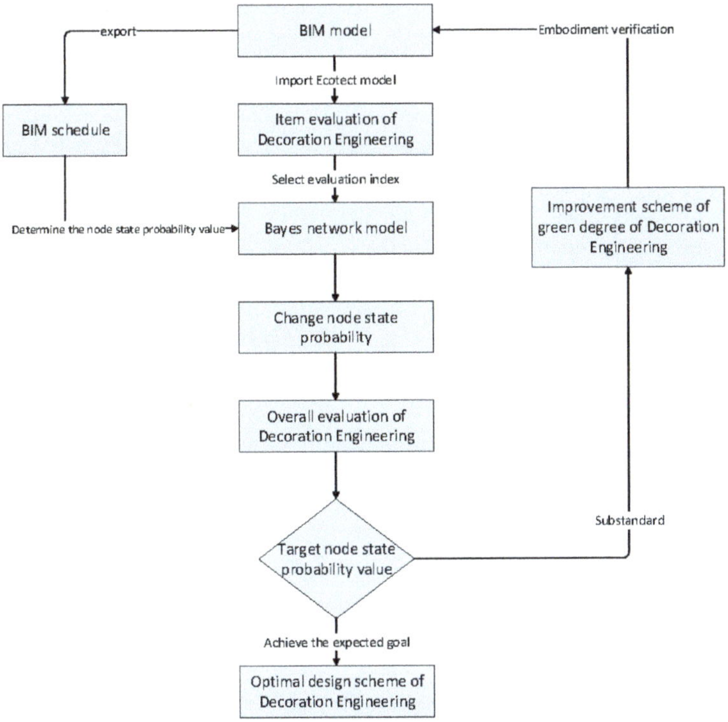


**Figure 4.** The evaluation process of building interior decoration is based on BIM-BN Technology.

## 3. Implementation of the Green Decoration Assessment Based on BIM-BN Technology

### *3.1. BIM Model Information for Green Decoration Assessment*

In the BIM model, the information can be divided into geometric information and nongeometric information. Geometric information refers to the information of shape, size, and location in the interior and exterior space of the building model, including the length, width, height, coordinate, area, volume, etc., of the components of the BIM model; nongeometric information refers to the general designation of other characteristic reflecting the interior and exterior space of the building model as well as geometric information, including the physical characteristics, technical information, product information and construction of the components Information, maintenance information, etc.

In addition, in order to make better use of BIM model data and information for green evaluation of building interior decoration, it is necessary to follow the corresponding decoration BIM creation standards and clarify the input and output rules of BIM information. According to "Implementation standard for BIM of building decoration engineering" (T/CBDA 3-2016) published in China on 1 December 2016, the additional BIM information includes component name, component code, and component attribute information [51]. The rule of component name is defined as model type—model element name, such as window—aluminum alloy window; the naming rule of component code is material code—Model—Specification—number, such as C-C0910-900 * 1000-01, which corresponds to the material ID number of the local material database; The attribute information can be input as required according to the real information of component material.

Nevertheless, some information and data for green decoration evaluation need to be calculated by BIM sustainability analysis software, such as Ecotect Analysis or Green Building Studio. Consequently, the green decoration evaluation information could be divided into three categories based on the evaluation criteria for green building interior decoration:

1. The information can be stored directly in BIM modeling software, such as "selecting a more water-efficient sanitary appliance." This sort of information can be viewed directly by the BIM visualization function or schedule.
2. The information needs to be analyzed in BIM sustainable software, such as the power density of indoor lighting reaches the target value of architectural lighting design standard of the "General Code for Energy Efficiency and Renewable Energy Application in Buildings" in China [52]; and the lighting of the main operation rooms meets the architectural lighting design standards of "General Code for Building Environment" in China [53]. The acquisition of such information also involves data exchange between different types of software.
3. The information cannot be output by BIM or BIM sustainable software. This kind of green evaluation information can only be obtained through actual measurements or traditional methods, which are sampling and testing the release of harmful substances from the main indoor pollution sources, including materials and furniture. These measurements and methods only aimed to fulfill the requirements of pre-evaluation of indoor pollution in the design stage without taking the comprehensive release rate of formaldehyde and TVOC pollution into consideration.

According to the above classifications, only the first two types of information are based on BIM technology. Both of them will be considered in the green evaluation index of building interior decoration. In addition, other information related to materials, furniture, and equipment should also take attribute information into consideration in BIM modelings, such as brand, model, manufacturer, and specification.

### *3.2. Quantification of Green Evaluation Index of Building Interior Decoration*

Compared with energy conservation and energy utilization, water conservation and water resource utilization, material conservation and material resource utilization, and land conservation and efficient space utilization, residents concentrated on the indoor environment quality which is closely related to their own health. Therefore, in order to ensure that the scope of the research is limited to energy and sustainability, in this paper, only five evaluation elements are selected to evaluate the indoor environment quality, which are the indoor thermal environment, indoor light environment, indoor acoustic environment, indoor wind environment, and indoor air environment.

According to the above five evaluation elements in the China national standard (T/CBDA 2-2016), combined with the analysis characteristics of BIM Technology, the indicators that have a great impact on the green degree of buildings are selected with careful considerations. The quantitative standard corresponding to each indicator in the design stage was further refined to obtain the quantitative level. In the subsequent evaluation process, as long as the quantitative standards are compared, the corresponding index evaluation grade can be determined quickly. Here, indoor environmental quality was chosen as an example to obtain the quantitative grade table of the evaluation indicators (Table 1).

**Table 1.** The quantitative grade for the index of indoor environmental quality assessment elements.

| | | Quantization Standard Stated in T/CBDA 2-2016 | Quantization Level |
|---|---|---|---|
| Indoor thermal environment | Indoor thermal comfort | $-1 \leq$ Predicted Mean Vote (PMV) $\leq 1$ and $9\% \leq$ Predicted Percentage Dissatisfied (PPD) $\leq 29\%$ | Yes |
| | | | No |
| | Independent control at the end of heating and air conditioning system | The proportion of main function rooms that can be independently started and stopped for adjustment is less than 70% | Bad |
| | | $70\% \leq$ the proportion of main function rooms with independent start stop adjustment $< 90\%$ | Medium |
| | | $90\% \leq$ proportion of main function rooms that can be independently started and stopped for adjustment | Good |
| | Outdoor view of room | The main function rooms are provided with external windows, which have good vision and avoid the interference of vision in the living space between households | Yes |
| | | | No |
| Indoor light environment | Window area ratio | The window area ratio of bedroom and living room is less than 1/6 | Bad |
| | | The window area ratio of bedroom and living room is 1/6 | Medium |
| | | The window area ratio of bedroom and living room is 1/5 | Good |
| | Lighting coefficient meets the proportion of required area | $60\% \leq$ area ratio | Yes |
| | | | No |
| | Room surface reflectance | Reflection ratio requirements—Ceiling: 0.60~0.90; wall: 0.30~0.80; floor: 0.10~0.50; desktop, worktop, equipment surface: 0.20~0.60 | Yes |
| | | | No |
| | Indoor lighting light source color | Determine the light source color according to the illuminance requirements and indoor surface conditions | Yes |
| | | | No |
| Indoor acoustic environment | Indoor noise level of main function room | Meet the low limit value and average value required by high standard in the current code for sound insulation design of civil buildings. | Yes |
| | | | No |
| Indoor wind environment | Ratio of ventilation opening area to room floor area | 10% in hot summer and warm winter; 8% in hot summer and cold winter; 5% in other areas | Yes |
| | | | No |
| | Ventilation and exhaust facilities | Ventilation and auxiliary smoke exhaust facilities shall be provided in kitchen and toilet without external windows | Yes |
| | | | No |
| | Air distribution design of main function rooms | Area ratio $< 60\%$ | Bad |
| | | $60\% \leq$ area ratio $< 70\%$ | Medium |
| | | $70\% \leq$ area ratio $< 80\%$ | Good |
| | | $80\% \leq$ area ratio | Excellent |
| Indoor Air Quality | Pre assessment of indoor air pollution | Prediction of indoor air pollutant concentration value at completion acceptance | Yes |
| | | | No |
| | | Predict and analyze the air pollutant concentration value of each construction node in the indoor construction process | Yes |
| | | | No |
| | | Prediction and analysis of indoor air pollutant concentration change and attenuation rate law | Yes |
| | | | No |
| | Control of hazardous materials | Put forward the pollution release characteristic parameters of main pollution source materials and furniture | Yes |
| | | | No |
| | | Prediction of comprehensive emission rate of formaldehyde and TVOC in materials and furniture | Yes |
| | | | No |
| | Control the type and quantity of main pollution source materials | Control the type and quantity of main pollution source materials in the room and replace at least one main pollution source material | Yes |
| | | | No |
| | Indoor air quality monitoring | Data collection and analysis of indoor CO2 concentration | Yes |
| | | | No |
| | | Real time alarm when indoor pollutant concentration exceeds | Yes |
| | | | No |

### 3.3. Formatting of Mathematical Components

The process of constructing a Bayesian network structure normally requires a lot of data for network structure training. However, it is difficult to obtain relatively complete green evaluation data of building interior decoration. Moreover, it is easy to lead to deviation when using incomplete data for research. On the basis of national standards and the relevant literature, the causal relationships between nodes were identified as the preliminary version. The final causal relationships between nodes were determined after conducting expert interviews with 20 design institutes. After that, the Bayesian network model of interior environmental quality assessment of building interior decoration is established, as shown in Figure 5.

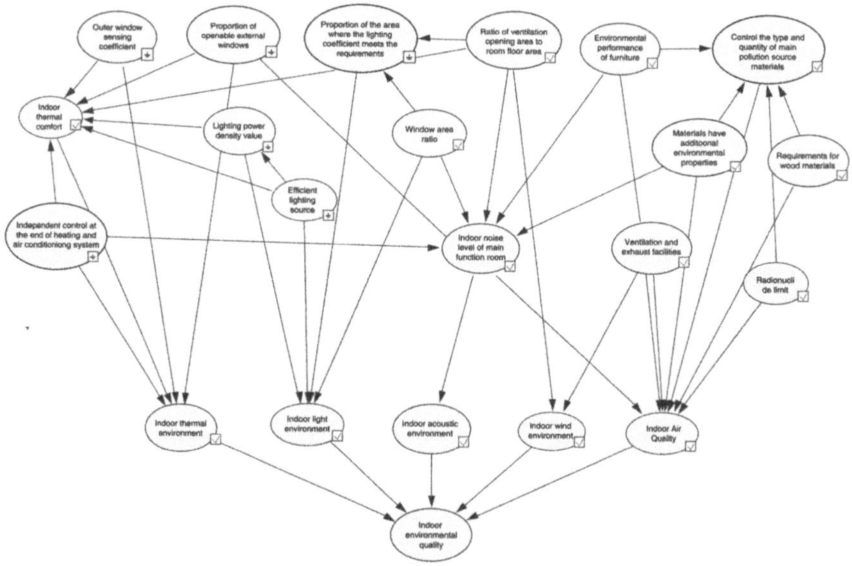


**Figure 5.** The Bayesian network structure of interior environmental quality assessment of building interior decoration.

In this BN model, the green evaluation index of building interior decoration based on BIM technology is regarded as a node variable, and the quantization level is determined according to the quantization standard of the index system to determine the status of the evidence node and intermediate node. According to the evaluation grade division principle of the evaluation standard for interior decoration of green buildings, each node and target node corresponding to the first level index are divided into four states, among which the worst state is corresponding.

In general, the indoor environmental quality of building interior decoration is affected by the indoor thermal environment, indoor light environment, indoor acoustic environment, indoor wind environment, and indoor air quality. The interaction between the secondary indicators can be clearly found from the constructed Bayesian network diagram of indoor environmental quality evaluation of building interior decoration. Among them, the ratio of window area to floor area and the ratio of ventilation opening area to room floor area affect the proportion of the area where the lighting coefficient meets the requirements; The indoor thermal comfort could be affected by many influencing factors, including the heat transfer coefficient of the external window, the independent control of the heating and air conditioning system end, the ratio of the area of the openable external window and the lighting power density value, the environmental performance of furniture, the additional environmental performance of materials, the ratio of the area of the window to the floor, as well as indoor thermal comfort, the indoor noise level of main functional rooms, types and quantity of main pollution sources.

## 4. Case Study

### 4.1. Item Evaluation of Green Degree of Building Interior Decoration Based on BIM

An interior decoration project in Xuzhou, China, was used as a case study. It is a sub-work of residential real estate. The case is one typical apartment on the 6th floor of this project. It is three bedrooms and two living rooms, which are shown in Figure 6. The Ecotect analysis software was selected to simulate and calculate indoor thermal comfort. As shown in Figure 7, it can be seen that the average value of PMV in the chosen building is 0.31. In the ISO 7730-2005 thermal comfort criteria, the recommended value for the PMV indicator is −0.5 to +0.5, while PPD is < 10% [54]. Combined with Chinese conditions and relevant studies, the generally acceptable range of thermal comfort is PMV: $-1.0 < \text{PMV} < 1.0$; the corresponding PPD < 26%. Therefore, for an ordinary room without heating and air conditioning, PMV = 0.3 represents a comfortable environment. in Figure 8, it can be seen that the average value of PPD is 17.24, which is also within a reasonable range [54,55]. In conclusion, the indoor thermal comfort index of the building meets the standard. Thus, the indoor thermal comfort value state1 is set to 100%.

**Figure 6.** The BIM plan layout of the apartment as a case study in Xuzhou, China.

The results can also be verified through the window list generated by Revit software (Figure 9): all windows of the case project are made of aluminum alloy glass, and the heat transfer coefficient is more than 2.0 W/($m^2$· k). Therefore, it is determined that the external window heat transfer coefficient does not meet the standard requirements. Hence, the value is set to 100%.

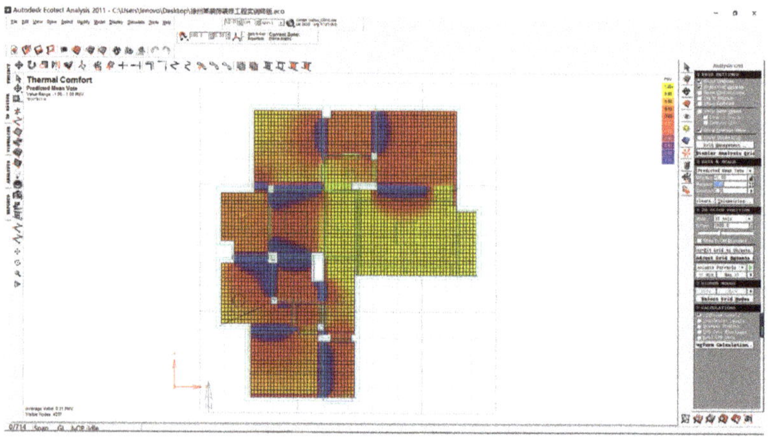


**Figure 7.** Thermal comfort PMV analysis results of indoor areas of a decoration project.

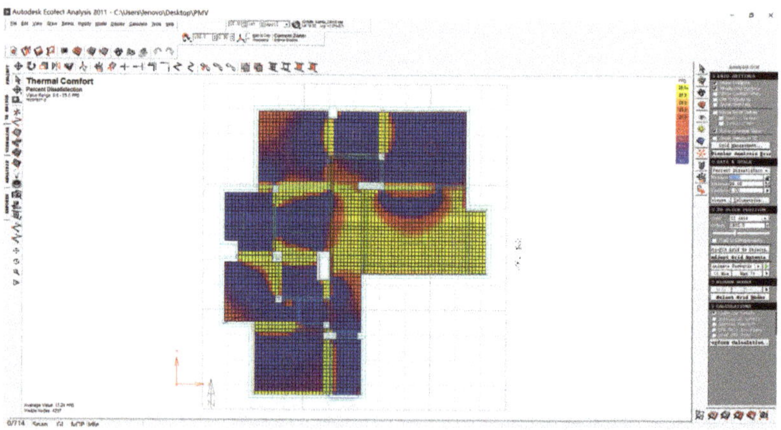


**Figure 8.** Analysis results of thermal comfort PPD in each area of a decoration project in Xuzhou.

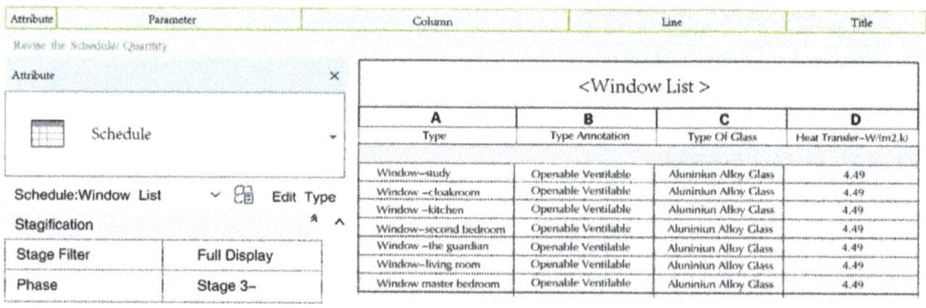


<Window List >

| A<br>Type | B<br>Type Annotation | C<br>Type Of Glass | D<br>Heat Transfer–W/(m2.k) |
|---|---|---|---|
| Window–study | Openable Ventilable | Aluninium Alloy Glass | 4.49 |
| Window –cloakroom | Openable Ventilable | Aluninium Alloy Glass | 4.49 |
| Window –kitchen | Openable Ventilable | Aluninium Alloy Glass | 4.49 |
| Window–second bedroom | Openable Ventilable | Aluninium Alloy Glass | 4.49 |
| Window –the guardian | Openable Ventilable | Aluninium Alloy Glass | 4.49 |
| Window–living room | Openable Ventilable | Aluninium Alloy Glass | 4.49 |
| Window master bedroom | Openable Ventilable | Aluninium Alloy Glass | 4.49 |

**Figure 9.** Analysis results of heat transfer coefficient of the exterior window of a decoration project in Xuzhou.

Through the above methods, the final indoor environmental quality sub-evaluation results were obtained (Table 2).

**Table 2.** Evaluation of indoor environmental quality.

| Evaluation Item | Valuation Index | Node Probability Value |
|---|---|---|
| Indoor thermal environment assessment | Indoor thermal comfort | State1 = 100% |
| | Heat transfer coefficient of outer window | State0 = 100% |
| | Area ratio of openable external window | State2 = 100% |
| | Independent control at the end of heating and air conditioning system | State2 = 100% |
| Indoor light environment evaluation | Proportion of the area where the lighting coefficient meets the requirements | State0 = 100% |
| | Window area ratio | State2 = 100% |
| | Lighting power density value | State2 = 100% |
| Indoor acoustic environment assessment | Indoor noise level of main function room | State1 = 100% |
| Indoor wind environment assessment | Ratio of ventilation opening area to room floor area | State1 = 100% |
| | Ventilation and exhaust facilities | State1 = 100% |
| Indoor air quality assessment | Environmental performance of furniture | State1 = 100% |
| | Materials have additional environmental functions | State2 = 100% |
| | Material radionuclide limit | State1 = 100% |
| | Requirements for wood materials | State0 = 100% |
| | Control the types and quantity of main pollution sources | State1 = 100% |

### *4.2. General Assessment of Indoor Environmental Quality*

On the basis of sub-item evaluation, the established Bayesian network model with node condition probability was used for the quantitative evaluation of the indoor environmental quality of the case project, and the evaluation results are shown in Figure 10.

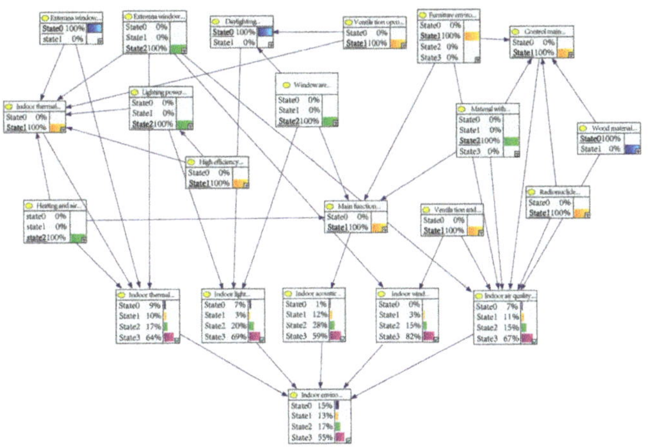

**Figure 10.** Indoor environmental quality evaluation model of a decoration project in Xuzhou.

It can be seen from Figure 10 that the probability value of the indoor thermal environment index of the case project: the probability value of reaching a three-star rating standard is 64%; the probability value of the indoor light environment index reaching a three-star rating standard is 69%; the probability value of the indoor acoustic environment index reaching three-star rating standard is 59%; the probability value of the indoor wind environment index reaching three-star rating standard is 82%; the indoor air quality index reaching the three-star rating standard is 64%. The probability value of the average standard is 67%. Finally, the probability that the indoor environmental quality of the target node does not meet the green evaluation standard of the decoration project is 15%, the probability of reaching the one-star rating standard is 13%, the probability of reaching the two-star rating standard is 17%, and the probability of reaching the three-star rating standard is 55%.

### *4.3. Optimization Analysis of Indoor Environment Quality*

Based on the current status, the node status of each secondary evaluation index is upgraded by one level. For example, the node value of "Heat transfer coefficient" is changed from "state0 = 100%" to "state1 = 100%". It means the designer requires to change the new window glass material, which has lower thermal conduction properties. According to this conception, the Bayesian network model is updated, and the evaluation results are shown in Figure 11. The indoor thermal environment state 3 value increased from 64% to 79%; the indoor light environment state 3 value increased from 69% to 78%; the indoor acoustic environment state 3 value increased from 59% to 68%; the indoor wind environment state 3 value increased from 82% to 85%; the indoor air quality environment state 3 value increased from 67% to 81%; the value of state 3 of target node indoor environment quality increased from 55% to 78%. The state 3, was 41.82% higher than the previous design scheme. The data demonstrates that the above optimization scheme is feasible for the indoor environmental quality of building interior decoration.

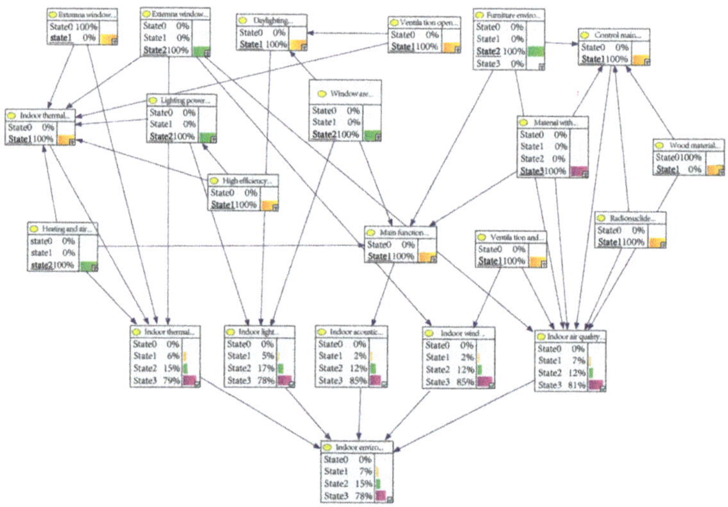


**Figure 11.** Bayesian network reasoning optimization diagram of indoor environmental quality assessment of a decoration project in Xuzhou.

## 5. Discussion

As an indispensable part of construction projects, interior decoration has an inevitable impact on resource conservation and public health. Nonetheless, the current green decoration faces many practical difficulties with less guarantee. As a big energy-consuming country, it is of great significance for China to research and develop new energy-saving technologies and enhance energy efficiency. Moreover, as the main theme of the development of China's construction industry in the next decade, a series of applications based on BIM has become an irreversible trend.

In the practice of the construction industry, there is still a lot of work for the green evaluation of building interior decoration based on BIM. The existing green evaluation methods that focus on qualitative evaluation seem to relatively rely on the subjective experience of experts. The relationship between the evaluation index and the contribution mechanism of greenness was still not clear. On the other hand, only BIM does not have the evaluation function of green buildings. It needs to add much of information and do some post-analysis work so that the designer can make simulations and optimization for the design.

Previous studies were often based on the original information type carried out by the BIM model. However, the results of this study show that, to a certain extent, the original information type cannot meet all the requirements of rule checking. For example, the parameter type attributes of BIM model components are basically physical attributes and generally do not include the function-based parameters proposed in the standard. Based on this finding, two methods, individual labeling, and centralized labeling were proposed, which can improve the information content of the model. In addition, this study not only analyzed the properties of the model components but also proposed two algorithms that be used to verify the spatial relationship between the model components. Therefore, this study expands the range of information that the model can respond to rule checking.

However, there are also some limitations to this study. This paper only focuses on the improvement of green evaluation methods for interior environmental quality evaluation indicators of decoration projects that are combined with BIM and the Bayesian network. In the future, continuous improvement of the green evaluation index system should be taken into consideration, as well as the range of research scope. On the other hand, this paper only studied the traditional decoration in green evaluation. However, China is encouraging assembly decoration to replace traditional decoration. In the future, the role of BIM-BN technology in the green evaluation of decoration projects can be discussed from the perspective of assembly and decoration so as to maximize the realization of the "green" level.

## 6. Conclusions

Considering the urgent need for green decoration and the great potential of BIM technology, this paper tries to build an index system for the green evaluation of interior decoration projects, with emphasis on the internal relationship between the indicators and the contribution mechanism to the final green degree of the project. In addition, this paper also aims to explore a green evaluation method based on BIM and Bayesian network so as to evaluate the design scheme green and evaluate the results. Which could feedback on the design scheme and provide technical support for the realization of green decoration. Thus, the improvement of green evaluation methods could be accomplished by the given references.

First of all, the green evaluation index system of interior decoration projects is established through an in-depth study of relevant standards, specifications, documents, and other information on green evaluation of interior decoration projects. Then, the corresponding quantitative standards of each index are further refined according to the order of five evaluation elements: energy saving and energy utilization, water saving and water resource utilization, materials and materials utilization, efficient utilization of land and space, and indoor environmental quality. Therefore, the quantitative grade is obtained. At the same time, information can be directly stored in BIM modeling software, and BIM sustainable analysis software data exchange functions are also obtained. On the basis of information extraction, the index system of decoration and fitment project based on BIM technology is determined, which provides convenience for the follow-up study of the contribution mechanism and green evaluation method based on the Bayesian network.

Secondly, the evaluation model of four states of target nodes in the green evaluation of building interior decoration based on BIM-BN technology is constructed on the basis of learning the basic theorem and elements of the Bayesian network through the structure learning and parameter learning of the Bayesian network. When using this evaluation model, the forward reasoning optimization and reverse reasoning optimization of the Bayesian network are carried out, respectively, and the results are sorted out. The Qualitative/quantitative relationship between evaluation indexes of the green degree of decoration projects based on BIM technology and the contribution mechanism of each index to the green degree is discovered.

Lastly, a green evaluation method of decoration projects based on BIM- BN technology is proposed by combining BIM technology and Bayesian network effectively with the guiding principle of results-oriented principle. In the design stage, the BIM model, BIM detailed list, and BIM sustainability analysis software are used to evaluate the greenness of interior decoration projects; on this basis, the contribution mechanism and influence path obtained by Bayesian network learning is used to evaluate the overall greenness of decoration and building interior decoration; and based on the evaluation results, the design scheme is reasoned and optimized to quickly lock in the best designer. So as to avoid the blindness of the design scheme and ensure the realization of green decoration.

**Author Contributions:** Conceptualization, J.Z. and W.F.; methodology, J.Z.; software, W.F.; validation, B.Y. and Y.Z.; formal analysis, W.F. and B.Y.; investigation, B.Y., Q.B. and Y.Z.; resources, J.Z. and W.F.; data curation, W.F. and B.Y.; writing—original draft preparation, W.F., B.Y. and Q.B.; writing review and editing, J.Z., B.Y. and Y.Z.; visualization, W.F., Q.B. and Y.Z.; supervision, J.Z.; project administration, J.Z. All authors have read and agreed to the published version of the manuscript.

**Funding:** This work was supported by the National Natural Science Foundation of China (Grant number 72171224) and The Humanities and Social Sciences Foundation of China's Education Ministry (Grant number 19YJAZH122).

**Data Availability Statement:** Not applicable.

**Acknowledgments:** The authors sincerely acknowledge the editors and anonymous reviewers for their valuable comments and constructive suggestions, which considerably improved the exposition of this work. The authors also gratefully acknowledge those who provided data and suggestions.

**Conflicts of Interest:** The authors declare no conflict of interest.

## References


1. Wu, Z.; Li, H.; Feng, Y.; Luo, X.; Chen, Q. Developing a green building evaluation standard for interior decoration: A case study of China. *Build. Environ.* **2019**, *152*, 50–58. [CrossRef]
2. Ding, Z.; Fan, Z.; Tam, V.W.Y.; Bian, Y.; Li, S.; Illankoon, I.M.C.S.; Moon, S. Green building evaluation system implementation. *Build Env.* **2018**, *133*, 32–40. [CrossRef]
3. Li, Y.; Chen, X.; Wang, X.; Xu, Y.; Chen, P. A review of studies on green building assessment methods by comparative analysis. *Energ Build.* **2017**, *146*, 152–159. [CrossRef]
4. He, Y.; Kvan, T.; Liu, M.; Li, B. How green building rating systems affect designing green. *Build Env.* **2018**, *133*, 19–31. [CrossRef]
5. *G. T. 50378–52006*; China, M.O.H.A. Evaluation Standard for Green Building. Architecture and Building Press: Beijing, China, 2006.
6. Gong, W. Indoor air quality is affected by construction materials and related evaluation analysis. *Fujian Constr. Mater.* **2015**, *1*, 82–83.
7. Fang, Y.; Zhang, W. Current Situation and Development Trend of Energy Resources in China. *Miner. Prot. Util.* **2018**, *38*, 34–42.
8. Qian, S.; Yu, Y.; Jian, Z.; Tao, Y. Objective conflicts in green buildings projects: A critical analysis. *Build. Environ.* **2016**, *96*, 107–117.
9. Cao, Y.; Xu, C.; Kamaruzzaman, S.N.; Aziz, N.M. A Systematic Review of Green Building Development in China: Advantages, Challenges and Future Directions. *Sustainability* **2022**, *14*, 12293. [CrossRef]
10. Zhao, T.; Cheng, D. Research on Green Degree Evaluation of Residential Buildings Based on IAHP. *Eng. Constr.* **2019**, *51*, 81–85.
11. Zhu, R.; Wang, Q. Comparative Analysis of the Weights of Green Standards for Typical Existing Buildings Abroad. *Constr. Technol.* **2014**, *43*, 14–17.
12. Huang, L. Application Research on the Evaluation Index System of Green Buildings in China. *J. Hebei North Univ. (Nat. Sci. Ed.)* **2017**, *33*, 32–37.
13. Zhang, M. Research on Evaluation System and Evaluation Method of Green House. Master's Thesis, Beijing Jiaotong University, Beijing, China, 2009.
14. Cai, X. Research on Low Carbon Building Evaluation Based on LCA. Master's Thesis, Jiangnan University, Wuxi, China, 2011.
15. Li, J. Research on Green Building Evaluation. Master's Thesis, Tianjin University, Tianjin, China, 2013.
16. Shen, L.; Li, X. BIM-LCA-based Method for EIA of Building Environment. *For. Eng.* **2015**, *31*, 145–155.
17. Huang, S. Research on Green Construction Evaluation System for Construction Projects. Master's Thesis, Tianjin University, Tianjin, China, 2016.
18. Lyu, C.; Hu, J.; Zhang, R.; Chen, W.; Xu, P. Optimizing the evaluation model of green building management based on the concept of urban ecology and environment. *Front Ecol. Evol.* **2023**, *10*, 1234. [CrossRef]
19. Li, L. Research on 6E Comprehensive Evaluation System for Greenness of Building Materials. Ph.D. Thesis, Dalian University of Technology, Dalian, China, 2012.

20. Li, X. Green Building Evaluation Based on BP Neural Network. *J. Saf. Environ.* **2012**, *12*, 103–106.
21. Xu, W.; Chang, X. Green Construction Evaluation Based on Variable Fuzzy Sets. *J. Shenyang Univ. Technol.* **2015**, *37*, 690–694.
22. Lan, J. Evaluation and Optimization of Green Buildings Based on BIM Technology. Ph.D. Thesis, Southeast University, Nanjing, China, 2016.
23. Cao, Z. Research on Greenness Evaluation of Prefabricated Concrete Public Buildings. Ph.D. Thesis, Beijing Jiaotong University, Beijing, China, 2019.
24. Zhao, W. Dynamic fusion mechanism of building information modeling technology and low-carbon green building evaluation index system. *Sci. Technol. Eng.* **2019**, *19*, 196–201.
25. Qin, Q. Optimization and Model Research of China's Green Building Evaluation Index System. Master's Thesis, East China Jiaotong University, Nanchang, China, 2014.
26. Xu, P. Research on Green Construction Evaluation System. Master's Thesis, Chongqing University, Chongqing, China, 2008.
27. Li, H.; Li, J. Evaluation of Green Construction Based on Grey Clustering Method. *J. Eng. Manag.* **2012**, *26*, 18–22.
28. Yan, J. Research on China's Green House Evaluation System. Master's Thesis, Xi'an University of Architecture and Technology, Xian, China, 2012.
29. Duan, X.; Bao, X. Evaluation of Structural Design of Green Residential Buildings Based on Grey Clustering Method. *J. Eng. Manag.* **2013**, *27*, 27–30.
30. Bao, X.; Wang, Q. Evaluation of Green Building Design Schemes Based on Grey Clustering Method. *Build. Econ.* **2014**, *375*, 84–86.
31. Jia, Y. Comprehensive Evaluation of Green Construction Based on Grey Clustering Method. Master's Thesis, Lanzhou Jiaotong University, Lanzhou, China, 2015.
32. Liu, X.; Zuo, Y.; Yin, Z.; Liang, C.; Feng, G.; Yang, X. Research on an evaluation system of the application effect of ground source heat pump systems for green buildings in China. *Energy* **2023**, *262*, 125374. [CrossRef]
33. Liu, H. Research on Construction and Comprehensive Evaluation of Green Building Index System. Master's Thesis, Tianjin University, Tianjin, China, 2013.
34. Xia, L.; Pan, J. Research on Evaluation Index System and Model of Green Buildings. *Build. Econ.* **2010**, *332*, 97–101.
35. Yang, M. Research on Green Degree Evaluation Model of Building Construction Based on Fuzzy Matter Element. Master's Thesis, Xi'an University of Architecture and Technology, Xian, China, 2014.
36. Wan, Y.; Wang, D. Green Construction Evaluation of Decoration Engineering Based on BP Artificial Neural Network. *Constr. Technol.* **2009**, *38*, 475–478.
37. Qi, B.; Niu, Y. Green Building Evaluation Method Based on BP Neural Network and Its Application. *J. Shenyang Jianzhu Univ. (Soc. Sci. Ed.)* **2012**, *14*, 152–156.
38. Li, H.; Wang, C. The Construction of Green Building Integrated Evaluation System Based on BIM Technology. *Mob. Inf. Syst.* **2022**, *2022*, 5906827. [CrossRef]
39. Wan, Y.; Zhai, Y.; Wang, X.; Cui, C. Evaluation of Indoor Energy-Saving Optimization Design of Green Buildings Based on the Intelligent GANN-BIM Model. *Math Probl. Eng.* **2022**, *2022*, 3130512. [CrossRef]
40. Li, N.; Liu, X. Application Prospect of BIM Technology in Architectural Lighting Design. *J. Light. Eng.* **2013**, *24*, 12–15.
41. Wei, H.; Li, C. Key points of BIM model applied to the evaluation of green residential buildings. In Proceedings of the 9th International Conference on Green Buildings and Building Energy Conservation, Beijing, China, 11–14 September 2013.
42. Feng, N. The Influence Mechanism of BIM on Green Building Engineering Project Management under the Background of Big Data. *Appl. Bionics. Biomech.* **2022**, *2022*, 8227930. [CrossRef] [PubMed]
43. Jalaei, F.; Jrade, A. Integrating building information modeling (BIM) and LEED system at the conceptual design stage of sustainable buildings. *Sustain. Cities Soc.* **2015**, *18*, 95–107. [CrossRef]
44. Wei, T.; Chen, Y. Green building design based on BIM and value engineering. *J. Amb. Intel. Hum. Comp.* **2020**, *11*, 3699–3706. [CrossRef]
45. Ilhan, B.; Yaman, H. Green building assessment tool (GBAT) for integrated BIM-based design decisions. *Automat. Constr.* **2016**, *70*, 26–37. [CrossRef]
46. Wang, T.; Liao, B. Bayesian network-based construction safety risk probability assessment method. *China Civ. Eng. J.* **2010**, *43*, 384–391.
47. Li, X.; Guo, Y.; Ge, F.; Yang, F. Human reliability assessment on building construction work at height: The case of scaffolding work. *Saf. Sci.* **2023**, *159*, 106021. [CrossRef]
48. Zhou, J.; Tang, C. Application of Bayesian network to dam risk analysis. *J. Hydroelectr. Eng.* **2010**, *29*, 192–196.
49. Hanea, A.M.; Christophersen, A.; Alday, S. Bayesian networks for risk analysis and decision support. *Risk Anal.* **2022**, *42*, 1149–1154. [CrossRef] [PubMed]
50. *T/CBDA 2-2016*; Assessment Standard for Green Interior Decoration. China Building Industry Press: Beijing, China, 2016; p. 158.
51. *T/CBDA 3-2016*; Implementation Standard for BIM of Building Decoration Engineering. China Building Industry Press: Beijing, China, 2016; p. 357.
52. *GB 55015-2021*; General Code for Energy Efficiency and Renewable Energy Application In Buildings. China Building Industry Press: Beijing, China, 2021; p. 188.
53. *GB 55016-2021*; General Code for Building Environment. China Building Industry Press: Beijing, China, 2021; p. 100.

54. *EN ISO 7730-2005*; Ergonomics of the Thermal Environment. Analytical Determination and Interpretation of Thermal Comfort Using Calculation of the PMV and PPD Indices and Local Thermal Comfort Criteria. British standards institution: London, UK, 2006; p. 64.
55. Yang, X.B.; Sun, D.F.; Zhou, X.J.; Lin, G.P. Literature survey on building rankings by indoor environment quality. In Proceedings of the International Conference on Civil Engineering and Transportation (ICCET 2011), Jinan, China, 14–16 October 2011; pp. 3047–3050.

*buildings*

MDPI

*Article*

# How Does the Government Policy Combination Prevents Greenwashing in Green Building Projects? An Evolutionary Game Perspective

**Yuqing Chen [1], Zhen Li [1], Jiaying Xu [1], Yingying Liu [2] and Qingfeng Meng [1,*]**

[1] School of Management, Jiangsu University, 301 Xuefu Road, Zhenjiang 212013, China
[2] AAC Technologies, Ruisheng Science Park, No. 3, Changcao Road, Changzhou 213100, China
* Correspondence: mqf@ujs.edu.cn

**Abstract:** Green buildings (GBs) can effectively reduce building energy consumption and alleviate energy problems. However, as green building projects swell, the development process of GBs in China has gradually exposed many problems, among which the greenwashing behavior of construction enterprises is the more serious. The government needs to adopt some appropriate policies to prevent problems in GBs. This paper uses the evolutionary game theory to construct models and proposes four policy combinations: static reward and static punishment, static reward and dynamic punishment, dynamic reward and static punishment, dynamic reward and dynamic punishment. We compare the impact of four combinations on construction strategy and analyze the inner mechanisms of the behavior evolution of government departments and construction enterprises. Our results revealed no evolutionary stabilization strategy in the game system under the static reward and punishment policy. Under the combination of dynamic subsidies and static penalty policy, the green construction effect of construction enterprises is the best. Furthermore, dynamic reward policy has a more obvious restraining effect on construction enterprises. The government should dynamically adjust rewards and punishments according to the construction quality and determine the appropriate upper limit of rewards and punishments to improve the policies' applicability and effectiveness. This study provides theoretical support for the healthy development of green buildings.

**Keywords:** green buildings; greenwashing; behavioral decision; policy combination; evolutionary game

## 1. Introduction

Green buildings (GBs) can reduce or eliminate adverse impacts during design, construction, or operation and positively affect climate and the natural environment [1]. It can mitigate phenomena such as global warming and climate change and also improve occupants' quality of life and provide economic benefits [2]. Thus, many counties have implanted GBs and enacted relevant laws [3,4], such as the Comprehensive Assessment System for Built Environment Efficiency (CASBEE) (Japan), the U.S. Leadership in Energy and Environmental Design (LEED), the UK's Building Research Establishment Environmental Assessment Method (BREEAM) [5,6], and the GB Evaluation Standard (GB/T50378-2019) [7]. GBs in this study are based on the GB Evaluation Standard because its index framework and evaluation method are more suitable for GBs in China. The standard comprehensively evaluates the life cycle of the building system in China.

However, some problems have been exposed in the development of GBs in China, such as high-cost premiums, unequal distribution of benefits [8], irrational development of the construction industry [9], greenwashing behavior [10], and generated waste at the different stages of construction [11], etc. Among these, the "greenwashing behavior" of construction enterprises (NBD, 2021. A-share environmental risk list | 360 companies on the list in the first half of the year, which is the highest risk, and who is deliberately "greenwashing"?

http://www.nbd.com.cn/articles/2021-07-31/1859595.html accessed on 31 July 2021) is the most common and severe [12]. Greenwashing refers to the behavior of an enterprise that advertises that its products, services and business activities meet the green standards but, in fact, does not live up to the name [10]. Therefore, this paper describes greenwashing behavior as an enterprise's deceptive "green construction" and non-green construction management made to pursue short-term interests. This conduct is fundamentally an opportunistic behavior of engineering management. For example, enterprises may claim to achieve GBs but do not take action because of their own interests [13]. In order to pursue short-term benefits by earning government subsidies, the construction project that calls "GB" did not implement the green construction scheme. This behavior led to the project failing to meet the GBs design standards and questioning government resource utilization efficiency [14]. GBs may not be truly "green," which hinders the sustainable development of GBs in China.

The government has formulated a series of policies and increased the supervision to regulate green construction behavior (The Central People's Government of the People's Republic of China, 2012. Opinions on the implementation of accelerating the development of green buildings in China. www.gov.cn/zwgk/2012-05/07/content_2131502.htm accessed on 7 May 2012; Department of Housing and Urban-Rural Development of Shanxi Province, 2018. Supplementary Notice on the Implementation of Green Construction to Accelerate the Construction of Transformation Projects. https://zjt.shanxi.gov.cn/zwgk/tfwj/202109/t20210907_1959322.shtml accessed on 19 September 2019). But green construction is different from general construction projects; the characteristics of high R&D costs, large initial investments, and low short-term profits make enterprises often reluctant to invest in GBs [15], and it is easy to have moral hazard problems in the construction process. New technologies, materials, and construction methods are involved, which creates a serious information asymmetry between enterprises and government. The government needs to adjust the original supervision methods, and the difficulty of supervision increases. The rapid development of GBs has made the government's regulatory forces (such as the scale and funding of departments at all levels) inadequate. Secondly, the particularity of GBs lies in requiring supervision of the whole process [16] (such as design, planning, construction, operation, transformation, demolition, etc.). The sharp rise in regulatory costs has also increased the difficulty of supervision to a certain extent. Under this circumstance, to ensure the standardization and effectiveness of GBs, the government needs to formulate reasonable reward and punishment policies, which is particularly important for regulating green construction behavior.

Therefore, to explore the influence of policies on the GBs' behavior, this study constructs an evolutionary game model of construction enterprises and government departments. Aiming to analyze whether a dynamic incentive mechanism is more effective than a static one, we consider four combinations: static reward and static punishment, static reward and dynamic punishment, dynamic reward and static punishment, and dynamic reward and dynamic punishment. This study needs to answer the following questions: Can the combination of dynamic rewards and punishments better motivate enterprises' green construction than static rewards and punishments? What combination of policies achieves overall optimality? By exploring the evolution process of the game, we can interpret the reasons for the behavioral tendencies of different subjects, analyze the internal mechanisms of evolution, and explore the effects of different policy combinations. The findings provide theoretical and decision-making support for the healthy development of GBs. The research flow of this paper is shown in Figure 1 below.

The chapters of this paper are structured as follows: Section 2, we provide a review and discuss the related literature to this study. Section 3 describes the model and hypothesis. Section 4 establishes the four policy combinations and analyzes evolutionary stabilization. Section 5 presents numerical simulations to explore the impact of different combinations on game subjects. Finally, Section 6 summarizes the main findings and provides managerial implications of this study.

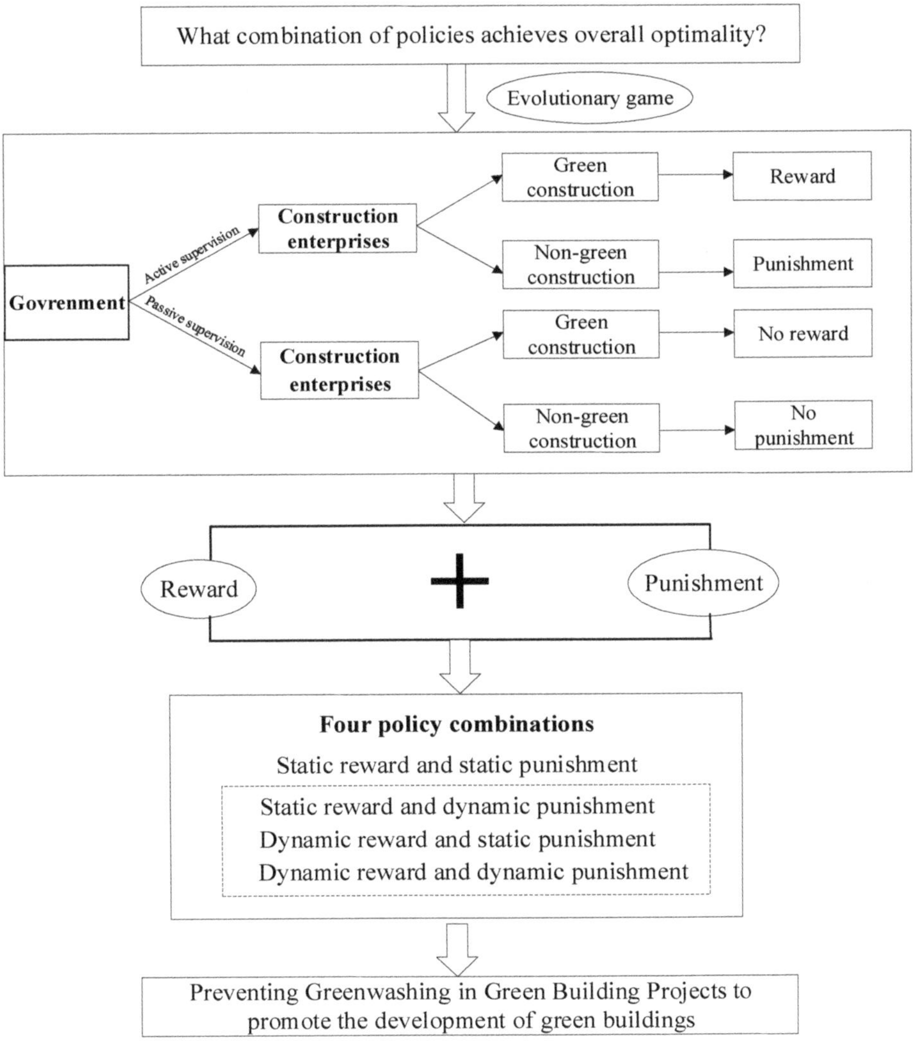


**Figure 1.** Research flow.

## 2. Related Literature

### *2.1. Research on Evolutionary Game Application*

The evolutionary game derives from the theory of biological evolution; since the evolutionary stabilization strategy was proposed [17], it has received widespread attention in construction engineering [18–20]. Wang et al. [21] established an evolutionary game model to analyze the effect of policy on real estate enterprises. They found that combining incentives and penalties was an effective way to create a sustainable real estate industry. Aiming to interpret diverse subjects' strategies and evolutionary behaviors, Jiang et al. [22] explore the construction safety supervisory mechanism in China from an evolutionary game view.

With the development of green concepts, many scholars use evolutionary games to study green building projects. Chen et al. [23] analyzed the effect of government policies on green building innovations through evolutionary game models and the results show that policy is the main factor affecting green procurement. Based on the evolution strategies of the construction enterprises under government supervision, Gao et al. [24] applied evolutionary game theory to analyze the path of government static and dynamic subsidies. To promote public acceptance of GBs, Jiang et al. [25] created a decision support model using evolutionary game theory to identify the optimal subsidy strategy and related factors.

### 2.2. Government Policies on Green Buildings

Academics have researched how governments develop strategies to regulate GBs' construction behavior [26–30]. Olubunmi et al. believe that GBs construction incentives can be divided into two categories: external incentives and internal incentive mechanisms. The external incentive mechanism mainly comes from the government's compulsory promotion, and the internal motivation comes from the beneficiary's income from GBs [31]. Cohen et al. used the prisoner's dilemma model to explain the barriers to energy conservation and emission reduction in the construction industry in Israel and demonstrated that government subsidies could help remove these barriers [32]. Liu and Li [33] reviewed the implementation of China's GBs incentive policy, indicating that China still needs to study establishing an incentive mechanism system further and evaluate the incentive effect. GBs construction enterprises in China are more sensitive to financial incentives than non-financial ones [34]. Monetary incentives and green construction technologies significantly impact the development of GBs [35]. Li et al. [36] constructed an evolutionary game model to analyze the impact of subsidy policies on construction units and believe that the government should reasonably adjust the subsidy amount, improve the penalty mechanism, and strengthen publicity to encourage construction units to apply actively. Kong and He [37] analyzed China-related GBs and divided these policies into supply-side and demand-side, arguing that supply-side policy incentives are much better than demand-side. Chen and Hong [38] studied the optimal subsidy problem in the green building market from the perspective of policy benefit. The research results show that the government's subsidy policy is not only affected by the government's subjective initiative but also by the construction cost and the preference of construction units for GBs. Song et al. [39] show that both mandatory regulations and incentive-based policies can help increase the adoption of GBs, with the former having a greater impact than the former.

In summary, the existing literature increasingly considers the factors of construction enterprises in the study of government policy formulation. However, there are still areas for improvement in the practice of GBs construction supervision, mainly manifested in two aspects: (1) Most of them assume that the policy strength is fixed, that is, the static reward and punishment mechanism. In the practice of GBs construction supervision, government regulatory authorities have greater discretion, and often dynamically adjust the intensity of rewards and punishments according to the construction quality of construction enterprises (i.e., dynamic reward and punishment mechanism). In construction safety [22], electric vehicle production [40], public health services [41], Platform E-commerce credit supervision [42], etc., DRPM has been proven to be practical and effective. (2) Existing studies still need to consider the bounded rationality of governments and construction enterprises fully. However, in the process of GBs, the government and enterprises are affected by various factors. Such as information asymmetry [43], social roles, cognition, economic interests, etc., which make the government and enterprises show limited rational behavior in construction [44,45], affecting the performance of green construction behavior and the choice of government regulatory departments.

## 3. Model Description and Hypothesis

### 3.1. Analysis of the Game Relationship between Government and Construction Enterprises

Although many factors prevent GBs from achieving the expected design standards, the passive behavior of government regulators and the non-green behavior of construction enterprises are undoubtedly the most critical factors. The construction quality and the government supervision during the construction phase directly affect the delivery of GB products. Government departments need to give necessary incentives and supervision to construction enterprises. Currently, China's incentives for GBs include financial subsidies, tax exemptions, and preferential use of land, and the punishment policy is mainly that the government fines construction enterprises to reduce moral risks [46]. In this paper, without considering external factors, whether GBs will be constructed according to standards can be regarded as the result of the game between construction companies and the government. Given this, this paper selects them as game subjects. Based on four policy combinations, the system evolution stabilization strategy is discussed.

### *3.2. Model Description and Hypothesis*

(1) This paper supposed that the game system contains two subjects: construction enterprises and government regulators, which play the roles of implementer and regulator. Both are finite and rational in the construction process and can continuously learn to achieve increased income. The final equilibrium is finding the optimal decision through continuous game and learning. There are only two strategic options for both sides of the game.

(2) Construction enterprise: Based on the reality of GBs' construction in China, the strategic choices of enterprises are abstracted into "green construction" and "non-green construction." The "green construction" strategy means that the enterprise can strictly follow the GBs' design standards, including using environmentally friendly construction materials, continuous improvement of construction technologies, etc., to ensure that the final GBs meet the design requirements. The "non-green construction" strategy refers to the moral risk behavior that enterprises' constructions do not meet the design requirements, resulting in a waste of energy and resources [47]. The "non-green construction" behavior will eventually harm the vital interests of the government and the public and cause environmental pollution to reduce social benefits.

(3) Government: As the subject of market regulation and the defender of public interest, government chooses the strategy of "positive supervision" and "passive supervision" in the abstract. With adequate information and analysis, government departments usually make decisions based on their experience and existing conditions. The "active supervision" strategy refers to the active performance of supervisory duties, adopting regular daily supervision and occasional non-daily supervision to focus on the construction process. Accordingly, the government regulator must bear the cost of supervision. Assuming an "active supervision" strategy, the government can decide to punish or reward construction companies as appropriate. However, in the actual supervision work, the government needs to invest a certain amount of human, material, and financial resources in supervising. In addition to being responsible for the supervision of GBs, it also shoulders many other tasks; Out of consideration of cost and benefit and based on the analysis of the construction information of existing construction enterprises, the government decides whether to actively supervise the construction enterprises.

(4) The probability of the behavior strategy of both sides of the game. In the game process between the government and the construction companies, assuming the probability of active supervision is $x$. Thus, passive supervision is $1-x$, the likelihood of construction companies complying with the law is $y$, and non-green construction is $1-y$.

(5) Cost-Benefit parameter assumptions and interpretation. The following parameter settings and instructions are given to construct the game's payment matrix further. Assuming that the cost that construction enterprises need to pay when choosing green construction is $S_1$, such as the adoption of new technologies, new materials, and training fees for construction personnel, etc., and the construction enterprise chooses non-green construction, the cost to be paid is $S_2$, obviously $S_1 > S_2$; Similarly, when the government regulatory department is actively supervised, the cost of supervision

is $S_3$, and the price paid when passive supervision is $S_4$, obviously $S_3 > S_4$; When the government actively supervises, it will carry out corresponding subsidies or penalties according to the green construction quality, and give subsidies $B$ to the construction enterprise; otherwise, a fine $P$ will be imposed, and the fine shall be owned by the government; The good green construction situation of the construction unit will be included in the integrity assessment system, and linked to the award evaluation, accumulating advantages for future undertaking projects, the potential future income is $R_1$, and the government department will obtain additional income $R_2$ due to "free riding" behavior. In addition, when the construction enterprise adopts non-green construction behavior, the government's credibility loss is caused by passive supervision by the government department $L$.

## 4. Model Building and Analysis

### *4.1. Static Reward and Static Punishment*

To further construct the game payment matrix, the following parameter settings and descriptions are given in Table 1.

**Table 1.** Parameter symbols and their meanings.

| Parameters | Meaning |
|---|---|
| $S_1$ | Green construction needs to pay the cost |
| $S_2$ | Non-green construction needs to pay the cost |
| $S_3$ | Costs when government regulators actively supervise |
| $S_4$ | Costs when government regulators passively supervise |
| $B$ | Subsidies given to green construction companies when actively supervision |
| $P$ | Fines for green construction companies when actively supervision |
| $R_1$ | Potential future benefits of green construction for construction companies |
| $R_2$ | Government will gain additional revenue from "free-riding" behavior |
| $L$ | The loss of government credibility and the cost of public opinion caused when construction companies adopt non-green construction practices |
| $x$ | Probability of active supervision |
| $y$ | Probability of green construction |

The assumptions and definitions based on the above, the game payment matrix of construction companies and government department, are obtained as shown in Table 2.

**Table 2.** The game payment matrix of government and construction enterprises under static reward and static punishment.

| | | Construction Enterprises | |
|---|---|---|---|
| | | **Green Construction** $y$ | **Non-Green Construction** $1-y$ |
| Government | Active supervision $x$ | $R_2 - B - S_3, R_1 + B - S_1$ | $P - S_3 - L, -P - S_2$ |
| | Passive supervision $1-x$ | $R_2 - S_4, R_1 - S_1$ | $-S_4 - L, -S_2$ |

4.1.1. Evolutionary Stabilization Analysis of Static Reward and Punishment Model

According to evolutionary game theory and the game payment matrix, the expected return and replication dynamic equations under the two strategies of construction enterprises and governments are expressed:

The expected benefit of government opting for active supervision is:

$$E_{1N} = (y-1)L - S_4 + yR_2 \quad (1)$$

The expected benefit of government opting for passive supervision is:

$$E_{1N} = (y-1)L - S_4 + yR_2 \quad (2)$$

According to evolutionary game theory, the replication dynamics equation is a system dynamics differential equation that describes the frequency or frequency of a particular strategy employed in a population [48], so the dynamic replication equation chosen by the government to actively regulate is:

$$F(x) = \frac{dx}{dt} = x(E_{1Y} - E_1) = x(1-x)(P - By - Py - S_3 + S_4) \tag{3}$$

Similar to this, the replication dynamic equation of GBs for companies can be obtained:

$$G(y) = \frac{dy}{dt} = y(1-y)(Bx + Px - S_1 + S_2 + R_1) \tag{4}$$

It is necessary to find the stable state of the replication dynamics and then discuss the influence of small perturbations on the stable state [49]. According to the replication dynamic Equations (3) and (4), let $F(x) = 0, G(y) = 0$, and discuss the evolutionary stabilization strategy of the system.

Evolutionary Stabilization Analysis of Government Strategies

Considering the evolutionary stabilization of government regulatory strategies, the derivative of Equation (3) is:

$$\frac{dF(x)}{dx} = (1-2x)(P + S_4 - S_3 - (P+B)y) \tag{5}$$

When $y = \frac{P+S_4-S_3}{P+B}$, $F(x) = 0$ is constant, and all $x$ are government stabilization strategies; when $y \neq \frac{P+S_4-S_3}{P+B}$, let $F(x) = 0$, then $x = 0$ and $x = 1$ are in equilibrium. According to the stability theorem of differential equations, $\left.\frac{dF(x^*)}{dx}\right|_{x=x^*} < 0$ and $x^*$ is an evolutionary stabilization strategy. Different scenarios of $P + S_4 - S_3$ need to be analyzed:

(1) When $P + S_4 - S_3 < 0, y > \frac{P+S_4-S_3}{P+B}$ is constant, and the evolutionarily stable strategy (ESS) is $x = 0$.
(2) When $P + S_4 - S_3 > 0$, there are two cases:
   (a) When $P + S_4 - S_3 > P + B, y < \frac{P+S_4-S_3}{P+B}$, and the ESS is $x = 1$.
   (b) When $P + S_4 - S_3 > P + B, 0 < \frac{P+S_4-S_3}{P+B} < 1$, The government's stabilization strategy depends on the size of $y$. When $y > \frac{P-S_3+S_4}{P+B}$, there is $\left.\frac{dF(x)}{dx}\right|_{x=0} < 0$, and $x = 0$ is the ESS. When $y < \frac{P-S_3+S_4}{P+B}$, there is $\left.\frac{dF(x)}{dx}\right|_{x=0} < 0$, and $x = 1$ is the ESS.

Evolutionary Stabilization Analysis of Construction Enterprise Strategies

Similarly, consider the evolutionary stability of the strategy of construction enterprises. For Equation (4), there is:

$$\frac{dG(y)}{dy} = (1-2y)((P+B)x - S_1 + S_2 + R_1) \tag{6}$$

When $x = \frac{S_1-S_2-R_1}{P+B}$, $G(y) = 0$ is constant, and all $y$ are enterprises' stabilization strategies; when $x \neq \frac{S_1-S_2-R_1}{P+B}$, let $G(y) = 0$, then $y = 0$ and $y = 1$ are in equilibrium. According to the stability theorem of differential equations, $\left.\frac{dG(y^*)}{dy}\right|_{y=y^*} < 0$ and $y^*$ is an evolutionary stabilization strategy. Different scenarios need to be analyzed:

(1) When $S_1 - S_2 - R_1 < 0, x > \frac{S_1-S_2-R_1}{P+B}$ is constant, and the evolutionary stabilization strategy is $y = 1$.
(2) When $S_1 - S_2 - R_1 > 0$, there are two cases:

(a) When $S_1 - S_2 - R_1 > P + B$, $x < \frac{S_1 - S_2 - R_1}{P+B}$, and the evolutionary stabilization strategy is $y = 0$.

(b) When $S_1 - S_2 - R_1 < P + B$, $0 < \frac{S_1 - S_2 - R_1}{P+B} < 1$, The government's stabilization strategy depends on the size of $x$. When $x > \frac{S_1 - S_2 - R_1}{P+B}$, and $y = 1$ is the evolutionary stabilization strategy. When $x < \frac{S_1 - S_2 - R_1}{P+B}$, and $y = 0$ is the evolutionary stabilization strategy.

Evolutionary Stabilization Analysis of Hybrid Strategies between Government and Construction Enterprises

The above analysis shows that under different initial conditions, the government and construction enterprises have different ESSs. However, in reality, when the additional benefits obtained while enterprises choose the "non-green" strategy are far greater than the total rewards and penalties, regardless of whether the government actively supervises, the construction enterprises will choose the "non-green" strategy. Suppose the government's reward and punishment mechanism is effective. In that case, it is necessary to ensure that when the government actively supervises, green construction benefits are greater than non-green. The government's punishment should be greater than the price paid by environmental governance, which is $S_1 - S_2 - R_1 < P + B$ and $P - S_3 - L > 0$. Therefore, this study focuses on the stability under these two constraints.

From Equations (3) and (4), if and only if $0 < \frac{P - S_3 + S_4}{B+P} < 1$, $0 < \frac{S_1 - S_2 - R_1}{B+P} < 1$ hold, the system contains five equilibrium points $A(0,0)$, $B(0,1)$, $C(1,0)$, $D(1,1)$, $E(x_1{}^*, y_1{}^*)$, where $x_1{}^* = \frac{S_1 - S_2 - R_1}{P+B}$, $y_1{}^* = \frac{P - S_3 + S_4}{P+B}$. According to Friedman [50], using the Jacobian matrix analyzes the stabilization of each equilibrium point, if $Det(J) > 0$ and trace $Tr(J) < 0$ in the matrix, it indicates that the equilibrium point is an ESS.

According to Equations (3) and (4), the Jacobian matrix of the system is as follows:

$$J = \begin{pmatrix} (1-2x)(P - By - Py - S_3 + S_4) & (P+B)(x-1)x \\ (P+B)(1-y)y & (1-2y)(Bx + Px - S_1 + S_2 + R_1) \end{pmatrix} \tag{7}$$

Therefore,

$$\begin{aligned} \det(J) = \ & (1-2x)(1-2y)(P - By - Py - S_3 + S_4)(Bx + Px - S_1 + S_2 + R_1) \\ & + xy(1-x)(1-y)(P+B)^2 \end{aligned} \tag{8}$$

$$tr(J) = (1-2x)(P - By - Py - S_3 + S_4) + (1-2y)(Bx + Px - S_1 + S_2 + R_1) \tag{9}$$

The expression of the determinant and trace of each equilibrium point is shown in Table 3.

**Table 3.** An expression of the determinant and trace of each equilibrium point.

| Equilibrium Point | *detJ* | *tr(J)* |
|---|---|---|
| $(0,0)$ | $(P - S_3 + S_4) * (-S_1 + S_2 + R_1)$ | $(P - S_3 + S_4) + (-S_1 + S_2 + R_1)$ |
| $(0,1)$ | $(B + S_3 - S_4) * (-S_1 + S_2 + R_1)$ | $(-B - S_3 + S_4) - (-S_1 + S_2 + R_1)$ |
| $(1,0)$ | $(P + S_3 - S_4) * (P + B - S_1 + S_2 + R_1)$ | $(P + B - S_1 + S_2 + R_1) - (P - S_3 + S_4)$ |
| $(1,1)$ | $(-B - S_3 + S_4) * (B + P - S_1 + S_2 + R_1)$ | $(B + S_3 - S_4) - (B + P - S_1 + S_2 + R_1)$ |
| $(x_1{}^*, y_1{}^*)$ | | $0$ |

When $0 < \frac{P - S_3 + S_4}{P+B} < 1$ and $0 < \frac{S_1 - S_2 - R_1}{P+B} < 1$, the local stabilization of the system is show in Table 4.

**Table 4.** Stability analysis results for fixed equilibrium points.

| Equilibrium Point | *detJ* | *trJ* | Stability |
|---|---|---|---|
| (0,0) | − | +/− | Saddle point |
| (0,1) | − | +/− | |
| (1,0) | − | +/− | |
| (1,1) | − | +/− | |

Now discuss the stability at $(x_1^*, y_1^*)$, the corresponding Jacobi matrix is given by:

$$J_1 = \begin{pmatrix} 0 & -\frac{(S_1-S_2-R_1)(B+P-S_1+S_2+R_1)}{B+P} \\ \frac{(B+S_3-S_4)(P-S_3+S_4)}{B+P} & 0 \end{pmatrix} \quad (10)$$

From the model solution, the characteristic root corresponding to the point $(x_1^*, y_1^*)$ is $\lambda_1, \lambda_2 = \pm\frac{i\sqrt{Z}}{B+P}$, and

$$Z = \sqrt{P - S_3 + S_4}\sqrt{S_1 - S_2 - R_1}\sqrt{(B + S_3 - S_4)(B + P - S_1 + S_2 + R_1)} \quad (11)$$

The system is not asymptotically stable. Construction firms' and government regulators' strategy is a closed-loop curve around the point $(x_1{}^*, y_1{}^*)$, which is infinitely near to that point but not automatically stable at that point, i.e., $(x_1{}^*, y_1{}^*)$ is not an ESS for the system.

**Proposition 1.** *The replica dynamic equations are not asymptotically stable at any of the five local equilibrium points, and the system does not have an ESS.*

**Proposition 2.** *During the game stage, there is no ESS, but a circular pattern is formed. The combination of static reward and static punishment policy has not achieved the effect of enhancing the motivation for green construction and reducing moral risks.*

In the case of $0 < \frac{P-S_3+S_4}{B+P} < 1$, $0 < \frac{S_1-S_2-R_1}{B+P} < 1$, the dynamic evolution of the game system is shown in Figure 2. There is no ESS in an evolutionary game between the government and construction enterprises, but it shows a cyclical characteristic. When the initial strategy choice of two sides of the game falls in region I, the final strategy choice of the government and construction enterprises will tend to be (positive supervision, non-green construction). When the original strategy choice of both main bodies falls in region II, the final strategy choice of government regulators and enterprises will tend to be (passive supervision, non-green construction). When the initial strategy choice falls in region III, the final strategy choice will tend to be (passive supervision, green construction); when the original strategy choice falls in region IV, the final strategy choice will be (positive supervision, green construction).

Since the government and construction enterprises pursue their own interests, which leads to a vicious circle between the two strategies and cannot achieve evolutionary stability, there is no ESS in the policy of static reward and static punishment. Thus, this paper proposes a study on dynamic policy combinations.

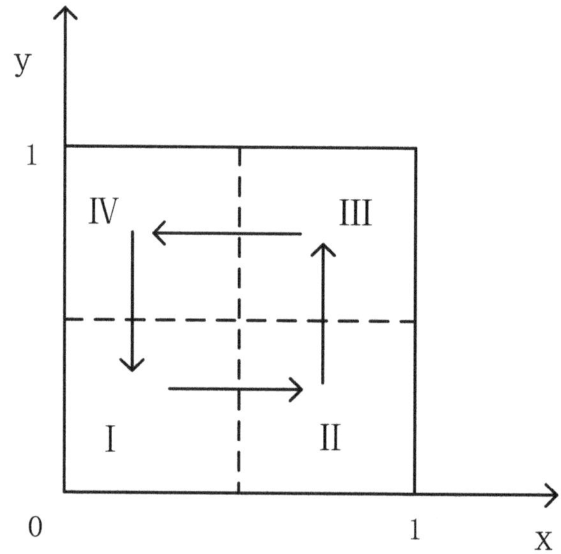


**Figure 2.** Evolution phase diagram of game system. (F2.I: When the initial state falls in region I, the stabilization strategy of the evolutionary game is (positive supervision, non-green construction); F2. II: When the initial state falls in region II, the stabilization strategy of the evolutionary game is (passive supervision, non-green construction); F2. III: When the initial state falls in region III, the stabilization strategy of the evolutionary game is (passive supervision, green construction); F2. IV When the initial state falls in region IV, the stabilization strategy of the evolutionary game is (positive supervision, green construction)).

### 4.2. Dynamic Reward and Static Punishment

It is assumed that the government's reward to construction enterprises is proportional to their "green construction" behavior, i.e., the reward from a fixed constant $B$ to $B(y) = yq$, where the reward is capped at $q$ and the punishment is still $P$. Then the game system is as shown in Table 5:

**Table 5.** The game payment matrix under dynamic reward and static punishment.

| | | Construction Enterprises | |
|---|---|---|---|
| | | Green Construction $y$ | Non-Green Construction $1-y$ |
| Government | Active supervision $x$ | $R_2 - yq - S_3, R_1 + yq - S_1$ | $P - S_3 - L, -P - S_2$ |
| | Passive supervision $1-x$ | $R_2 - S_4, R_1 - S_1$ | $-S_4 - L, -S_2$ |

The dynamic replication equations can be obtained:

$$F(x) = \frac{dx}{dt} = x(1-x)(P - y^2q - Py - S_3 + S_4) \quad (12)$$

$$G(y) = \frac{dy}{dt} = y(1-y)(xyq + Px - S_1 + S_2 + R_1) \quad (13)$$

**Proposition 3.** *The equilibrium points that exist for replicating the system of dynamic equations are as follows:*

1. *The system with four fixed equilibrium points always, namely* $(0,0), (0,1), (1,0), (1,1)$.
2. *When* $0 < \frac{S_1-S_2-R_1}{B(y)+P} < 1$ *and* $0 < \frac{P-S_3+S_4}{B(y)+P} < 1$, *there is an equilibrium point* $(x_2{}^*, y_2{}^*)$, *where* $x_2{}^* = \frac{S_1-S_2-R_1}{B(y)+P}$, $y_2{}^* = \frac{P-S_3+S_4}{B(y)+P} = \frac{-P+\sqrt{4qP+P^2-4qS_3+4qS_4}}{2q}$.

**Proposition 4.** *The system of replicated dynamic equations has asymptotic stability at* $(x_2{}^*, y_2{}^*)$ *and exists a stabilization strategy.*

**Proof.** When $S_1 - S_2 - R_1 < B(y) + P$, $P - S_3 - L > 0$, we can obtain the stability analysis results for each equilibrium point, as shown in Table 6. □

**Table 6.** Stability analysis of equilibrium points under dynamic reward and static punishment mechanism.

| Equilibrium Point | *detJ* | *trJ* | Stability |
|---|---|---|---|
| (0,0) | − | +/− | Saddle point |
| (0,1) | − | +/− | |
| (1,0) | − | +/− | |
| (1,1) | − | +/− | |
| $(x_2{}^*, y_2{}^*)$ | + | 0 | ESS |

Now consider the stability at $(x_2{}^*, y_2{}^*)$, the corresponding Jacobi matrix is given by:

$$J_2 = \begin{pmatrix} 0 & a_2 \\ b_2 & c_2 \end{pmatrix} \tag{14}$$

where $a_2 = (1 - x_2{}^*)x_2{}^*(-P - 2qy)$, $b_2 = (1 - y_2{}^*)y_2{}^*(P + qy_2{}^*)$, $c_2 = qx_2{}^*(1 - y_2{}^*)y_2{}^* + (1 - y_2{}^*)(Px_2{}^* + qx_2{}^*y_2{}^* - S_1 + S_2 + R_1) - y_2{}^*(Px_2{}^* + qx_2{}^*y_2{}^* - S_1 + S_2 + R_1)$.

According to the stability theory of differential equations, $(x_2{}^*, y_2{}^*)$ is the center point of stability. Therefore, the system has asymptotic stability under the policy combination of dynamic reward and static punishment, and $(x_2{}^*, y_2{}^*)$ is the ESS of the system.

It is not easy to envision the evolutionary curve as a spiral, which tends toward equilibrium $(x_2{}^*, y_2{}^*)$. Proposition 4 proves the existence of an ESS for the system with a dynamic reward and static punishment policy. The government actively regulates with probability $x_2{}^*$, and the construction companies embrace GBs with probability $y_2{}^*$.

**Proposition 5.** *The probability of active supervision and green construction is related to costs, benefits, rewards and punishments as follows:*

(1) $\frac{\partial x_2{}^*}{\partial R_1} < 0, \frac{\partial x_2{}^*}{\partial P} > 0, \frac{\partial x_2{}^*}{\partial q} < 0, \frac{\partial x_2{}^*}{\partial S_1} > 0, \frac{\partial x_2{}^*}{\partial S_2} < 0, \frac{\partial x_2{}^*}{\partial S_3} < 0, \frac{\partial x_2{}^*}{\partial S_4} > 0$

(2) $\frac{\partial y_2{}^*}{\partial P} > 0, \frac{\partial y_2{}^*}{\partial q} < 0, \frac{\partial y_2{}^*}{\partial S_3} < 0, \frac{\partial y_2{}^*}{\partial S_4} > 0.$

Proposition 5 illustrates that:

(i) Active supervision's probability is positively correlated with the cost of green construction $S_1$. When the cost of GBs is higher, construction enterprises will likely have a fluke mentality and believe the government will not detect non-compliance. Based on which government authorities will generally strengthen supervision. Active supervision's probability is inversely correlated with the cost of non-green construction. When non-green construction $S_2$ cost is higher, governmental thinks that construction enterprises will choose green construction and loosen supervision.

(ii) As the cost of active supervision increases, the cost of human and financial resources will also rise accordingly. This phenomenon will put pressure on the government departments, so the probability of active supervision will decrease, and the probability of GBs will also reduce. On the contrary, with the increase in the cost of passive regulation by governmental departments, the government is more willing to take the initiative to effectively regulate and improve the green construction enthusiasm of construction enterprises.

(iii) The higher the fine $P$ is, the greater the expense construction enterprises pay for non-green construction and the higher the subjective consciousness of conducting green construction behaviors. The higher the governmental supervision will raise the fine to prevent construction enterprises from unethical risk behaviors.

(iv) The higher the reward $q$ means more of the financial burden, the lower the probability of active supervision. Nevertheless, construction enterprises think the government may loosen regulations after giving higher subsidies. Enterprises may adopt non-green construction, leading to a lower probability of their green construction.

(v) In addition, regulatory motivation is negatively correlated with the potential future benefits of green construction. The government believes that the higher the foreseeable benefits of construction enterprises, the more they will choose green construction, leading to decreased regulatory motivation.

### 4.3. Static Reward and Dynamic Punishment

Supposing that the punishment strength is proportional to the non-green construction behavior, i.e., the punishment strength changes from a fixed constant $P$ to $P(y) = (1 - y)r$, where the punishment is capped at $r$, and the reward level remains $B$. Similarly, Propositions 6–8 can be drawn.

**Proposition 6.** *The equilibrium points that exist for replicating the system of dynamic equations are as follows:*

1. *The system with four fixed equilibrium points always, namely* $(0,0), (0,1), (1,0), (1,1)$.
2. *When* $0 < \frac{S_1 - S_2 - R_1}{B + P(y)} < 1$ *and* $0 < \frac{P(y) - S_3 + S_4}{B + P(y)} < 1$, *there is another equilibrium point* $(x_3{}^*, y_3{}^*)$, *where* $x_3{}^* = \frac{S_1 - S_2 - R_1}{B + P(y)}$, $y_3{}^* = \frac{P(y) - S_3 + S_4}{B + P(y)}$.

**Proposition 7.** *The system of replicated dynamic equations exists asymptotic stability at* $(x_3{}^*, y_3{}^*)$ *and an ESS. (The proof process is shown in Appendix A.1.)*

**Proposition 8.** *The probability of active supervision and corporates' green construction are related to costs, benefits, subsidies and fines as follows:*

(1) $\frac{\partial x_3{}^*}{\partial S_1} > 0, \frac{\partial x_3{}^*}{\partial S_2} < 0, \frac{\partial x_3{}^*}{\partial S_3} < 0, \frac{\partial x_3{}^*}{\partial S_4} > 0, \frac{\partial x_3{}^*}{\partial R_1} < 0, \frac{\partial x_3{}^*}{\partial r} > 0, \frac{\partial x_3{}^*}{\partial B} < 0.$

(2) $\frac{\partial y_3{}^*}{\partial S_3} < 0, \frac{\partial y_3{}^*}{\partial S_4} > 0, \frac{\partial y_3{}^*}{\partial r} > 0, \frac{\partial y_3{}^*}{\partial B} < 0.$

Proposition 8 is similar to Proposition 5. It is worth paying attention to the higher fine $r$, the enterprise that fails to meet the green standard will face a high expense, which means that the government's regulatory behavior may face close attention from all walks of life, so the probability of active supervision will increase, and the probability of green construction will also increase with the increase in fines.

### 4.4. Dynamic Reward and Dynamic Punishment

It is assumed that the government's policy formulation is related to the choices of construction enterprises, i.e., the subsidies are related to the green construction choice, and the punishments are associated with non-green construction behavior. Suppose the reward is $B(y) = yq$ and the punishment is $P(y) = (1 - y)r$, where $r$ is the maximum of the penalty and $q$ is the maximum of the reward. Propositions 9–11 can be obtained similarly.

**Proposition 9.** *The equilibrium points that exist for replicating the system of dynamic equations are as follows:*

1. *The system always has four fixed equilibrium points, namely* $(0,0), (0,1), (1,0), (1,1)$.
2. *When* $0 < \frac{S_1 - S_2 - R_1}{B(y)+P(y)} < 1, 0 < \frac{P(y) - S_3 + S_4}{B(y)+P(y)} < 1$, *there is another equilibrium point* $({x_4}^*, {y_4}^*)$, *where* ${x_4}^* = \frac{S_1 - S_2 - R_1}{B(y)+P(y)}$, ${y_4}^* = \frac{P(y) - S_3 + S_4}{B(y)+P(y)}$.

**Proposition 10.** *The system of replicated dynamic equations has asymptotic stability at* $({x_4}^*, {y_4}^*)$ *and the system has an ESS. (The proof process is shown in Appendix A.2.)*

The system has asymptotic stability under the dynamic reward and dynamic punishment mechanisms, and $({x_4}^*, {y_4}^*)$ is the ESS. The evolutionary trajectory is a spiral curve tending to the equilibrium point $({x_4}^*, {y_4}^*)$. There is an ESS under the dynamic reward and dynamic punishment policy, the active supervision's probability is ${x_4}^*$ and the probability of construction companies choosing green construction is ${y_4}^*$.

**Proposition 11.** *The active supervision's probability and corporates' green construction is related to costs, benefits, subsidies and fines as follows:*

*(1)* $\frac{\partial {x_4}^*}{\partial S_1} > 0, \frac{\partial {x_4}^*}{\partial S_2} < 0, \frac{\partial {x_4}^*}{\partial S_3} < 0, \frac{\partial {x_4}^*}{\partial S_4} > 0, \frac{\partial {x_4}^*}{\partial R_1} < 0, \frac{\partial {x_4}^*}{\partial r} > 0, \frac{\partial {x_4}^*}{\partial q} < 0.$

*(2)* $\frac{\partial {y_4}^*}{\partial S_3} < 0, \frac{\partial {y_4}^*}{\partial S_4} > 0, \frac{\partial {y_4}^*}{\partial r} > 0, \frac{\partial {y_4}^*}{\partial q} < 0.$

Proposition 11 analysis is similar to Proposition 5 and 8.

## 5. Results

### 5.1. Simulation Analysis

To more intuitively describe the impact of the behaviors of two game subjects on each other and compare the differences in the evolutionary stability under four policy combinations, this chapter uses Matlab software and validates specific numerical examples to verify the established model and the conclusions obtained. To more clearly display the superiority of the dynamic incentive model compared with the static one, we study the effects of evolving choices under a change of parameters.

Under the premise of meeting the model requirements, combined with the specific reality of the GB construction stage, the relevant initial parameters were assigned. Considering that GBs projects are different from traditional engineering projects, construction projects have novelty, high complexity, and long-term goals, and the understanding of government and construction enterprises is gradually increasing. This paper draws on the study of Meng et al. [51] and Liang et.al. [52]. The number of games is expressed in time to observe parameter changes that influence the evolutionary results. The initial assignment of each parameter is shown in Table 7.

**Table 7.** The initial value of the game model.

| Parameters | Value | Parameters | Value |
|---|---|---|---|
| $S_1$ | 1.5 | $R_2$ | 0.85 |
| $S_2$ | 1 | $q$ | 0.85 |
| $S_3$ | 0.2 | $r$ | 0.8 |
| $S_4$ | 0.1 | $L$ | 0.6 |
| $B$ | 0.5 | $x$ | 0.2 |
| $P$ | 0.8 | $y$ | 0.5 |
| $R_1$ | 0.4 | | |

In this part, the evolution path diagrams will be compared. The evolutionary path diagrams of government regulators and construction enterprises under four policy combinations are shown in Figure 3a–d.

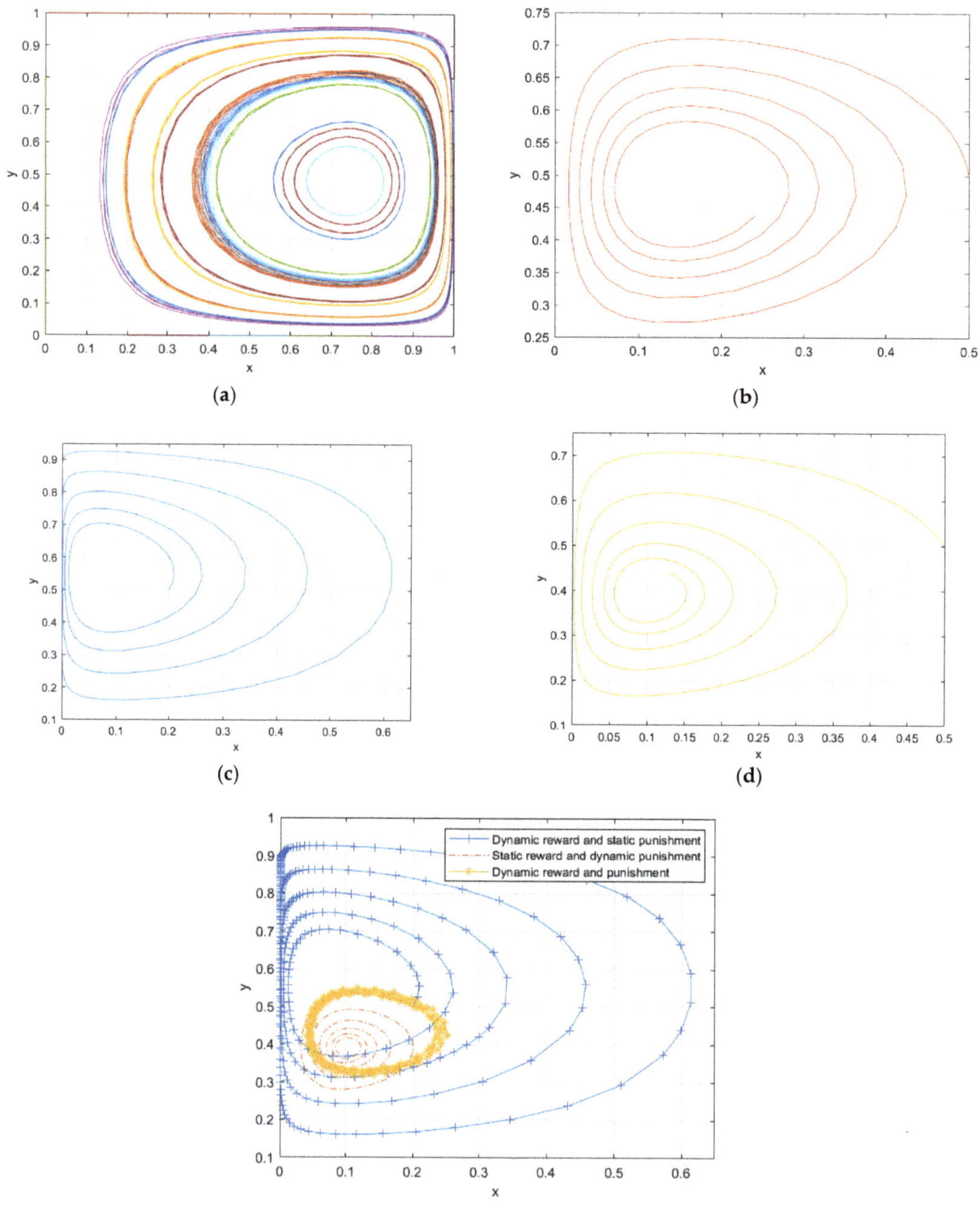


**Figure 3.** (**a**) Evolutionary path diagram under static reward and punishment (When the time $t$ is from 0 to 100, the evolution trajectory of the evolutionary game). (**b**) Evolutionary path diagram under dynamic reward and punishment. (**c**) Evolutionary path diagram under dynamic reward and static punishment. (**d**) Evolutionary path diagram under static reward and dynamic punishment. (**e**) Evolutionary path diagram under three policies.

### 5.2. Discussion

As can be seen from Figure 3a, when $0 < \frac{P-S_3+S_4}{B+P} < 1$, $0 < \frac{S_1-S_2-R_1}{B+P} < 1$, the evolutionary process moves in a closed-loop, wave-like motion. Both sides' evolutionary trajectories swing endlessly about this point $(x_1{}^*, y_1{}^*)$ and cannot establish equilibrium. This figure indicates that under static incentives, the government and construction companies exhibit a cyclical behavior pattern and no ESS. For the reason that the strategy choice impacts both parties' benefits, the finite rational government and construction enterprises will adjust their strategies according to the changes, leading to an infinite cycle of both parties' choices. This condition is consistent with the realistic situation that the government's incentive policies are challenging to achieve long-lasting effects in the current GBs construction process.

Second, after verifying that the system cannot reach evolutionary stability under the static reward and punishment mechanism, the evolutionary stability results are investigated based on the dynamic incentive mechanisms. When the government adopts three dynamic policy combinations, the evolutionary path diagrams are shown in Figure 3b–d. From the above three diagrams, it can be seen that with the three dynamic reward and punishment policy combinations, the evolutionary trajectories converge in a spiral around their respective equilibrium points and finally reach a stable state. These figures show that the method of dynamic incentives is reasonable and practical. The market situation largely influences policies' formulation and implementation, and since it is complex and changeable in reality. Applicability and timeliness will be the first choices for the government. For this reason, three dynamic policy combinations are more in line with objective reality than a static policy combination.

Finally, the effects of the three dynamic incentive policies are compared to judge which policy combination works best. By comparing the locations of the equilibrium points in Figure 3e, when dynamic reward and dynamic punishment mechanisms are adopted, the probability of active supervision is slightly higher than for the other two policy combinations. When dynamic reward and static punishment mechanisms are adopted, construction enterprises are willing to choose green construction. When a static reward and dynamic punishment mechanism is adopted, the probability of active supervision and green construction is lower than the other two policies. These findings show that the government can get better results by adopting dynamic reward policies. Since the fundamental purpose of government regulation is to ensure green construction, it is more important to improve the green construction enthusiasm of construction enterprises. Therefore, through comprehensive comparison, it is concluded that the dynamic reward and static punishment policy combination is the most effective in terms of green construction behavior, followed by the dynamic reward and punishment policy combination, and finally, static reward and dynamic punishment.

Under the dynamic reward and static punishment policy, we further investigate the influence of the maximum value of government reward $q$ on government and construction enterprises. The evolutionary trajectory of both subjects can be obtained by setting the upper limit of reward $q \in (0.3, 1.6)$. At the same time, keep other relevant parameters unchanged, as shown in Figure 4.

Keep other parameters constant and set the reward $q$ to vary between the interval $(0.3, 1.6)$. Figure 4 shows that the probability $x$ of active supervision converges to 0 as the reward $q$ increases. For construction companies, the probability $y$ of green construction decreases with the $q$ increases and tends to converge to 0. It verifies the conclusion obtained in the previous section that the probability of active supervision and green construction are both inversely proportional to the upper limit of subsidies.

Based on the dynamic reward and punishment policy, we further investigate the influence of the maximum value of punishment $r$ on the strategy choice. The evolutionary track of two subjects can be obtained by setting the upper limit of penalty $r \in (0.5, 2)$. Meanwhile, other relevant parameters are fixed, as presented in Figure 5.

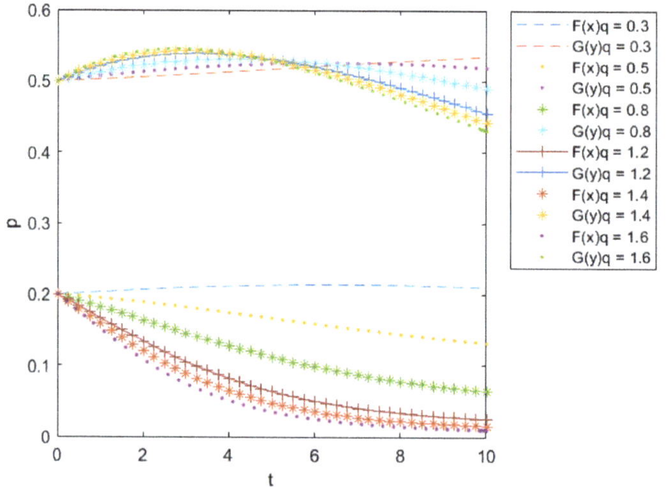


**Figure 4.** Effect of reward $q$ changes on evolutionary outcomes.

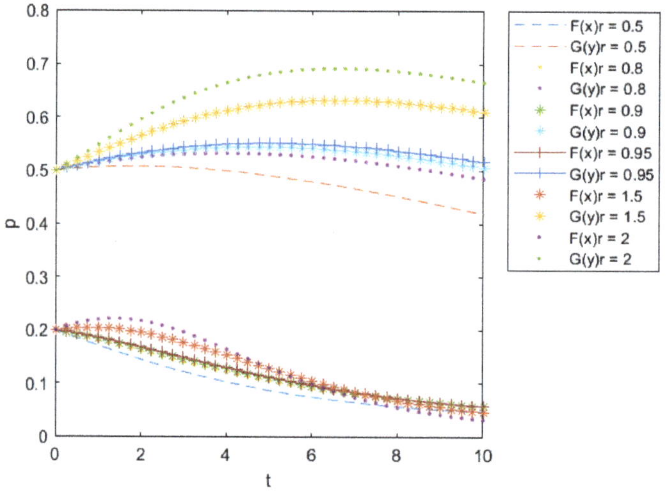


**Figure 5.** Effect of penalty $r$ changes on evolutionary outcomes.

Figure 5 describes that the probability $x$ of active supervision increases with the increase of the maximum value of the fine $r$. For construction companies, the probability $y$ of green construction also rises with increasing the fine $r$. It is verified that the probability of both governmental supervision and green construction are proportional to the maximum value of fines.

## 6. Conclusions

From the perspective of GBs supervision, this paper studies the impact of government policy combinations on green construction behavior. We constructed the evolutionary game models under static and dynamic incentive mechanisms. The reasons for the behavioral tendencies of both subjects and the effects of different policy combinations were explained, and the relevant essential parameters were analyzed. The model's effectiveness was verified through example analysis. The main conclusions of this paper are as follows:

(1) Under the static reward and punishment policy, the behavioral of the government and construction enterprises evolves in a cyclical cycle and does not exist ESS. The reward and punishment are set as fixed constants in the game process according to the construction quality and green construction management level. The evolution game is in a cyclical change process, which cannot achieve a stable equilibrium state under the static incentive mechanism.
(2) With the dynamic policy combinations, the strategy choice of the two subjects will eventually stabilize at the system equilibrium point, independent of the initial strategy choices of both sides. If the government can adjust the reward and punishment in real-time according to the construction quality, impose more severe punishment on the non-green construction behaviors of construction enterprises and give higher subsidies to the GB construction. Regardless of the initial strategy choice, when the dynamic incentive mechanism is adopted, the evolutionary trajectories of game systems are spirally approaching the stable equilibrium point, which is infinitely close to the evolutionary stability point.
(3) As the main body of GBs implementation, the choice of behavior strategy of construction enterprises is mainly affected by the government's strategies. Adopting a dynamic reward policy has a more obvious restraining effect on construction enterprises. When considering the bounded rationality of government and companies, the numerical simulation shows that the adoption of dynamic reward and static punishment, as well as dynamic reward and dynamic punishment policy combinations, can better promote the development of GBs.
(4) Under the dynamic incentive mechanisms, the probability of active supervision is inversely proportional to the reward ceiling value. The government tends to adopt a passive regulatory attitude with the reward ceiling value increase. At the same time, the probability of active supervision is proportional to the upper limit of punishment. The government is willing to choose an active supervision strategy as its value increases.
(5) Under the dynamic incentive mechanisms, the probability of green construction by construction enterprises is inversely proportional to the upper limit of subsidies and positively proportional to the fine. Unlike traditional intuition, this shows that enterprises, as a bounded rational decision-making subject, have a certain risk aversion and are more sensitive to losses than subsidies. Accordingly, when the government implements GB supervision, it is necessary to increase the punishment of "greenwashing," thereby increasing the probability of enterprises constructing according to regulations. This conclusion provides a theoretical reference for formulating policies and the green construction of construction enterprises.

Based on the conclusions, the following recommendations are made:

(1) The policy combination should be adjusted from static to dynamic. When formulating a dynamic policy combination, the government should adjust the intensity of rewards and punishments according to the construction quality to improve the applicability and effectiveness of the policies. At the same time, relevant departments should set up red lines for GBs construction supervision, accelerate the formulation of relevant laws and regulations, continuously improve supervision mechanisms, strengthen law enforcement, and adopt diversified punishment models.
(2) Construction enterprises should strictly regulate their construction behaviors. High construction costs are the main reason construction enterprises choose illegal construction. Construction enterprises should realize that the potential benefits, such as government reward and reputation accumulation, outweigh the benefits of non-green construction. Companies should regularly conduct self-inspection and self-correction activities and build a sound self-restraint system. In addition, construction companies should also retrain and update the knowledge reserve of on-site construction personnel to master the new construction processes and procedures of GBs and achieve them as a long-term goal.

This study uses the evolutionary game to study the problem of GB construction supervision, analyzes and discusses the behavior strategies in the process of GBs construction, and proposes dynamic incentive policy combinations that can improve the applicability and accuracy of policies. However, due to the large and complex content involved in the supervision of GBs, there are still certain areas for improvement in the model assumptions in this paper. We only study government and construction enterprises as game participants. Still, many participants in the GBs construction process, such as the public and owners, are also essential for the constraint of construction enterprises. Future research can build a multi-agent game model from this perspective to make the game model more realistic, targeted, and effective.

**Author Contributions:** Conceptualization, Y.L. and Y.C.; methodology, Y.C. and Q.M.; software, Z.L.; validation, Y.C., J.X. and Z.L.; formal analysis, Y.C. and Q.M.; investigation, Y.L.; resources, J.X.; writing—original draft preparation, Y.C.; writing—review and editing, Q.M.; visualization, Z.L.; supervision, J.X.; project administration, Y.L.; funding acquisition, Z.L. All authors have read and agreed to the published version of the manuscript.

**Funding:** This research was funded by the National Natural Science Foundation of China (Nos. 72071096, 71971100).

**Data Availability Statement:** Not applicable.

**Acknowledgments:** The author thank the editors and reviewers for their hard work.

**Conflicts of Interest:** The authors declare no conflict of interest.

## Appendix A

### *Appendix A.1. The Proof of Proposition 7*

**Proof.** When $S_1 - S_2 - R_1 < B + P(y)$, $P(y) - S_3 - L > 0$, the stability at $({x_3}^*, {y_3}^*)$, the corresponding Jacobi matrix is given by:

$$J_2 = \begin{pmatrix} 0 & a_3 \\ b_3 & c_3 \end{pmatrix}$$

where $a_3 = ({x_3}^* - 1){x_3}^*(B - 2r({y_3}^* - 1)), b_3 = (B + r(1 - {y_3}^*))(1 - {y_3}^*){y_3}^*, c_3 = {x_3}^*(B + r - 2B{y_3}^* + r{y_3}^*(-4 + 3{y_3}^*)) + (-1 + 2{y_3}^*)S_1 + S_2 + R_1 - 2{y_3}^*(S_2 + R_1)$. □

The system has asymptotic stability under the static reward and dynamic punishment mechanisms, and $({x_3}^*, {y_3}^*)$ is the ESS of the system.

### *Appendix A.2. The Proof of Proposition 10*

**Proof.** When $S_1 - S_2 - R_1 < B(y) + P(y)$,$P(y) - S_3 - L > 0$, the stability at $({x_4}^*, {y_4}^*)$, the corresponding Jacobi matrix is given by:

$$J_2 = \begin{pmatrix} 0 & a_4 \\ b_4 & c_4 \end{pmatrix}$$

where $a_4 = -2(-1 + {x_4}^*){x_4}^*(r(-1 + {y_4}^*) - q{y_4}^*), b_4 = (1 - {y_4}^*){y_4}^*(r(1 - {y_4}^*) + q{y_4}^*)$, $c_4 = q{x_4}^*(2 - 3{y_4}^*){y_4}^* + r{x_4}^*(-1 + {y_4}^*)(-1 + 3{y_4}^*) + (-1 + 2{y_4}^*)S_1 + S_2 + R_1 - 2{y_4}^*(S_2 + R_1)$. □

## References

1. Shen, Y.; Faure, M. Green building in China. *Int. Environ. Agreem. Politics Law Econ.* **2021**, *21*, 183–199. [CrossRef]
2. Liu, Y.; Sun, X.; Sun, T.; Shi, X.; Liu, J. Promoting green residential buildings by increasing homebuyers' willingness to pay: Evidence from Sino-Singapore Tianjin Eco-city in China. *J. Clean. Prod.* **2019**, *238*, 117884. [CrossRef]

3. Illankoon, I.; Tam, V.; Le, K.N.; Tran, C.; Ma, M. Review on Green Building Rating Tools Worldwide: Recommendations for Australia. *J. Civ. Eng. Manag.* **2019**, *25*, 831–847. [CrossRef]
4. Zainine, M.A.; Mezni, T.; Baeshen, Y.; Rahmoun, M.; Guizani, A. The assessment of buildings and constructions sector of economy proposal: An environmental perspective. *Environ. Sci. Pollut. Res.* **2021**, *28*, 22510–22521. [CrossRef] [PubMed]
5. Lei, M.; Cui, T. A Scientometric Analysis and Visualization of Global LEED Research. *Buildings* **2022**, *12*, 1099. [CrossRef]
6. Jiang, S.; Wang, N.; Wu, J. Combining BIM and Ontology to Facilitate Intelligent Green Building Evaluation. *J. Comput. Civ. Eng.* **2018**, *32*, 04018039. [CrossRef]
7. Ye, L.; Cheng, Z.; Wang, Q.; Lin, H.; Lin, C.; Liu, B. Developments of Green Building Standards in China. *Renew. Energy* **2015**, *73*, 115–122. [CrossRef]
8. Hwang, B.; Tan, J. Green building project management: Obstacles and solutions for sustainable development. *Sustain. Dev.* **2012**, *20*, 335–349. [CrossRef]
9. Jin, J. Problem Analysis and Countermeasures of the Construction Enterprises in the Green Construction. In *Applied Mechanics and Materials*; Tang, X., Chen, X., Dong, Y., Wei, X., Yang, Q., Eds.; Trans Tech Publications Ltd.: Cham, Switzerland, 2013; Volume 291–294, pp. 1016–1020.
10. He, Q.; Wang, Z.; Wang, G.; Zuo, J.; Wu, G.; Liu, B. To be green or not to be: How environmental regulations shape contractor greenwashing behaviors in construction projects. *Sustain. Cities Soc.* **2020**, *63*, 102462. [CrossRef]
11. Rehman, U.; Iatisam, M.; Younas, W.; Tariq, N. Review on the Role of Sustainability Techniques in Development of Green Building. In Proceedings of the 2017 International Conference on Energy Conservation and Efficiency (ICECE), Lahore, Pakistan, 22–23 November 2017; IEEE: Piscataway, NJ, USA, 2017; pp. 74–80.
12. Chen, Y.; Wang, G.; He, Y.; Zhang, H. Greenwashing behaviors in construction projects: There is an elephant in the room! *Environ. Sci. Pollut. Res.* **2022**, *29*, 64597–64621. [CrossRef]
13. Du, X. How the Market Values Greenwashing? Evidence from China. *J. Bus. Ethics* **2015**, *128*, 547–574. [CrossRef]
14. Huang, N.; Bai, L.; Wang, H.; Du, Q.; Shao, L.; Li, J. Social Network Analysis of Factors Influencing Green Building Development in China. *Int. J. Environ. Res. Public Health* **2018**, *15*, 2684. [CrossRef] [PubMed]
15. Liu, H.; Li, J.; Sun, Y.; Wang, Y.; Li, W. Government incentives to promote the development of green building based on games theory. In *Advanced Materials Research*; Xu, Q., Li, H., Li, Q., Eds.; Trans Tech Publications Ltd.: Cham, Switzerland, 2014; Volume 869–870, pp. 381–384.
16. Liu, Y. Legal construction of green building standards:realistic problems and improvement path. *China Popul. Resour. Environ.* **2020**, *30*, 170–178.
17. Smith, J.M.; Price, G.R. The Logic of Animal Conflict. *Nature* **1973**, *246*, 15–18. [CrossRef]
18. Tao, Z.; Wang, B.; Shu, L. Analysis on the Procurement Cost of Construction Supply Chain based on Evolutionary Game Theory. *Arab. J. Sci. Eng.* **2021**, *46*, 1925–1940. [CrossRef]
19. Xu, X.; Li, Y.; Hou, X.; Zeng, S. Evolutionary Game Analysis of Dangerous-Waste Disposal of Major Engineering Enterprises and Government Supervision. In Proceedings of the International Conference on Construction and Real Estate Management 2021, Beijing, China, 16–17 October 2021; Wang, Y., Zhang, D., Shen, G., Eds.; ASCE: Reston, VA, USA, 2021; pp. 554–560.
20. Qin, Z.; Wei, W.; Qin, M. Study on the Green Construction Supervision Based on Evolutionary Game Theory. In *Advances in Computer Science Research*; Jing, W., Guiran, C., Huiyu, Z., Eds.; Springer: Berlin/Heidelberg, Germany, 2017; Volume 59, pp. 1034–1040.
21. Wang, J.; Qin, Y.; Zhou, J. Incentive policies for prefabrication implementation of real estate enterprises: An evolutionary game theory-based analysis. *Energy Policy* **2021**, *156*, 112434. [CrossRef]
22. Jiang, X.; Sun, H.; Lu, K.; Lyu, S.; Skitmore, M. Using evolutionary game theory to study construction safety supervisory mechanism in China. *Eng. Constr. Archit. Manag.* **2023**, *30*, 514–537. [CrossRef]
23. Chen, L.; Gao, X.; Hua, C.; Gong, S.; Yue, A. Evolutionary process of promoting green building technologies adoption in China: A perspective of government. *J. Clean. Prod.* **2021**, *279*, 123607. [CrossRef]
24. Gao, Y.; Jia, R.; Yao, Y.; Xu, J. Evolutionary Game Theory and the Simulation of Green Building Development Based on Dynamic Government Subsidies. *Sustainability* **2022**, *14*, 7294. [CrossRef]
25. Jiang, Y.; Hu, T.; Zhao, D.; Liu, B.; Zhang, H.; Zhang, Y.; Xu, Z. Decision model to optimize long-term subsidy strategy for green building promotion. *Sustain. Cities Soc.* **2022**, *86*, 104126. [CrossRef]
26. Gang, L.; Cui, L. Analysis of urban green building and sustainable development. *Agro Food Ind. Hi-Tech* **2017**, *28*, 1031–1035.
27. Hwang, B.-G.; Shan, M.; Phuah, S.L. Safety in green building construction projects in Singapore: Performance, critical issues, and improvement solutions. *KSCE J. Civ. Eng.* **2018**, *22*, 447–458. [CrossRef]
28. Xing, Z.; Cao, X. Promoting Strategy of Chinese Green Building Industry: An Evolutionary Analysis Based on the Social Network Theory. *IEEE Access* **2019**, *7*, 67213–67221. [CrossRef]
29. Zeng, X.; Li, S.; Yin, S.; Xing, Z. How Does the Government Promote the Collaborative Innovation of Green Building Projects? An Evolutionary Game Perspective. *Buildings* **2022**, *12*, 1179. [CrossRef]
30. Assylbekov, D.; Nadeem, A.; Hossain, M.A.; Akhanova, G.; Khalfan, M. Factors Influencing Green Building Development in Kazakhstan. *Buildings* **2021**, *11*, 634. [CrossRef]
31. Olubunmi, O.A.; Xia, P.B.; Skitmore, M. Green building incentives: A review. *Renew. Sustain. Energy Rev.* **2016**, *59*, 1611–1621. [CrossRef]

32. Cohen, C.; Pearlmutter, D.; Schwartz, M. A game theory-based assessment of the implementation of green building in Israel. *Build. Environ.* **2017**, *125*, 122–128. [CrossRef]
33. Liu, G.; Li, X. Literature Review on the Incentive Policies to Promote the Development of Green Building. In *Applied Mechanics and Materials*; Hou, H., Tian, L., Eds.; Trans Tech Publications Ltd.: Cham, Switzerland, 2013; Volume 357–360, pp. 2785–2792.
34. He, C.; Wang, X.; Zhao, G. Developer's willingness to construct green dwellings in china: Factors and stimulating policies. *J. Civ. Eng. Manag.* **2018**, *24*, 378–389. [CrossRef]
35. Onuoha, I.J.; Aliagha, G.U.; Rahman, M.S.A. Modelling the effects of green building incentives and green building skills on supply factors affecting green commercial property investment. *Renew. Sustain. Energy Rev.* **2018**, *90*, 814–823. [CrossRef]
36. Li, X.; Wang, C.; Kassem, M.A.; Liu, Y.; Ali, K.N. Study on Green Building Promotion Incentive Strategy Based on Evolutionary Game between Government and Construction Unit. *Sustainability* **2022**, *14*, 10155. [CrossRef]
37. Kong, F.; He, L. Impacts of supply-sided and demand-sided policies on innovation in green building technologies: A case study of China. *J. Clean. Prod.* **2021**, *294*, 126279. [CrossRef]
38. Chen, W.; Hong, X. Design of effective subsidy policy to develop green buildings: From the perspective of policy benefit. *Clean Technol. Environ. Policy* **2015**, *17*, 1029–1038. [CrossRef]
39. Song, Y.; Li, C.; Zhou, L.; Huang, X.; Chen, Y.; Zhang, H. Factors affecting green building development at the municipal level: A cross-sectional study in China. *Energy Build.* **2021**, *231*, 110560. [CrossRef]
40. Liu, C.; Huang, W.; Yang, C. The evolutionary dynamics of China's electric vehicle industry–Taxes vs. subsidies. *Comput. Ind. Eng.* **2017**, *113*, 103–122. [CrossRef]
41. Wang, C.; Cui, W. Supervision for the Public Health Services for Older Adults Under the Background of Government Purchasing: An Evolutionary Game Analysis Framework. *Front. Public Health* **2022**, *10*, 881330. [CrossRef]
42. Wang, X.; Ren, X. Research on Dynamic Evolutionary Game of Platform E-commerce Credit Supervision from the Perspective of Government Governance. *Chin. J. Manag. Sci.* **2021**, *29*, 29–41.
43. Collin-Dufresne, P.; Fos, V. Do Prices Reveal the Presence of Informed Trading? *J. Financ.* **2015**, *70*, 1555–1582. [CrossRef]
44. Pi, Z.; Gao, X.; Chen, L.; Liu, J. The New Path to Improve Construction Safety Performance in China: An Evolutionary Game Theoretic Approach. *Int. J. Environ. Res. Public Health* **2019**, *16*, 2443. [CrossRef]
45. Ma, Y.; Wan, Z.; Jin, C. Evolutionary Game Analysis of Green Production Supervision Considering Limited Resources of the Enterprise. *Pol. J. Environ. Stud.* **2021**, *30*, 1715–1724. [CrossRef]
46. Cai, D.; Guo, C.; Bai, L.; Huang, H. Multi-Attribute Procurement Auctions Considering Green Degree of Buildings Under Different Subsidy Objects. *J. Syst. Manag.* **2022**, 1–16. Available online: https://kns.cnki.net/kcms/detail/31.1977.N.20221130.1611.001.html (accessed on 11 February 2023).
47. Kim, S.; Lim, B.T.H. Analysing the Characteristics of Green and Non-Green Buildings: From the Real Estate Perspective. In *AIP Conference Proceedings*; Chieh, D., Ed.; AIP Publishing LLC: Melville, NY, USA, 2019; Volume 2124.
48. Chen, W.; Hu, Z.-H. Using evolutionary game theory to study governments and manufacturers' behavioral strategies under various carbon taxes and subsidies. *J. Clean. Prod.* **2018**, *201*, 123–141. [CrossRef]
49. Wu, H.; Ying, M. An Analyzes on Complexity and Evolutionary Stability in Co-petition Games. *Syst. Eng.-Theory Pract.* **2004**, *2*, 90–94.
50. Dan, F. Evolutionary Games in Economics. *Econom. Econom. Society* **1991**, *59*, 637–666.
51. Meng, Q.; Liu, Y.; Li, Z.; Wu, C. Dynamic reward and penalty strategies of green building construction incentive: An evolutionary game theory-based analysis. *Environ. Sci. Pollut. Res.* **2021**, *28*, 44902–44915. [CrossRef] [PubMed]
52. Liang, X.; Fu, Y. Study on the Supply Side Evolutionary Game of Green Building under the Mechanism of Government Dynamic Reward and Punishment. *Chin. J. Manag. Sci.* **2021**, *29*, 184–194.

buildings

MDPI

*Article*

# Research on Cooperative Evolutionary Game of Design and Construction Consortium of Green Building Project under Design Change

**Yingchen Wang [1], Ling Lv [1], Xiaoxiao Geng [2,*], Liyuan Ren [1] and Ran Sun [1]**

[1] School of Management Engineering and Business, Hebei University of Engineering, Handan 056038, China; wangyingchen@hebeu.edu.cn (Y.W.); l15031960826@163.com (L.L.); renliyuan@hebeu.edu.cn (L.R.); sunran1110@163.com (R.S.)

[2] School of Architecture and Art, Hebei University of Engineering, Handan 056038, China

* Correspondence: gengxiaoxiao@hebeu.edu.cn

**Abstract:** In recent years, with the country's vigorous promotion of green buildings and the increasingly complex and large-scale engineering projects, the design-construction consortium model can better meet the needs of the organization and implementation of large-scale green projects and become a realistic choice for enterprises in project implementation. Therefore, the formation of a good and stable cooperative relationship between consortium members is increasingly important in improving project revenue and quality. The issue of maintaining the stability of consortium relationships is an urgent problem to be solved at this stage. As such, a three-party evolutionary game model, based on evolutionary game theory, comprising the developer unit, design unit, and construction unit, is constructed here. Then, strategies for ensuring evolutionary stability under different design modalities are discussed. Finally, the influence of relevant parameters under changing design conditions on the stability of the design and construction consortium of green building projects is analyzed through numerical simulation. The research results show the following: (1) If the additional revenue distribution coefficient within the consortium members is closer to 0.5, the influence on the stability of the design and construction consortium will be smaller; in contrast, if the influence on the design and construction consortium is increased, the cooperative relationship within the consortium will be more unstable. (2) The presence of additional revenue $\Delta\pi_1$ can increase the stability of the design and construction consortium. An increase in the additional revenue $\Delta\pi_1$ will inhibit the instability of the consortium on the one hand and strengthen the stability of the consortium on the other but will also lead to the occurrence of opportunistic behavior. (3) The construction unit's payment of a subsidy to the cooperative members can help promote the stability of the design and construction consortium to a certain extent and can also weaken the effects of other factors on the stability of the consortium, but there is a threshold value for the amount of said subsidy. (4) On the one hand, the cooperation members actively cooperate with each other to maximize the cooperation benefits of the design and construction consortium, while on the other hand, the construction unit actively promotes the implementation of the green building project, strictly monitors the implementation of the green design and green construction approach by the design and construction units in the early and implementation stages of the project, prevents the design changes caused by the final product failing to meet the green building standard, and actively solves design change problems in a manner that benefits the sustainable development of the green building, so that the cooperative relationship among the members of the consortium can develop steadily, which is beneficial to the green and ecological development of architectural design and construction.

**Keywords:** green building; design and construction consortium; design changes; evolutionary game

## 1. Introduction

Eco-city and green building are global development strategies and promoting the development of green building is the trend of building development in the new period. With the continuous development of construction projects, the project is becoming more and more complex, large-scale, and green, which determines that it has higher requirements for design and construction, which requires not only coordinated management within each enterprise [1] but also the cooperation between enterprises, so the application of the design-construction consortium model in the future construction industry is an inevitable trend. The design-construction consortium model can bring more benefits to large-scale green building projects, and it is superior to the traditional model in terms of cost, schedule, and quality [2–4]. Some studies have also pointed out that this team cooperation mode is also helpful to avoid contract disputes in the construction process [5]. Therefore, the cooperative relationship between the design and construction consortium is an important factor affecting the success or failure of large-scale green building projects.

However, compared with the traditional model, the design and construction consortium model needs the information transmission of the project and the coordination of the stakeholders. In the process of project implementation, the interests of individual members may conflict with those of other partners or even the whole project. In order to safeguard their own interests, the consortium partners may be reluctant to share the project information involving their related interests, or even hide information that is unfavorable to them from the partners in order to maximize their own interests. In the end, it will not only lead to poor implementation of the project in terms of time, cost, and quality but also benefit the consortium [6]. Some scholars have also confirmed this point, and the lag of information transmission and uncoordinated cooperation in the design and construction process of the design and construction consortium will affect the quality of the project [7]. Some studies also believe that the excessive intervention and decision-making mistakes of the owners are also important factors affecting the design and construction consortium [8]. In addition, the stability of the cooperative relationship within the consortium will also affect the number of design changes and the cost of changes [9,10]. In other words, when the cooperation within the consortium is unstable, it is easy to cause design changes. However, there is little research on the influence of design change on consortium relations.

During the implementation of large-scale green projects, design changes are inevitable. Design changes have an important impact on project schedule management, quality management, and investment control, and also have an important impact on the coordination among participants [11]. From extensive research, it can be seen that design change is one of the main factors of cost overrun, and in some cases, it may lead to cost overrun accounting for 5% to 40% of project cost [12,13]. In addition, design change is also an important influencing factor in project delay [14]. Therefore, this is a very important and serious problem. Good design change management is of great significance for improving the construction quality of building projects, enhancing the green degree of building products, and promoting the sustainable development of buildings and the environment.

At present, the research on design change management and cooperation in the design and construction consortium is relatively mature, but most of the research on influencing factors is limited to static analysis. In reality, the influence of a certain factor on the main body is a dynamic process. In addition, many stakeholders are involved in the implementation of large-scale green building projects, and they have different preferences when making decisions, but other stakeholders cannot predict them accurately. Therefore, based on evolutionary game theory, this paper analyzes the cooperative relationship of the design and construction consortium under design change. Evolutionary game theory can not only fully reflect the interaction between participants and describe the specific process of strategy selection, but also combine the research thinking of traditional game theory with the dynamic evolution process and put forward that the game subjects can refer to, imitate, and learn from other people's strategy selection ideas in the game process to make the best judgment, and finally tend to a stable and balanced state. Therefore, it is of certain reference value to use evolutionary game theory to analyze the stability of design and construction consortium under design change.

Therefore, in order to solve the problem of a complex dynamic game and multi-party participation in the cooperation process of the design and construction consortium in large-scale green projects, this paper will analyze the evolutionary game relationship of consortium cooperation stability under design change, build an evolutionary game model composed of a constructer, designer, and developer, combine the computer simulation means of Matlab with the dynamic evolution idea, and describe the influence of equilibrium point stability and related parameters on the strategic evolution of participants and the stability of the design and construction consortium through numerical simulation. This paper aims to reveal the complex dynamics of the game process of consortium cooperation in large-scale green building projects and put forward constructive suggestions for promoting the establishment of harmonious and stable consortium cooperation and the sustainable development of green buildings.

The rest of the paper is structured as follows. The second part summarizes the existing literature, and the third part introduces the game relationship among the constructer, designer, and developer and introduces the construction of the model. The fourth part analyzes the stability of the equilibrium point. In the fifth part, Matlab (2019b) is used to simulate the evolution of the model. The sixth part contains the conclusions and recommendations.

## 2. Literature Review

This paper reviews the relevant literature in terms of three aspects: The design and construction consortium, design changes in construction projects, and the application of the evolutionary game in the green building, as shown in Table 1.

**Table 1.** Research related to the design and construction consortium, design changes in construction projects, and the application of evolutionary game in green buildings.

| Research Topics | Dimensions | Source Papers |
|---|---|---|
| Design and construction consortium | Selection of consortium members and formation of cooperation | [1,15] |
| | Advantages of consortium model in cost, time and quality | [5,16,17] |
| | Identification and evaluation of factors affecting consortium cooperation | [18,19] |
| Design changes in construction projects | Causes of design changes | [20–22] |
| | Impact of design changes | [23–25] |
| | In terms of research methods | [26–28] |
| The application of evolutionary game in green buildings | Save energy and cut emissions | [29] |
| | Green construction | [30] |
| | Government dynamic reward and punishment policy | [31–33] |
| | Construction waste recycling | [34,35] |

### 2.1. Design and Construction Consortium

The development of the consortium mode has become an inevitable trend in the world. As an international intensive form, the international engineering construction consortium has gained popularity in international engineering contracting projects. In the 1980s and 1990s, two cooperative project management modes emerged in developed countries, such as the United States and Britain, namely, Partnering proposed by Charles Cava and Alliancing proposed by BP. The American Building Industry Association (CII1995) pointed out that Partnering can achieve cultural sharing among organizations and establish a relationship of understanding, trust, and common goals among members of organizations. This is undoubtedly the ultimate goal of the consortium. In the formation of the consortium, Wang Xuetong [1] put forward a development model of integrated and collaborative management of design and construction, which made all organizations and departments more integrated and collaborative. In terms of research methods, Park H [15] uses social network analysis (SNA) to study the formation of consortium members' selection and cooperation. In the application of consortium mode, Nasrun M [5] thinks that team cooperation is helpful to avoid delay in the construction process, and Okunlola Ojo S [16] compares the traditional contracting and design-build procurement modes, and the research shows that the design-build mode is superior to the traditional mode in cost, time, and quality standards. Other scholars have suggested that the consortium model can give full play to the advantages of various stakeholders and expand the design depth, which is conducive to the success of construction projects [17]. In addition, Khalef R [18] identifies the influencing factors of the consortium through the analysis of social networks and association rules, and Dogan S Z [19] uses the social network analysis on the e-mail communication network between participants to quantitatively measure the coordination of the consortium cooperation. The research shows that the coordination score is highly correlated with the centrality index.

### 2.2. Design Changes in Construction Projects

Construction projects are easily influenced by various factors in the implementation process, on the one hand, due to the characteristics of large-scale, complex, and green construction projects themselves, and on the other hand, due to external factors such as the competent decision-making of stakeholders or environmental changes [26]. When a building project is affected to some extent, engineering changes usually occur. Research shows that the standard to measure the success of a construction project is primarily to finish it within time, cost budget, and quality restraints [36], and it is undeniable that design changes will have an impact on these three indicators more or less [37]. Some scholars directly show that frequent design changes are one of the reasons leading to project failure [38]. Therefore, many scholars have carried out extensive research on the management of design change. In terms of the causes of design change, Gharaibeh et al. [20] investigated the factors leading to design change according to the opinions of 252 professionals in Jordan's construction industry and found that the owner's requirements, design errors and omissions, and value engineering are the most important main influencing factors. Yap and Skitmore [14] showed that design changes will cause 5–20% time and cost overruns for construction projects in Malaysia. A lack of coordination among various professional consultants, changes in requirements/specifications, increases/omissions of scope, errors/differences in design documents, and unpredictable ground conditions are the five most important reasons. Yap et al. [21] investigated 39 reasons leading to design changes. The research shows that eight factors, such as the project team's ability, quality and technology, site restrictions and safety considerations, legislation and regulations, active rework, project communication, end-user requirements, and risk management, are the basic reasons leading to design changes. On this basis, Shahab Shoar [22] detailed 23 key reasons and analyzed the relationship among them. The research showed that factors such as "unfamiliar with new construction methods", "design errors", "value engineering", "uncertainty of scope", "change orders", and "neglected constructability in design stage" are easily influenced by other factors, among which is "customer's attitude and experience". In terms of the impact of design changes, Aslam et al. [23] studied the impact of design changes on the project cost, and the research showed that design changes were one of the main factors leading to cost overruns, which might even account for 19% of the project cost. Matusala Bassa [24], through an analysis of questionnaire results, showed that the biggest impact of design changes is the delay in completion schedule, the increase in project cost, the waste of materials during rework, the decline in productivity, and overtime to meet the project deadline. Rahman [25] also confirmed that design changes in Malaysia are the main reason for project schedule and cost overruns. Saad et al. [27] put forward a system dynamics model to capture the factors that may lead to engineering and medical-related design changes in healthcare projects and investigated the impact of these design changes on project performance. Afsharghotli and Yitmen [28] used the artificial neural network (ANN) model to evaluate the quantitative measurement of the time and cost performance of petrochemical projects caused by the interactive design change. AAGA Yana [26] used partial least squares (PLS) to analyze the factors that affect the design change and divided the factors that affect the design change into internal factors and external factors.

*2.3. The Application of Evolutionary Game in the Green Building*

Evolutionary game theory comes from the theory of biological evolution. In 1973, Smith and Price [39] put forward the strategy of evolutionary stability. Since then, evolutionary game theory has been widely used in the construction industry. Scholars have also studied green buildings from various perspectives. Cohen et al. [29] used game theory to explain the obstacles to energy saving and emission reduction in Israel's construction industry and proved that government subsidies can help eliminate these obstacles. Geng X [30] established an evolutionary game model among developers, contractors, and the government from the perspective of consumers' different green preferences, studied the optimal strategy of the green construction system in different situations, and put forward effective suggestions for each subject. Li X [31], based on the evolutionary game theory, analyzed the influence of a local government subsidy policy on the application strategy of the construction unit. The research showed that in the long run, government subsidies cannot improve the willingness of the construction unit to promote green buildings, so controlling the amount of subsidies and strengthening publicity is conducive to encouraging the construction unit to participate in the transformation and upgrading of the construction industry. Based on evolutionary game theory, Chen Y [32] comparatively analyzed the internal mechanism of behavior evolution of government and construction enterprises under four policies: Static reward and static punishment, static reward and dynamic punishment, dynamic reward and static punishment, and dynamic reward and dynamic punishment. The research results provide theoretical support for the formulation of government policies, and the government should dynamically adjust the intensity of rewards and punishments and determine the upper limit of rewards and punishments. Meng Q [33], based on evolutionary game theory, studied the role of the government incentive mechanism in promoting green building construction. The research shows that dynamic reward and static punishment are the best strategies to promote green building construction. Chen J [34] studied the management of building demolition waste based on evolutionary game theory. The research showed that supervision intensity, supervision cost, punishment, garbage disposal cost, and illegal dumping income are the main factors affecting the decision-making behavior of contractors and government departments. In addition, encouraging the public to participate in supervision can effectively promote the recycling of waste. Su Y [35] studied the strategic changes of local governments, contractors, and recycling factories in the construction waste recycling market based on evolutionary game theory, and analyzed the behaviors, demands, and synergies of stakeholders in the construction waste market, providing management suggestions for policy makers. Evolutionary game theory can aptly study the problem of green buildings, which is the basis of this research model.

To summarize, although domestic and foreign research has achieved certain results related to design and construction consortia, design changes during construction projects, and the application of an evolutionary game in the construction industry, certain limitations still arise, which manifest as follows: (1) Research on the cooperative relationship amongst consortia is mostly limited to the static analysis of influencing factors, while research on the dynamic mechanisms of factors influencing the cooperative relationship of consortia is still rare. (2) On the one hand, although many scholars have recognized the impact of design change management on the smooth implementation of construction projects, most of the research focuses on the causes of design changes and their countermeasures, but the existing research results do not generally consider that design changes are unpredictable, meaning that the relevant enterprises cannot implement the necessary countermeasures in a timely manner. On the other hand, little of the literature has studied design changes during green building projects. (3) Previous studies have shown that the application of evolutionary games to the construction industry is applicable to a certain degree. Evolutionary games can help reveal the long-term dynamic game process, but few have applied evolutionary games to solve the problems related to the stability of the design and construction consortium's cooperative relationship. Therefore, the research focus of this paper is on how to maintain a cooperative relationship within the design and construction consortium so as to manage the impact of design changes during the implementation of large-scale green building projects. This paper addresses the stability of a consortium's cooperative relationship and the influence of relevant factors by constructing a three-party evolutionary game model comprising a construction unit, a design unit, and a construction unit. In relation to improving design change management, this paper aims to provide theoretical support and practical guidance for establishing a harmonious and stable cooperative relationship within a consortium, and further promote the sustainable development of green buildings.

## 3. Model

The abovementioned design and construction consortium refers to the consortium formed by designers and constructors, in which the lead party bids for the general contracting project. After winning the bid, relying on their respective strengths, they separately complete the design and construction tasks, and both answer to the developer. In this way, the members of the consortium share resources and risks, i.e., maintain individual independence while serving the project together in a stable partnership. After the project is completed, internal liquidation is undertaken, and the organization is dissolved [40].

This paper primarily studies three important groups: The constructors, designers, and developers. In the process of consortium cooperation under the design change of large-scale green projects, the strategic choices of these three groups are both different and changing. In real society, the constructer, designer, and developer are all bounded by rationality, and the information of the three parties in the same construction project is incomplete and unequal. Therefore, each group observes the change in each other's strategies, constantly trying and imitating to improve their own strategies, so that the evolution of all group strategies will eventually become dry and stable. The research logic conforms to the bounded rational conditions, mutation, and selection ideas of evolutionary games. Therefore, evolutionary game theory is suitable for studying the strategy choice of the constructer, designer, and developer.

Under the conditions of bounded rationality and cooperation between the two parties, the developer will initiate a design change, after which the designer and constructer will start to comprehensively consider the costs and benefits and formulate two different strategies, namely "cooperation" and "non-cooperation". In a long-term evolutionary game, an uncooperative designer or constructer may not cooperate, or may withdraw, due to the prioritization of their individual interests, the potential risks related to joint contracting, or the uneven distribution of benefits derived from project implementation. However, when higher or additional revenue is generated via the optimization of cooperation between the two parties, and the revenue obtained comes to be greater than that yielded by separate contracts, the designer and constructer will select "cooperation". When the designer actively cooperates, green, environmentally friendly, and safe products can be made as a result of their professional abilities. When a designer chooses not to cooperate, the efforts made by both parties under the previously pertaining cooperative relationship will be rendered void, and the achievements will be greatly reduced. A designer who makes strategic changes in the middle and later stages of the project will also cause huge losses to be suffered during project implementation. On the other hand, when a constructor pursues cooperation, the product can be made to reach the green quality standard, and its safety will be improved. At the same time, they can help to strictly control the costs associated with the construction process and ensure the maximization of overall interests. However, a constructer who abandons cooperation and withdraws halfway through the project will cause serious losses to be suffered by the developer, resulting in delays in construction. If the design and construction consortium becomes extremely unstable, not only will great economic losses be suffered by the developer but design change behaviors will be triggered [41]. If a consortium member commits a breach of contract when the developer initiates a design change, the developer may seek a new partner to complete future work. If this occurs, the potential benefits brought by the new member will compensate the developer for the losses already incurred, but the developer can also choose whether to initiate further design changes based on the principle of maximizing their own interests.

In a green building project, the foundation of the design and construction consortium is the maximization of personal interests, and design changes can be beneficial to the stable and sustainable development of internal relations. Therefore, this paper focuses on the stability of a design and construction consortium following design changes proposed by a developer.

### 3.1. Hypotheses and Descriptions

The parameter settings in this paper are shown in Table 2.

**Table 2.** Main body parameters.

| Main Body | Parameters | Explanatory Notes |
|---|---|---|
| constructer | $\pi_1$ | Normal revenue of enterprise with non-cooperation of the constructer |
| | $\Delta\pi_1$ | Additional project benefits derived from cooperation between designer and constructer |
| | a | Additional income distribution coefficient of the constructer |
| | $D_1$ | When the developer initiates a design change, the loss of reputation suffered by the constructer within the industry as a result of choosing the "non-cooperation" strategy |
| | $E_1$ | The additional income obtained by the constructer when the designer cooperates after the default mode on the basis of the original average normal income |
| | F | Liquidated damages to be paid by the constructer when choosing the non-cooperation strategy |
| designer | $\pi_2$ | Normal revenue of an enterprise with the non-cooperation of the designer |
| | $\Delta\pi_1$ | Additional project benefits derived from cooperation between the designer and constructer |
| | $1 - a$ | Additional income distribution coefficient of the designer |
| | $D_2$ | The loss of reputation suffered by the designer within the industry as a result of choosing the "non-cooperation" strategy when the developer initiates a design change |
| | $E_2$ | The additional income obtained by the designer when the constructer cooperates after the default mode on the basis of the original average normal income |
| | F | Liquidated damages to be paid by the designer when choosing the non-cooperation strategy |
| developer | $\Delta\pi_2$ | Additional benefits generated by the developer due to design changes |
| | B | Loss of normal project income received by the developer when the non-cooperation strategy is adopted by the designer or constructer |
| | b | Distribution coefficient of developer's profits and losses due to the constructer's non-cooperation |
| | $1 - b$ | Distribution coefficient of the developer's profits and losses due to the designer's non-cooperation |
| | C | Additional costs incurred by the developer due to design changes |
| | $F_1$ | When the constructer cooperates, the developer subsidizes them for making design changes |
| | $F_2$ | When the designer cooperates, the developer subsidizes them for making design changes |
| | $G_1$ | When the constructer breaches the contract, the potential benefits that the developer will receive after making design changes by seeking a new enterprise |
| | $G_2$ | When the designer breaches the contract, the potential benefits that the developer will gain after making design changes by seeking a new enterprise |

In order to further clarify the game relationship between the research subjects, we have made the following assumptions:

**Hypothesis 1 (H1).** *Whether or not the designer and the constructer cooperate with the design changes proposed by the developer is denoted as {cooperation, non-cooperation}. Under the "cooperation" strategy, both parties make design changes according to the requirements of the construction unit. Under the "non-cooperation" strategy, one party or both parties break the contract during project implementation and withdraws from the consortium, thus terminating the cooperation relationship. The strategy set of the developer is {initiated design change, not initiated design change}.*

**Hypothesis 2 (H2).** *When the members of the design and construction consortium all choose the "cooperation" strategy, the increases in revenue received by the constructer and the designer when they cooperate is $\Delta\pi_1$, and the distribution coefficient of $\Delta\pi_1$* within the design and construction consortium is *a*.

**Hypothesis 3 (H3).** *When one or both parties of the design and construction consortium choose the "non-cooperation" strategy, the average normal incomes of the constructer and the designer are* $\pi_1$ and $\pi_2$, *respectively. When one party chooses to cooperate and the other chooses the non-cooperation strategy, resulting in reputational losses,* $D_1$ and $D_2$, *are felt, but additional income (*$E_1$ and $E_2$*) is obtained by the defaulting party on the basis of the original average normal income. If "asymmetric information" regarding one party's enterprise is obtained after a breach of contract, the breaching party shall pay liquidated damages L to the other in accordance with the contract and treaty, but if both parties choose the "non-cooperation" strategy, the liquidated damages will be offset against each other and ignored.*

**Hypothesis 4 (H4).** *When the developer chooses to initiate a change, additional revenue* $\Delta\pi_2$ *will be derived from the optimization of output resulting from cooperation with the green design change, but the additional cost C will be incurred by the design change, and the subsidies* $S_1$ *and* $S_2$ *must be paid to the cooperative members. When a member chooses not to cooperate, the construction unit will look for a new cooperative enterprise due to the design change. The potential revenue is* $G_1$ *and* $G_2$.

**Hypothesis 5 (H5).** *When the developer chooses not to initiate a change, its basic revenue is* $\pi_3$. *Regardless of whether the developer initiates a design change or not, when there are uncooperative members in the design and construction consortium, the developer will bear the corresponding loss B, wherein the profit and loss distribution coefficient among all members is b.*

### 3.2. Construction of the Model

Assuming bounded rationality, the probability of the constructer choosing the cooperative strategy is $x$, and the probability of them choosing the non-cooperative strategy is $1 - x$. The probability that the designer chooses the cooperation strategy is $y$, and the probability that they choose not to cooperate is $1 - y$. The probability that the developer chooses to initiate a design change is $z$, and the probability that they choose not to initiate a design change is $1 - z$. Based on these assumptions, the revenue matrix is shown in Table 3. In each cell, the first column is the revenue contributed by the constructor, the second is that contributed by the designer, and the third is that contributed by the developer.

**Table 3.** Matrix of benefits for game subjects.

| | Active Supervision ($z$) | | Negative Supervision ($1 - z$) | |
|---|---|---|---|---|
| | Cooperation ($y$) | Non-Cooperation ($1 - y$) | Cooperation ($y$) | Non-Cooperation ($1 - y$) |
| Cooperation ($x$) | $\pi_1 + a\Delta\pi_1 + S_1$<br>$\pi_3 - C + \Delta\pi_2 - S_1 - S_2$<br>$\pi_2 + (1 - a)\pi_1 + S_2$ | $\pi_1 + L + S_1$<br>$\pi_3 - C + \Delta\pi_2 + G_2 - (1 - b)B - S_1$<br>$\pi_2 - D_2 + E_2 - L$ | $\pi_1 + a\Delta\pi_1$<br>$\pi_3$<br>$\pi_2 + (1 - a)\pi_1$ | $\pi_1 + L$<br>$\pi_3 - (1 - b)B$<br>$\pi_2 + E_2 - L$ |
| Non-Cooperation ($1 - x$) | $\pi_1 - D_1 + E_1 - L$<br>$\pi_3 - C + \Delta\pi_2 + G_1 - bB - S_2$<br>$\pi_2 + L + S_2$ | $\pi_1 - D_1$<br>$\pi_3 - C + \Delta\pi_2 + G_1 + G_2 - B$<br>$\pi_2 - D_2$ | $\pi_1 + E_1 - L$<br>$\pi_3 - bB$<br>$\pi_2 + L$ | $\pi_1$<br>$\pi_3 - B$<br>$\pi_2$ |

## 4. Evolutionary Game Model Analysis

### 4.1. Calculation of Stable Points

The revenue when the constructor chooses to cooperate is denoted as $U_{11}$, that when it chooses not to cooperate is denoted as $U_{12}$, the average revenue of the strategic choice is $\overline{U_1}$, and the replication dynamic equation is given by $F(x)$. The results are as follows:

$$U_{11} = zS_1 + y(a\Delta\pi_1 + L) + \pi_1 + L \quad (1)$$

$$U_{12} = y(E_1 - L) - zD_1 + \pi_1 \quad (2)$$

$$\bar{U_1} = xU_{11} + (1-x)U_{12} = xy[\pi_1 - E_1] + xz(D_1 + S_1) + xL + y(E_1 - L) - zD_1 + \pi_1 \quad (3)$$

According to the Malthusain principle, the dynamic equation regarding the constructor can be written as follows:

$$F(x) = \frac{dx}{dt} = x(1-x)(U_{11} - U_{12}) = x(1-x)[y(a\Delta\pi_1 - E_1) + z(D_1 + S_1) + L] \quad (4)$$

When $y = \frac{z(D_1+S_1)+L}{-(a\Delta\pi_1 - E_1)}$ and $F(x) = 0$, regardless of the value of $x$, both of the constructor's potential strategies are ESS [39,42]. When $y > \frac{z(D_1+S_1)+L}{-(a\Delta\pi_1 - E_1)}$, let $F(x) = 0$, and then $x= 0$ and $x= 1$ can be obtained. $F(0) > 0$, and $F(1) < 0, x= 1$ is the stability point. That is, the constructor chooses the cooperative strategy as the equilibrium point. When $y < \frac{z(D_1+S_1)+L}{-(a\Delta\pi_1 - E_1)}$, let $F(x) = 0$, and then $x= 0$ and $x= 1$ can be obtained; $F(0) < 0$, and $F(1) > 0, x= 0$ is the stability point. That is, the constructor chooses the non-cooperative strategy as the equilibrium point.

The revenue when the designer chooses to cooperate is denoted as $U_{21}$, that when it chooses not to cooperate is denoted as $U_{22}$, the average revenue of the strategic choice is $\bar{U_2}$, and the replication dynamic equation is given by $G(y)$.

$$U_{21} = x[(1-a)\pi_1 - L] + zS_2 + \pi_2 + L \quad (5)$$

$$U_{22} = x(E_2 - L) - zD_2 + \pi_2 \quad (6)$$

$$\bar{U_2} = yU_{21} + (1-y)U_{22} = xy[(1-a)\Delta\pi_1 - E_2] + yz(S_2 + D_2) + yL + x(E_2 - L) - zD_2 + \pi_2 \quad (7)$$

$$G(y) = \frac{dy}{dt} = y(1-y)(U_{21} - U_{22}) = y(1-y)[x[(1-a)\Delta\pi_1 - E_2] + z(S_2 + D_2) + L] \quad (8)$$

When $x = \frac{z(S_2+D_2)+L}{-[(1-a)\Delta\pi_1 - E_2]}$, regardless of the value of y, both of the designer's strategies are ESS. When $x > \frac{z(S_2+D_2)+L}{-[(1-a)\Delta\pi_1 - E_2]}$, let $G(y) = 0$, and then $y= 0$ and $y= 1$ can be obtained; $G(0) > 0$, and $G(1) < 0$, $y= 1$ is the stability point. That is, the designer chooses cooperation as the equilibrium point. When $x < \frac{z(S_2+D_2)+L}{-[(1-a)\Delta\pi_1 - E_2]}$, let $G(y) = 0$, and then $y= 0$ and $y= 1$ will be obtained; $G(0) < 0$, and $G(1) > 0$, $y= 0$ is the stability point. That is, the designer chooses non-cooperation as the equilibrium point.

The revenue when the developer chooses to initiate the design change strategy is recorded as $U_{31}$, the revenue when he chooses not to initiate the design change strategy is recorded as $U_{32}$, the average revenue of the strategy selection is $\bar{U_3}$, and the replication dynamic equation is $H(z)$. The results are as follows:

$$U_{31} = x(bB - S_1 - G_1) + y[(1-b)B - S_2 - G_2] + \pi_3 - C + \Delta\pi_2 + G_1 + G_2 - B \quad (9)$$

$$U_{32} = xbB + y(1-b)B + \pi_3 - B \quad (10)$$

$$\bar{U_3} = zU_{31} + (1-z)U_{32} = xz(-S_1 - G_1) + yz(-S_2 - G_2) + z(-C + \Delta\pi_2 + G_1 + G_2) + xbB + y(1-b)B + \pi_3 - B \quad (11)$$

$$H(z) = \frac{dz}{dt} = z(1-z)(U_{31} - U_{32}) = z(1-z)[x(-S_1 - G_1) + y[-S_2 - G_2] + \Delta\pi_2 + G_1 + G_2 - C] \quad (12)$$

When $x = \frac{y[-S_2-G_2]+\Delta\pi_2+G_1+G_2-C}{S_1+G_1}$, regardless of the value of $z$, both of the developer's potential strategies are ESS. When $x > \frac{y[-S_2-G_2]+\Delta\pi_2+G_1+G_2-C}{S_1+G_1}$, let $H(z) = 0$, and then $z= 0$ and $z= 1$ can be obtained; $H(0) > 0$, and $H(1) < 0, z= 1$ is the stability point. That is, the developer initiating the design change is taken as the equilibrium point. When $x < \frac{y[-S_2-G_2]+\Delta\pi_2+G_1+G_2-C}{S_1+G_1}$, let $H(z) = 0$, and then $z= 0$ and $z= 1$ will be obtained; $H(0) <0$, and $H(1) > 0, z= 0$ is the stability point. That is, the developer not initiating design changes is the equilibrium point.

### *4.2. Evolutionary Equilibrium Stability Analysis*

According to the method of analysis proposed by Friedman, the stability of the equilibrium point of the game can be determined by the local stability of the Jacobian matrix. The partial derivatives of x, y, and z within the replicated dynamic equation are obtained individually, and in this way, the Jacobian matrix J of the system is obtained.

$$J = \begin{bmatrix} \frac{\partial F(x)}{x} & \frac{\partial F(x)}{y} & \frac{\partial F(x)}{z} \\ \frac{\partial G(y)}{x} & \frac{\partial G(y)}{y} & \frac{\partial G(y)}{z} \\ \frac{\partial H(z)}{x} & \frac{\partial H(z)}{y} & \frac{\partial H(z)}{z} \end{bmatrix} \tag{13}$$

Based on the Jacobian matrix [43], we let $F(x) = 0$, $G(y) = 0$, and $H(z) = 0$, and in this way, eight pure strategy equilibrium solutions can be obtained: A1 = (0, 0, 0), A2 = (1, 0, 0), A3 = (0, 1, 0), A4 = (0, 0, 1), A5 = (1, 0, 1), A6 = (0, 1, 1), A7 = (1, 1, 0), and A8 = (1, 1, 1). According to Lyapunov's stability criterion, when the eigenvalue of the Jacobian matrix is non-positive, the equilibrium point will be evolutionarily stable.

$$J = \begin{bmatrix} \lambda_1 & & \\ & \lambda_2 & \\ & & \lambda_3 \end{bmatrix}. \tag{14}$$

where $\lambda_1 = (1-2x)[y(a\Delta\pi_1 - E_1) + z(D_1 + S_1) + L]$, $\lambda_2 = (1-2y)[x[(1-a)\Delta\pi_1 - E_2] + z(S_2 + D_2) + L]$, and $\lambda_3 = (1-2z)[x(-S_1 - G_1) + y[-S_2 - G_2] + \Delta\pi_2 + G_1 + G_2 - C]$.

In relation to the equilibrium point, the possible eigenvalues are shown in Table 4.

**Table 4.** Equilibria points and their characteristic values.

| | $\lambda_1$ | $\lambda_2$ | $\lambda_3$ |
|---|---|---|---|
| (0, 0, 0) | $L$ | $L$ | $\Delta\pi_2 + G_1 + G_2 - C$ |
| (1, 0, 0) | $D_1 + S_1 + L$ | $S_2 + D_2 + L$ | $-\Delta\pi_2 - G_1 - G_2 + C$ |
| (0, 1, 0) | $a\Delta\pi_1 - E_1 + L$ | $-L$ | $-S_2 + \Delta\pi_2 + G_1 - C$ |
| (0, 0, 1) | $-L$ | $(1-a)\Delta\pi_1 - E_2 + L$ | $-S_1 + \Delta\pi_2 + G_2 - C$ |
| (1, 0, 1) | $a\Delta\pi_1 - E_1 + D_1 + S_1 + L$ | $-S_2 - D_2 - L$ | $S_2 - \Delta\pi_2 - G_1 + C$ |
| (0, 1, 1) | $-D_1 - S_1 - L$ | $(1-a)\Delta\pi_1 - E_2 + S_2 + D_2 + L$ | $S_1 - \Delta\pi_2 - G_2 + C$ |
| (1, 1, 0) | $-a\Delta\pi_1 + E_1 - L$ | $-(1-a)\Delta\pi_1 + E_2 - L$ | $-S_1 - S_2 + \Delta\pi_2 - C$ |
| (1, 1, 1) | $-a\Delta\pi_1 + E_1 - D_1 - S_1 - L$ | $-(1-a)\Delta\pi_1 + E_2 - S_2 - D_2 - L$ | $S_1 + S_2 - \Delta\pi_2 + C$ |

Following the assumptions of parameter size in the model, $\Delta\pi_2 + G_1 + G_2 - C > 0$. Therefore, (0, 0, 0) and (0, 0, 1) are not asymptotically stable points.

In this game system, the more stable the design–construction consortium is, the less likely it is that design changes will occur, and the more unstable it is, the more likely they become. On the one hand, when the project fails to meet green building standards due to the speculative behavior of the members of the consortium or their failure to build according to the contract, and the required standards can only be met after a change, if it is possible for the standards to be met, the developer will not initiate design changes. However, when the design and construction consortium chooses the "non-cooperation" strategy, the developer tends to favor "initiating design change", meaning we can select

(1, 1, 0) as the optimal asymptotic stability point; that is, (cooperation, cooperation, not initiating design change). On the other hand, due to design changes initiated as a result of an unreasonable design or the irregular construction approaches of members in the consortium, the project will be more capable of meeting the green building standards, thus increasing the developer's income. Further, if one of the consortium members does not adopt the "cooperation" strategy, the developer may seek new members who will cooperate and can carry out the design changes, at which point we can select (1, 1, 1) as the asymptotic stability point.

(1) Stability analysis of the evolutionary game following the first design change.

When the first design change initiated by the developer is greater,$-S_2 + \Delta\pi_2 + G_1 - C_3 < 0, -S_1 + \Delta\pi_2 + G_2 - C_3 < 0$ holds, which indicates that for the developer, but not for the members of the design and construction consortium, all other strategies will help make the project green. Here, (1, 0, 0), (0, 1, 0), and (1, 1, 0) are asymptotically stable points, and $(1, 1, 0)$ is the optimal stable point.

① When a $a\Delta\pi_1 - E_1 + L < 0$ and $(1-a)\Delta\pi_1 - E_2 + L < 0$ holds—that is, when the added value of the design and construction consortium's members due to cooperation is low—(1, 0, 0) or (0, 1, 0) will be the stable point of system evolution. When this occurs, both sides may show opportunistic behavior, potentially resulting in one party cooperating and the other party not cooperating. When this happens, the final result of the evolution of the system is determined by the position of the saddle point and the initial point of the system. As shown in the evolutionary phase diagram (Figure 1), if the chosen initial strategy falls within the region of A, the system will eventually converge to (1, 0, 0); that is, the constructer chooses a cooperative strategy, and the designer chooses a green, non-cooperative strategy. If the initial strategy choice falls within the area of B, the system converges to (0, 1, 0); that is, the constructer chooses an uncooperative strategy, and the designer chooses a cooperative strategy.

$$S_A = \frac{1}{2}\left(\frac{-L}{a\Delta\pi_1 - E_1} + \frac{(1-a)\Delta\pi_1 - E_2 + L}{(1-a)\Delta\pi_1 - E_2}\right) \tag{15}$$

According to Equation (15), the size of area $S_A$ is determined by five factors, namely, $a, \Delta\pi_1, E_1, E_1,$ and $L$. The influence of one such factor, such as $\Delta\pi_1$ or $L$, on the area of area A is uncertain. The other three parameters will show a monotonically increasing or decreasing relationship with the area of A. The specific effects of these three parameters on the choices of the developer and constructor are shown in the table.

As shown in Table 5, the sizes of parameters $a, E_1$ and $E_2$ will affect the saddle point and the area of A. With increases in parameters $a$ and $E_2$, the saddle point will move to the upper left, and the area of A will increase. At this point, when the probability of the system converging to (1, 0, 0) is greater, the saddle point will move to the upper left, and the area of A will increase with the reduction in parameter $E_1$. On the other hand, when the probability of the system converging to (0, 1, 0) is greater, the designer will be more willing to implement a cooperative strategy, and the constructor will be more willing to implement a non-cooperative strategy.

② When $a\Delta\pi_1 - E_1 + L > 0$ and $(1-a)\Delta\pi_1 - E_2 + L < 0$, the system's equilibrium point will be stable at (1, 0, 0). This indicates that the income distribution coefficient a has increased. For the constructor, the income created by choosing the "cooperation" strategy will be greater than that generated as a result of the "private information" obtained by choosing "non-cooperation", so regardless of which strategy the designer chooses, the constructor's optimal strategy will be "cooperation". For the designer, the benefits yielded by "private information" will be much higher than those from choosing "cooperation" and will be sufficient to make up for the loss of liquidated damages. Therefore, the designer will choose an "uncooperative" strategy.

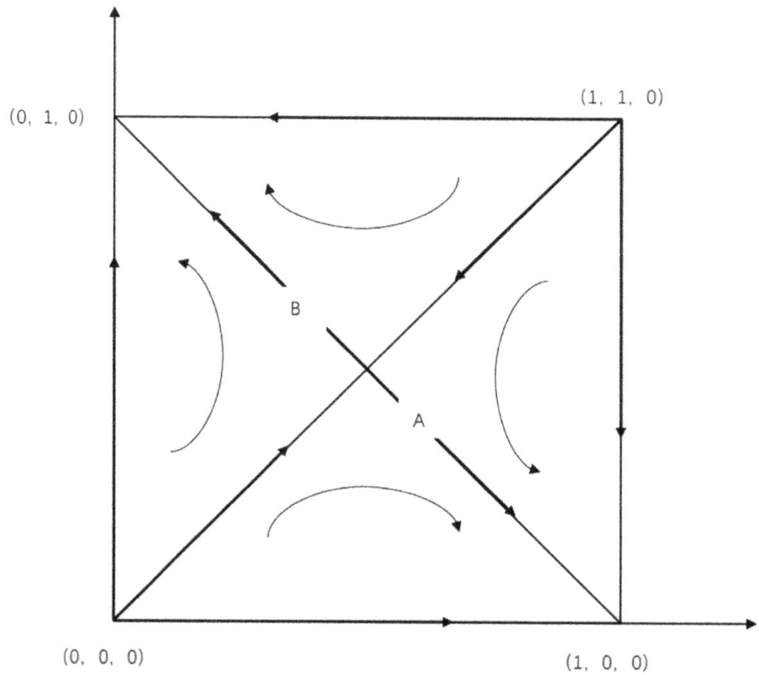


**Figure 1.** Evolutionary phase diagram (a).

**Table 5.** Correlation analysis of system parameters (a).

| Parametric Variation | $S_M$ Change | Evolutionary Direction |
|---|---|---|
| $a \uparrow$ | $S_A \uparrow$ | (1, 0, 0) |
| $E_1 \downarrow$ | $S_A \uparrow$ | (1, 0, 0) |
| $E_2 \uparrow$ | $S_A \uparrow$ | (1, 0, 0) |

③ When $a\Delta\pi_1 - E_1 + L < 0$ and $(1-a)\Delta\pi_1 - E_2 + L > 0$, the system's equilibrium point will be stable at (0, 1, 0). This will indicate that the income distribution coefficient a has decreased. For the designer, the income created as a result of choosing a "cooperation" strategy will be higher than the income generated as a result of the "private information" obtained by choosing "non-cooperation", so regardless of which strategy the constructer chooses, the designer's optimal strategy will be "cooperation". For the constructer, the benefits yielded by "private information" are much greater than those from choosing "cooperation" and will be sufficient to make up for the loss of liquidated damages. As such, the constructor will choose an "uncooperative" strategy.

④ When $a\Delta\pi_1 - E_1 + L > 0$ and $(1-a)\Delta\pi_1 - E_2 + L > 0$, the system's equilibrium point will be stable at (1, 1, 0). This indicates that the added value of the benefits generated by cooperation or the payment of liquidated damages by the consortium is increased, so members of the consortium will tend to choose the "cooperation" strategy regardless of what the other party chooses, at which time they will obtain the greatest benefit.

(2) Stability analysis of the evolutionary game under the second design change.

In the second case of the possible design change initiated by the developer, $-S_2 + \Delta\pi_2 + G_1 - C > 0, -S_1 + \Delta\pi_2 + G_2 - C > 0$ holds, which indicates that for the developer, all strategy combinations result in the project reaching the green building standard, but if a design change is made, the design will be more reasonable and the construction process will be more convenient. Here, (1, 0, 1), (0, 1, 1), (1, 1, 0), and (1, 1, 1) are asymptotically stable points, and (1, 1, 1) is the most stable point.

① When $a\Delta\pi_1 - E_1 + L + D_1 + S_1 < 0$ and $(1-a)\Delta\pi_1 - E_2 + L + D_2 + S_2 < 0$—that is, when the added value produced by the members of the design and construction consortium is very low—the reputational loss suffered and the design change subsidy that must be paid by the developer will still be ignored. Under these conditions, there is a stable strategy, but it is not unique; that is, (1, 0, 1) or (0, 1, 1) can be taken as the stable point of the system's evolution. Both sides may choose to undertake opportunistic behaviors, resulting in one party cooperating and the other party not cooperating. At this point, the final result of the evolution of the system will be determined by the positions of the saddle point and the initial point of the system. As shown in the evolutionary phase diagram (Figure 2), if the initial strategy choice falls within the region of C, the system will eventually converge to (1, 0, 1); that is, the constructer will choose a cooperative strategy and the designer will choose a green, non-cooperative strategy. If the initial strategy choice falls within the area of D, the system will converge to (0, 1, 1); that is, the constructer will choose an uncooperative strategy and the designer will choose a cooperative strategy.

$$S_C = \frac{1}{2}\left(-\frac{D_1 + S_1 + L}{a\Delta\pi_1 - E_1} + \frac{(1-a)\Delta\pi_1 - E_2 + D_2 + S_2 + L}{(1-a)\Delta\pi_1 - E_2}\right) \tag{16}$$

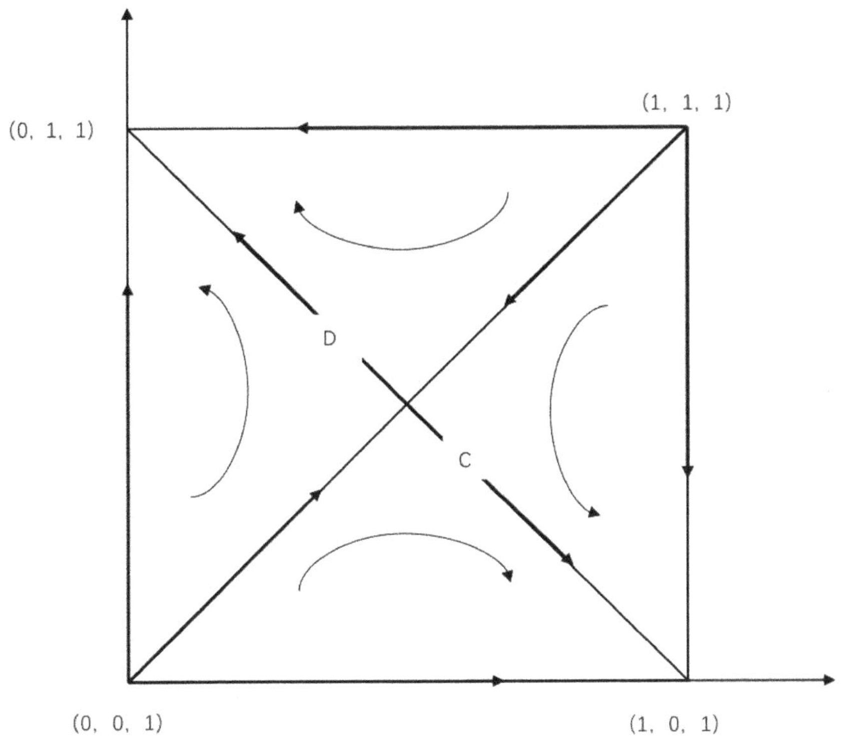


**Figure 2.** Evolutionary phase diagram (b).

According to Equation (16), the size of area $S_C$ is affected by nine factors: $a, \Delta\pi_1, E_1, E_1, D_1, D_2, S_1, S_2$, and $L$. Here, the influences of factors $\Delta\pi_1$ and $L$ on the area of C are uncertain, but the other seven parameters show a monotonically increasing or decreasing relationship with the area of C. The specific influences of these seven parameters on the choices of the developer and constructor are shown in Table 6.

**Table 6.** Correlation analysis of system parameters (b).

| Parametric Variation | $S_C$ Change | Evolutionary Direction |
|---|---|---|
| $a\uparrow$ | $S_C\uparrow$ | $(1, 0, 1)$ |
| $E_1\downarrow$ | $S_C\uparrow$ | $(1, 0, 1)$ |
| $E_2\uparrow$ | $S_C\uparrow$ | $(1, 0, 1)$ |
| $D_1\downarrow$ | $S_C\uparrow$ | $(1, 0, 1)$ |
| $D_2\uparrow$ | $S_C\uparrow$ | $(1, 0, 1)$ |
| $S_1\downarrow$ | $S_C\uparrow$ | $(1, 0, 1)$ |
| $S_2\uparrow$ | $S_C\uparrow$ | $(1, 0, 1)$ |

As shown in Table 6, the sizes of parameters $a, \Delta\pi_1, E_1, E_1, D_1, D_2, S_1, S_2$, and $L$ will all affect the saddle point and the area of C. With increases in parameters $a, E_1, D_1$, and $S_1$, the saddle point will move to the upper left. With reductions in parameters $E_2, D_2$, and $S_2$, the saddle point will move to the upper left, and the area of region C will increase. Under these conditions, the probability that the system converges to $(1, 0, 1)$ will be greater; that is, the constructer will be more willing to undertake a cooperative strategy, and the designer will be more willing to implement a non-cooperative strategy. On the other hand, when the opposite pertains, the probability that the system converges to $(0, 1, 1)$ will be greater; the designer will be more willing to implement a cooperative strategy, and the constructer will be more willing to implement a non-cooperative strategy.

② When $-D_1 - S_1 < a\Delta\pi_1 - E_1 + L < 0$ and $(1-a)\Delta\pi_1 - E_2 + L + D_2 + S_2 < 0$, the system's equilibrium point will be stable at $(1, 0, 1)$. This means that the income distribution coefficient a will have increased. For the constructor, due to the necessity of the payment of developer subsidies, the benefits induced by choosing a "cooperation" strategy will be sufficient to make up for the reputation loss caused by choosing "non-cooperation", so for the constructor, the optimal strategy is "cooperation". For the designer, the profit yielded by choosing "cooperation" will remain very low, and they would most likely rather bear the reputational loss and compensate for the liquidated damages. Therefore, the designer will most likely choose an "uncooperative" strategy.

③ When $a\Delta\pi_1 - E_1 + L + D_1 + S_1 < 0$ and $-D_2 - S_2 < (1-a)\Delta\pi_1 - E_2 + L < 0$, the system's equilibrium point will be stable at $(0, 1, 1)$. This means that the income distribution coefficient a will have decreased. For the designer, because of the necessity of paying a developer subsidy, the income derived from choosing a "cooperation" strategy will be sufficient to make up for the reputational loss brought about by choosing "non-cooperation", so the optimal strategy for the designer is "cooperation". For the constructor, the benefits resulting from choosing "cooperation" remain very low, meaning they would most likely rather bear the reputational loss and compensate for the liquidated damages. As such, the constructor will most likely choose an "uncooperative" strategy.

④ When $-D_1 - S_1 < a\Delta\pi_1 - E_1 + L < 0$ and $-D_2 - S_2 < (1-a)\Delta\pi_1 - E_2 + L < 0$, if $-S_1 - S_2 + \Delta\pi_2 - C < 0$, the system has no stable equilibrium point, but if $-S_1 - S_2 + \Delta\pi_2 - C > 0$, the system's equilibrium point will be stable at $(1, 1, 1)$. If the developer's net income is insufficient to pay the required subsidies to the cooperative enterprise, one or perhaps both parties in the consortium will choose not to cooperate; however, this situation is not optimal for the developer, so they will tend to choose to initiate a design change. When this happens, the consortium members tend to cooperate. However, when the developer recognizes the willingness of some members to cooperate, they will be more likely to not initiate any changes, meaning there will be no equilibrium and stability point in the system. When the developer can use the net income derived from design changes to pay the subsidy to the cooperative enterprise, they will most likely choose to initiate such changes, as long as the consortium members do not choose the strategy of (non-cooperation, non-cooperation); under these conditions, the consortium members will choose the cooperation strategy, and the system will show equilibrium and stability points $(1, 1, 1)$.

⑤ When $a\Delta\pi_1 - E_1 + L > 0$ and $(1-a)\Delta\pi_1 - E_2 + L > 0$, if $-S_1 - S_2 + \Delta\pi_2 - C < 0$, the system's equilibrium point will be stable at (1, 1, 0), but when $-S_1 - S_2 + \Delta\pi_2 - C > 0$, the system's equilibrium point will be stable at (1, 1, 1). This means that the added value of the benefits generated by cooperation or the liquidated damages paid by the design and construction consortium is increased, meaning the consortium members will tend to choose the "cooperation" strategy regardless of whether the developer gives subsidies or not. If the developer's net income is insufficient to pay the subsidy to the cooperative enterprise, the system's equilibrium point will be stable at (1, 1, 0). If the developer's net income can be used to pay the subsidy, the system's equilibrium point will be stable at (1, 1, 1).

## 5. Simulation Analysis and System Optimization

From the previous analysis, we can see that the evolutionary stability strategy and evolutionary path are different in different scenarios. In order to analyze the evolution trajectory and final stable state of the game process of related agents more intuitively, this paper uses Matlab2019 software to carry out numerical simulation analysis, and the specific simulation process is divided into two parts. In the first part, the evolutionary trajectory of ESS is numerically simulated by the equilibrium point of the tripartite evolutionary game. The second part simulates the influence of the additional income distribution coefficient, additional income, default cost, and subsidy on the evolution results.

Because there are many parameters involved in this study, the interaction between parameters is extremely complicated. In order to accurately reflect the actual situation, we primarily refer to relevant literature to determine some basic data. We also consulted experts from relevant enterprises to discuss and improve some missing data. Finally, the parameter values shown in Tables 7 and 8 are sorted.

**Table 7.** Parameter value setting—the first design change.

| Parameter | $a$ | $\Delta\pi_1$ | $\Delta\pi_2$ | $C$ | $D_1$ | $D_1$ | $E_1$ | $E_2$ | $S_1$ | $S_2$ | $G_1$ | $G_2$ | $F$ |
|---|---|---|---|---|---|---|---|---|---|---|---|---|---|
| numeric value | 0.5 | 3.5 | 3 | 3.5 | 0.4 | 0.5 | 3 | 2.5 | 0.3 | 0.2 | 0.6 | 0.6 | 1 |

**Table 8.** Parameter value setting—the second design change.

| Parameter | $a$ | $\Delta\pi_1$ | $\Delta\pi_2$ | $C$ | $D_1$ | $D_1$ | $E_1$ | $E_2$ | $S_1$ | $S_2$ | $G_1$ | $G_2$ | $F$ |
|---|---|---|---|---|---|---|---|---|---|---|---|---|---|
| numeric value | 0.5 | 3.5 | 3.8 | 3.5 | 0.4 | 0.5 | 3 | 2.5 | 0.3 | 0.2 | 0.2 | 0.1 | 1 |

### *5.1. Stability Analysis of the Equilibrium Point*

Here, Matlab is used to simulate an evolutionary game within a tripartite system, and numerical simulations are carried out for A3 = (0, 1, 0), A4 = (1, 0, 0), A5 = (0, 1, 1), A6 = (1, 0, 1), A7 = (1, 1, 0), and A8 = (1, 1, 1) in order to more clearly characterize the behaviors within the three-party system and verify the correctness of the game model.

(1) The developer-initiated design is set as the first case, during which there are three asymptotic stability points, A3 = (0, 1, 0), A4 = (1, 0, 0), and A7 = (1, 1, 0). The initial values in the design are set to $a = 0.5, \Delta\pi_1 = 3.5, \Delta\pi_2 = 3, C = 3.5, E_1 = 3, E_2 = 2.5$, $D_1 = 0.4, D_2 = 0.5, S_1 = 0.3, S_2 = 0.2, G_1 = 0.6, G_2 = 0.6$, and $F = 1$. The results of the evolution are shown in Figure 3.

As shown in Figure 3, the initial values described above mean that $\Delta\pi_1 - E_1 + L < 0$, $(1-a)\Delta\pi_1 - E_2 + L < 0$ is true, and so the initial strategy that is selected falls into the area of N, meaning the equilibrium stability point is (0, 1, 0).

If we reduce the size of parameter a such that $a\Delta\pi_1 - E_1 + L < 0$ and $(1-a)\Delta\pi_1 - E_2 + L > 0$ is true, then the equilibrium stability point is (0, 1, 0); on the other hand, if we increase the size of a, $a\Delta\pi_1 - E_1 + L > 0$ and $(1-a)\Delta\pi_1 - E_2 + L < 0$ will hold, and the equilibrium stability point will be (1, 0, 0).

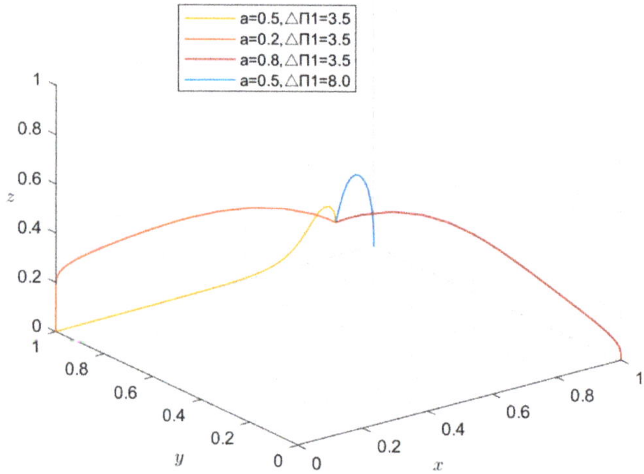


**Figure 3.** Evolution results in different situations—the first design change.

If we adjust parameter $\Delta\pi_1$ such that $a\Delta\pi_1 - E_1 + L > 0$ and $(1-a)\Delta\pi_1 - E_2 + L > 0$ holds, the equilibrium stability point will be (1, 1, 0).

(2) Under the second developer-initiated design, there are four asymptotic stability points: A5 = (0, 1, 1), A6 = (1, 0, 1), A7 = (1, 1, 0), and A8 = (1, 1, 1). The initial values that we use are set to $a = 0.5, \Delta\pi_1 = 3.5, \Delta\pi_2 = 3.8, C = 3.5, E_1 = 3, E_2 = 2.5, D_1 = 0.4, D_2 = 0.5, S_1 = 0.3, S_2 = 0.2, G_1 = 0.2, G_2 = 0.1$ and $F = 1$, and the results of the evolution are shown in Figure 4.

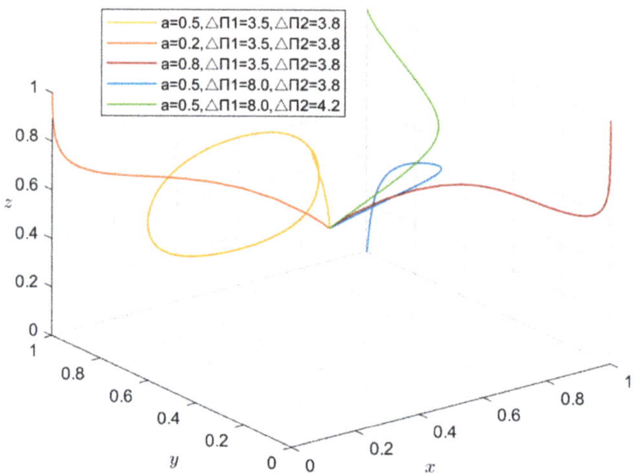


**Figure 4.** Evolution results in different situations—the second design change.

As shown in Figure 4, The initial value setting makes $-D_1 - S_1 < a\Delta\pi_1 - E_1 + L < 0$, $-D_2 - S_2 < (1-a)\Delta\pi_1 - E_2 + L < 0$ hold, but at this time, the benefits of the developer's choice to initiate design changes are not enough to subsidize the cooperative members, so the system is in a periodic state.

If we reduce the size of parameter a such that $a\Delta\pi_1 - E_1 + L + D_1 + S_1 < 0$ and $-D_2 - S_2 < (1-a)\Delta\pi_1 - E_2 + L < 0$ are true, then the equilibrium stability point will be (0,1,1); on the other hand, if the size of a is increased, such that $-D_1 - S_1 < a\Delta\pi_1 - E_1 + L < 0$

and $(1-a)\Delta\pi_1 - E_2 + L + D_2 + S_2 < 0$ are true, then the equilibrium stability point will be $(1, 0, 1)$.

If we adjust the size of parameter $\Delta\pi_1$ so that $a\Delta\pi_1 - E_1 + L > 0$ and $(1-a)\Delta\pi_1 - E_2 + L > 0$ are true, the equilibrium stability point will be $(1, 1, 0)$; on this basis, $\Delta\pi_2 - C - S_1 - S_2 > 0$ holds, and as a result, the equilibrium point will change from $(1, 1, 0)$ to $(1, 1, 1)$ and system optimization will be achieved.

### *5.2. Correlation Parameter Analysis*

Based on the above theoretical analyses performed under different circumstances, we next used Matlab2019 simulations to study the impacts of the benefit distribution coefficient, the revenue added value, and the default cost on the stability of the design–construction consortium.

#### 5.2.1. The First Case of Contractor-Initiated Design Changes

Assume that the initial states of the constructor, designer, and developer are equal to 0.5. Table 7 shows a breakdown of the parameters.

(1) The influence of income distribution coefficient A on the stability of the design and construction consortium.

We have adjusted the size of the additional income distribution coefficient to observe its effects on the stability of the design–construction consortium; we set a = 0.2, a = 0.5, and a = 0.8 in respective simulations, and the results are shown in Figure 5.

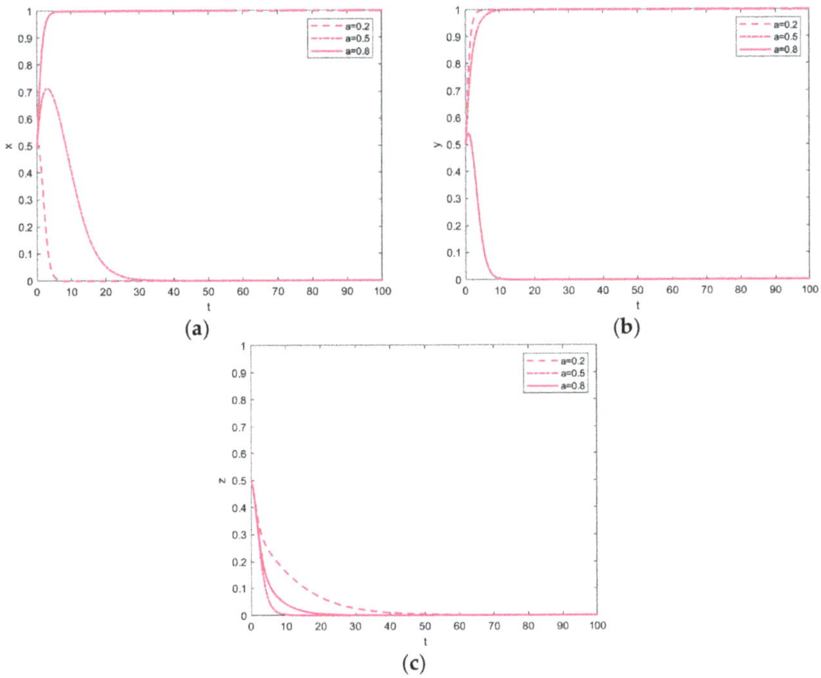


**Figure 5.** (**a**) Evolutionary path of the constructer with different additional income distribution coefficients a—The first design change; (**b**) evolutionary path of the designer with different additional income distribution coefficients a—The first design change; (**c**) evolutionary path of the developer with different additional income distribution coefficients a—The first design change.

Figure 5 shows that when the constructer's additional income distribution coefficient a is low, they will choose a cooperative strategy, but they will be relatively more inclined to choose a non-cooperation strategy so as to obtain income related to obtaining "private information" and the payment of liquidated damages. On the other hand, the designer's additional revenue distribution coefficient is $1 - a$, which is higher. This means that the designer's return as a result of choosing a cooperative strategy will be higher—sufficiently so to exceed the benefits derived from "private information". This means that regardless of which strategy the construction company chooses, the designer will choose a cooperation strategy. When the absolute value of the difference between the constructer's additional income distribution coefficient a and the designer's additional income distribution coefficient is $1 - a$—that is, when coefficient a is closer to 0.5—the constructer and designer's strategies tend to stabilize more slowly. Further, the additional income distribution coefficient a has little effect on the stability of the design and construction consortium, which is primarily affected by other factors. When the constructer's additional income distribution coefficient a is high, their income derived from the choice of a cooperative strategy will be higher—sufficiently so to exceed the returns generated by "private information", so they will tend to cooperate. However, the designer's additional income distribution coefficient at this time will be low, making them more inclined to choose a non-cooperation strategy in order to obtain the benefits of "private information" and the payment of liquidated damages.

In general, when the absolute value of the difference between a and $1 - a$ is smaller, the impact on the stability of the design and construction consortium will be smaller, and when the inverse is true, the impact on the design and construction consortium will be greater, making the design and construction consortium more unstable. For the developer, the smaller the absolute value of the difference between a and $1 - a$, the faster they will converge on the choice of not initiating design changes.

(2) The impact of revenue added value $\Delta\pi_1$ on the stability of the design and construction consortium.

We have adjusted the size of the revenue added value in order to observe its effect on the stability of the design and construction consortium; we set values of $\Delta\pi_1 = 2, \Delta\pi_1 = 3.5$ and $\Delta\pi_1 = 5$ in the simulations, and the results are shown in Figure 6.

As can be seen from Figure 6, for the constructor, although the extra revenue $\Delta\pi_1$ generated from cooperation is lower, making cooperation strategies less profitable, the increase in extra revenue $\Delta\pi_1$ will still direct their evolutionary path towards cooperation at the beginning, but the difference between the revenue derived from obtaining "private information" and the payment of liquidated damages remains higher than the benefits generated by cooperation, so the result of long-term evolution will still tend towards non-cooperation. However, with increases in the additional revenue $\Delta\pi_1$, the inhibition of the evolutionary trend's movement towards non-cooperative strategies will be greater, and this will ultimately tend towards a cooperative strategy. For the designer, the revenue derived from choosing a cooperation strategy will be sufficient to exceed the revenue derived from "private information", so no matter which strategy the construction company chooses, the designer will choose a cooperation strategy. However, when the final result of the evolution of the construction company is the choice of an uncooperative strategy, the speed at which the designer converges towards choosing a cooperative strategy will slow with the increase in the additional revenue $\Delta\pi_1$. When the final result of the evolution of the construction company is the choice of cooperation, as the additional revenue $\Delta\pi_1$ increases, the convergence of the designer towards a cooperation strategy will be accelerated.

In general, when the benefits related to the "private information" held by members of the design and construction consortium and the payment of liquidated damages are smaller than the benefits generated by cooperation, with the increase in additional revenue $\Delta\pi_1$, the speed of convergence towards an uncooperative strategy by members ultimately choosing this strategy will be inhibited, and the speed of convergence towards a cooperative strategy by members ultimately choosing this strategy will also be inhibited. When the benefits related to the "private information" held by members of the consortium and to the payment of liquidated damages are greater than the benefits generated by cooperation, the speed of convergence of the members towards a cooperation strategy will be increased with increases in the additional gain $\Delta\pi_1$. Therefore, the higher the additional gain $\Delta\pi_1$, the greater the impact on the stability of the design and construction consortium, and the design and construction consortium will be made relatively more stable. The developer's strategy will converge more quickly towards not initiating a design change as the additional revenue $\Delta\pi_1$ increases.

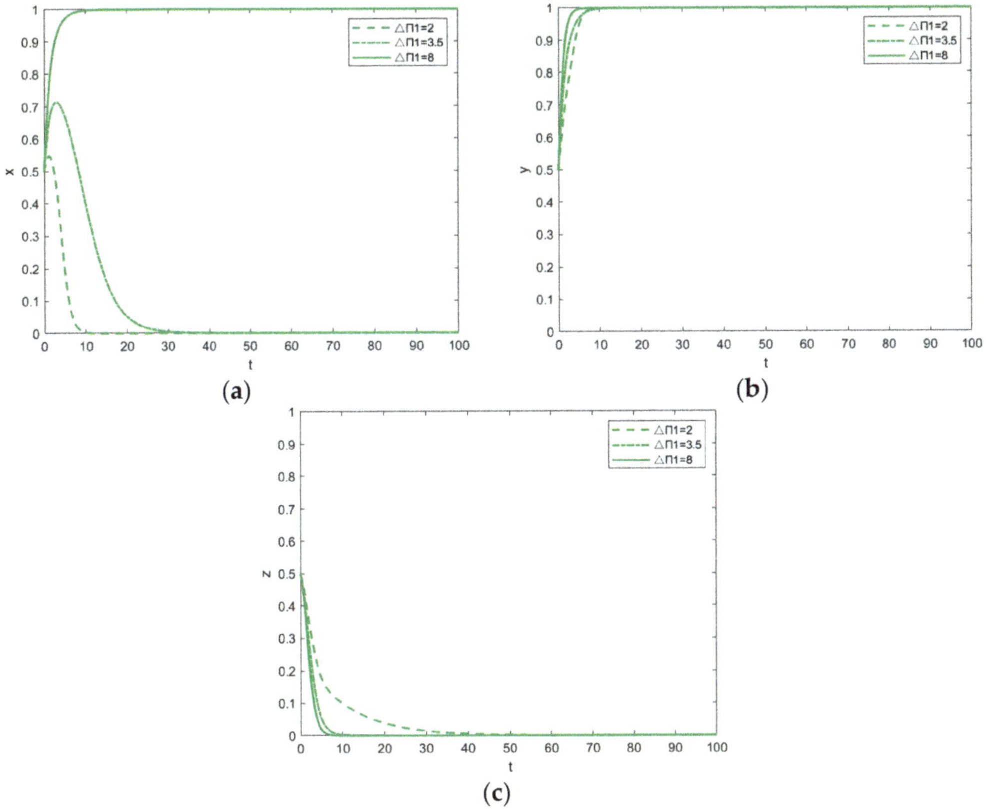


**Figure 6.** (**a**) Evolutionary path of the constructer under different values of additional revenue $\Delta\pi_1$—The first design change; (**b**) evolutionary path of the designer under different values of additional revenue $\Delta\pi_1$—The first design change; (**c**) evolutionary path of the developer under different values of additional revenue $\Delta\pi_1$—The first design change.

(3) The impact of default cost F on the stability of the design and construction consortium.

By adjusting the default cost, we can observe its effects on the stability of the design–construction consortium, and so we set the values of F = 0.5, F = 1, and F = 1.5 in the simulations. The results are shown in Figure 7.

As Figure 7 shows, with increases in the default cost f, the convergence of non-cooperative members of the consortium will be restrained, and the convergence of cooperative members will be accelerated. Therefore, the higher the default cost f, the more highly the cooperation within the design and construction consortium will be promoted, and the more stable it will be. In contrast, the lower the default cost F, the smaller the impact on the stability of the construction consortium. For a developer, with an increase in default cost F, their strategy will more quickly converge towards not initiating design changes.

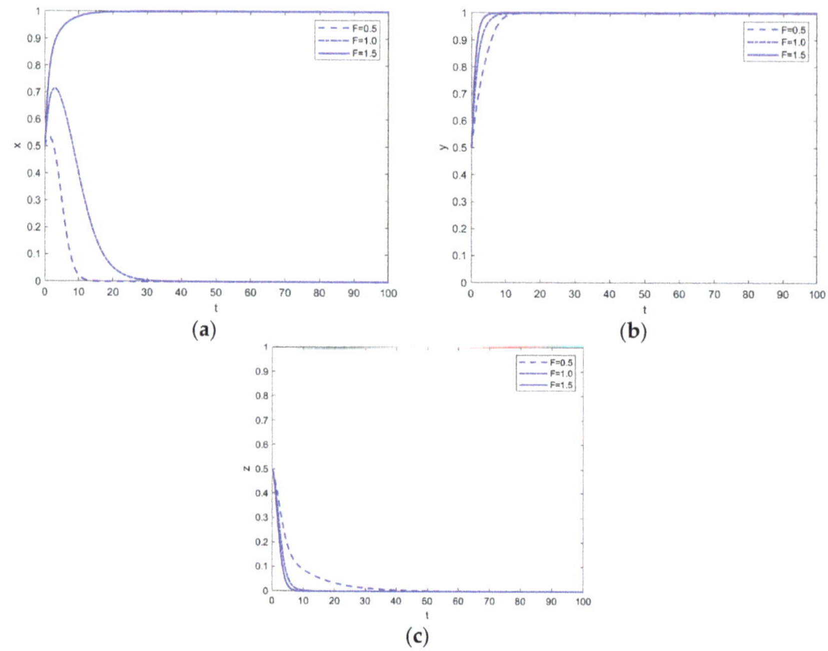

**Figure 7.** (**a**) Evolutionary path of the constructer with different default costs F—The first design change; (**b**) evolutionary path of the designer with different default costs F—The first design change; (**c**) evolutionary path of the developer with different default costs F—The first design change.

#### 5.2.2. Developer—Initiated Design Changes in the Second Case

We assume that the initial states of the constructor, designer, and developer equal 0.5. Table 8 shows the parameter details.

(1) The influence of income distribution coefficient A on the stability of the design and construction consortium.

We have adjusted the size of the additional income distribution coefficient to observe its effects on the stability of the design–construction consortium; we set values of a = 0.2, a = 0.5, and a = 0.8 for the simulations, and the results are shown in Figure 8.

As Figure 8 shows, when the additional income distribution coefficient a of the constructor is low, they will quickly tend towards being "uncooperative". At this time, when the designer chooses a cooperative strategy, their income will be higher, so when the additional income distribution coefficient a approaches 0, the system will quickly converge towards (0, 1, 1). When the constructer's additional income distribution coefficient a approaches 0.5, this factor will have little influence on the stability of the consortium, of which stability will primarily be affected by opportunistic behavior, subsidies, and other factors. When their net income derived from a design change is greater than that derived from a subsidy paid to the enterprise, if the design and construction consortium chooses the (cooperation, cooperation) strategy, the developer will more likely choose to initiate said change. On the other hand, when the consortium chooses (cooperation, cooperation), the developer will not choose to initiate such changes; as such, in a certain case, if the developer tends to initiate design changes, the design and construction consortium tends to cooperate. However, when the design and construction consortium chooses (cooperation, cooperation), the developer's optimal strategy is not to initiate design changes, meaning their evolutionary direction changes from initiating design changes to not initiating design changes. As such, in certain cases, when the additional income distribution coefficient a of the constructor is high, their strategy selection will quickly tend towards "cooperation".

Under this circumstance, the designer's income when choosing a cooperation strategy will be low, and so when the additional income distribution coefficient a approaches 1, the system will quickly converge to (1, 0, 1).

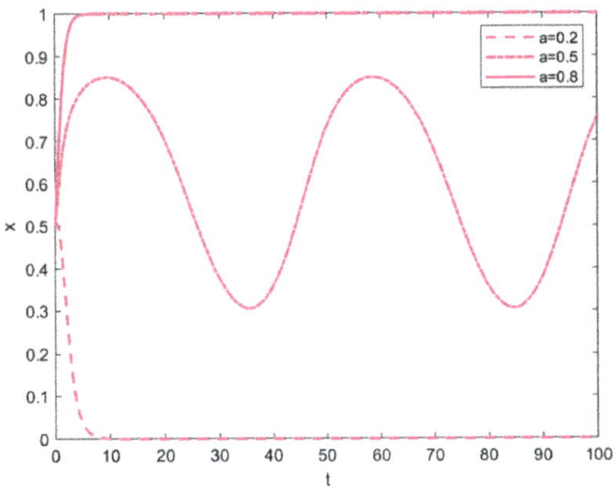


**(a)**

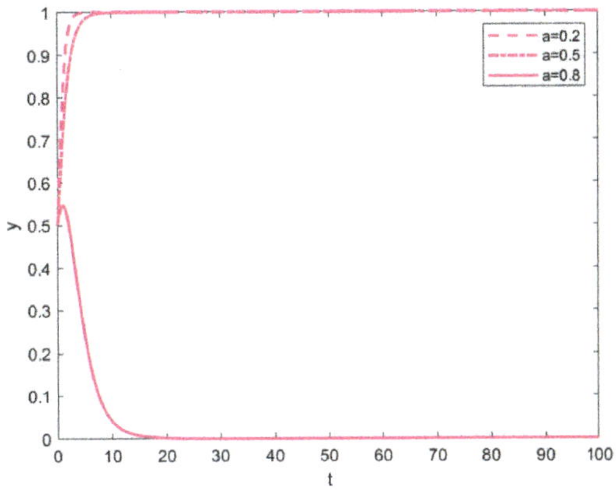


**(b)**

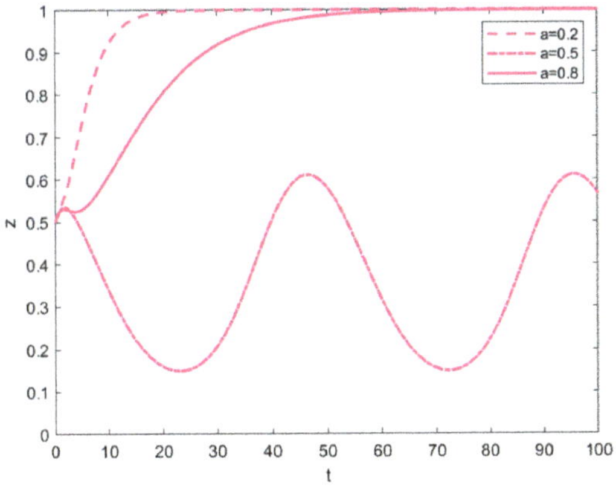


**(c)**

**Figure 8.** (**a**) Evolutionary path of the constructer with different additional income distribution coefficients a—The second design change; (**b**) evolutionary path of the designer with different additional income distribution coefficients a—The second design change; (**c**) evolutionary path of the developer with different additional income distribution coefficients a—The second design change.

Generally speaking, when the absolute value of the difference between a and $1 - a$ is smaller, its influence on the stability of the design and construction consortium is reduced, and in the converse case, its influence is increased, meaning the design and construction consortium will be more unstable. The greater the absolute value of the difference between a and $1 - a$, the faster the constructor's strategy will converge towards initiating design changes, while when the absolute value of the difference between a and $1 - a$ is smaller, the extent to which it inhibits convergence in the direction of not initiating design changes is altered.

(2) The impact of revenue added value $\Delta\pi_1$ on the stability of the design and construction consortium.

By adjusting the size of the revenue added value, we have observed its influence on the stability of the design and construction consortium; values of $\Delta\pi_1 = 2, \Delta\pi_1 = 3.5$ and $\Delta\pi_1 = 5$ have been set for the simulations, and the results are shown in Figure 9.

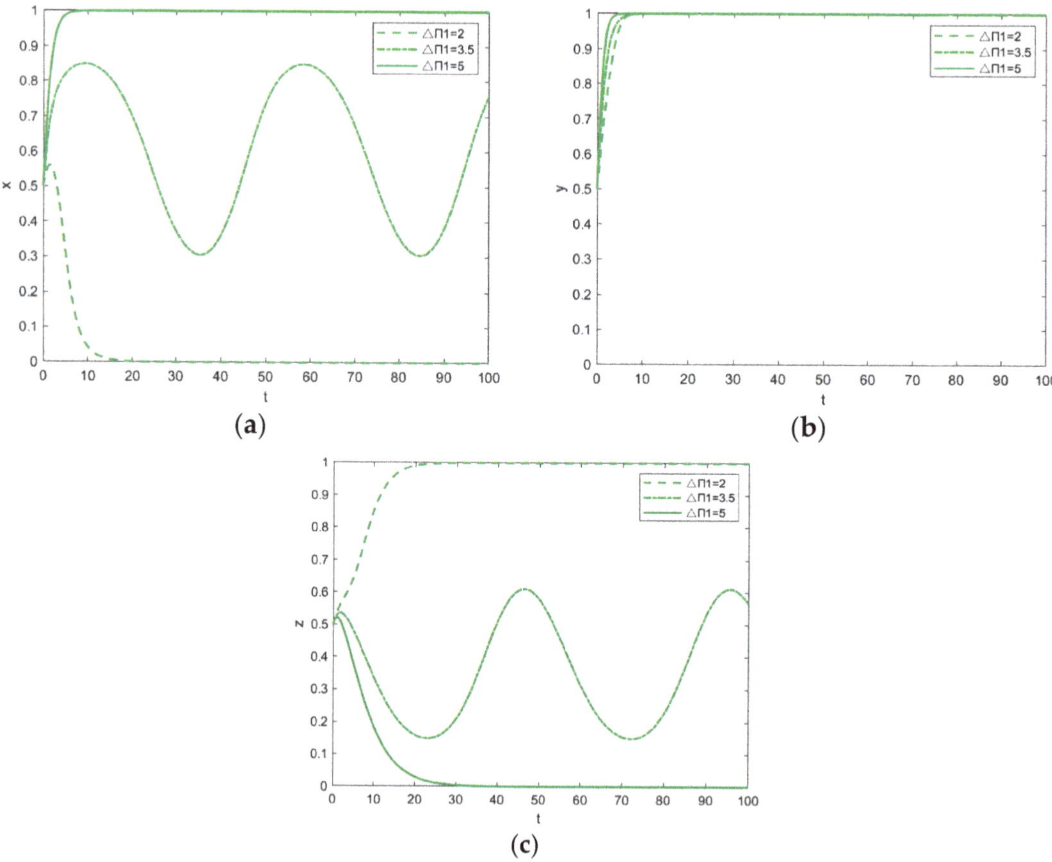


**Figure 9.** (**a**) Evolutionary path of the constructer with different additional revenue $\Delta\pi_1$—The second design change; (**b**) evolutionary path of the designer with different additional revenue $\Delta\pi_1$—The second design change; (**c**) evolutionary path of the developer with different additional revenue $\Delta\pi_1$—The second design change.

Generally speaking, when the constructor's $a\Delta\pi_1 - E_1 + L + D_1 + S_1 < 0$, as the additional income $\Delta\pi_1$ is increased, their convergence towards an uncooperative strategy will be inhibited, and their convergence towards a cooperative strategy will also be inhibited. When the designer's $(1-a)\Delta\pi_1 - E_2 + L + D_2 + S_2 < 0$, as the additional income $\Delta\pi_1$ is increased, the evolution path of their choice will be the same. When $-D_1 - S_1 < a\Delta\pi_1 - E_1 + L < 0$ and $-D_2 - S_2 < (1-a)\Delta\pi_1 - E_2 + L < 0$, the system will adopt a periodic state, but with the increase in additional income $\Delta\pi_1$, the volatility of the system will be enhanced. When $a\Delta\pi_1 - E_1 + L > 0$ and $(1-a)\Delta\pi_1 - E_2 + L > 0$, as the additional income $\Delta\pi_1$ is increased, its influence on the stability of the design and construction consortium will be increased, and the design and construction consortium will be made more stable. For the developer, when the additional income $\Delta\pi_1$ is increased, the convergence of its strategy towards not initiating design changes will be more restricted.

(3) The impact of default cost F on the stability of the design and construction consortium.

By adjusting the default cost, we can observe its influence on the stability of the design–construction consortium; as such, we set values of F = 0.5, F = 1, and F = 1.5 for the simulations. The results are shown in Figure 10.

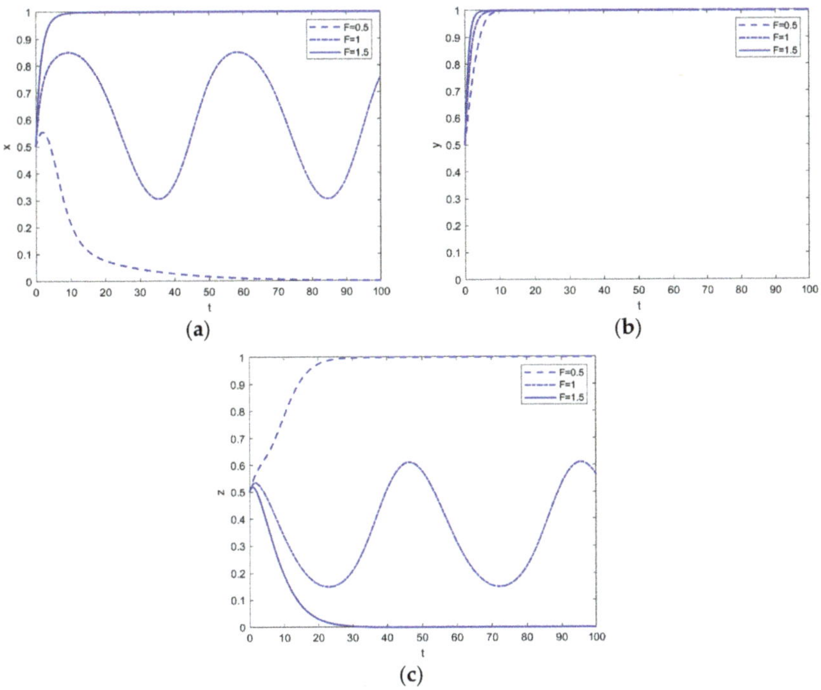


**Figure 10.** (**a**) Evolutionary path of the constructer with different default costs F—The second design change; (**b**) evolutionary path of the designer with different default costs F—The second design change; (**c**) evolutionary path of the developer with different default costs F—The second design change.

As Figure 10 shows, within the design and construction consortium, with increases in default cost F, the convergence of the non-cooperative members will be restrained, and that of cooperative members will be accelerated. Therefore, the higher the default cost F, the more strongly the cooperation amongst the design and construction consortium will be promoted, and the more stable the consortium will be. In contrast, the lower the default cost F, the lower the impact will be on the stability of the consortium. For the developer, as default cost F increases, it will restrain convergence in the direction of not initiating design changes, and the opposite case will accelerate convergence in the direction of initiating such change.

(4) The influence of the developer's subsidy S on the stability of the design and construction consortium.

By adjusting the size of the subsidy S paid to the cooperative members, we can observe its influence on the stability of the design and construction consortium; as such, four subsidy scenarios are established, namely, $S_1 = 0.1, S_2 = 0.1, S_1 = 0.1, S_2 = 0.3, S_1 = 0.3, S_2 = 0.1$, $S_1 = 0.3$, and $S_2 = 0.3$. The results are shown in Figure 11.

As can be seen from Figure 11, for the members of the design and construction consortium, if the subsidy paid to the other members is fixed, with an increase in the subsidy paid to one member, this member's evolutionary path will converge towards cooperation. If the subsidy paid to one member is fixed, increases in the subsidy paid to the other members will inhibit the convergence of its evolutionary path towards cooperation. For a developer, if the subsidies paid to the design and construction consortium are increased, the speed of convergence towards initiating design changes will slow down. If the system resides in a periodic state, increases in subsidies paid to the design and construction consortium will make the evolutionary path of the developer more volatile.

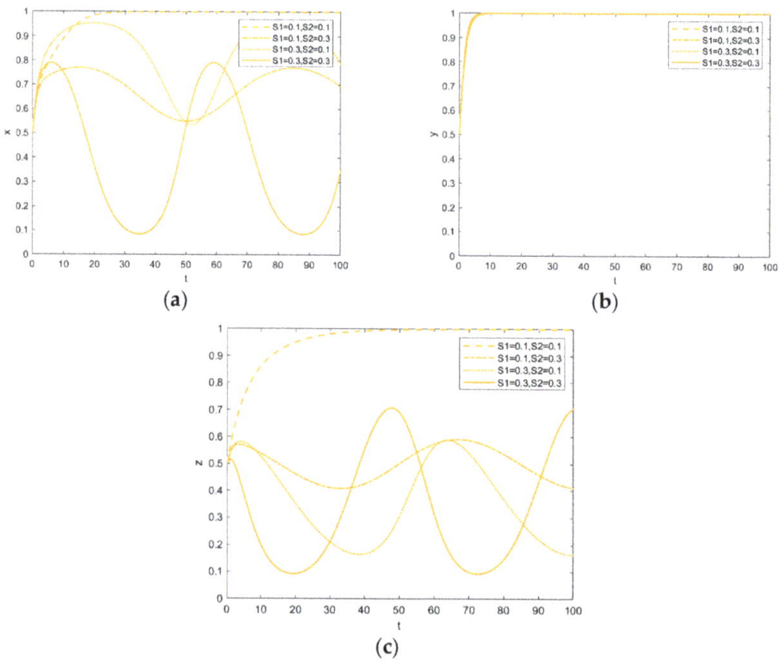


**Figure 11.** (**a**) Evolutionary path of the constructer with different change subsidies $S_1$,$S_2$—The second design change; (**b**) evolutionary path of the designer with different change subsidies $S_1$,$S_2$—The second design change; (**c**) evolutionary path of the developer with different change subsidies $S_1$,$S_2$—The second design change.

Generally speaking, if one member of the design and construction consortium is stable in their choice of a cooperative strategy, then as the subsidy paid by the developer to the other member increases, the opportunistic behavior of the other member will be inhibited, and the speed of their convergence towards non-cooperation will too. That is to say that, with increases in the subsidy paid by the developer to certain members, it may not be possible to stabilize the member's strategy selection at (1, 1), but the member's strategy selection may be prevented from stabilizing at (0, 0), (1, 0) or (0, 1). Therefore, the subsidy paid by the developer to the members of the design and construction consortium will promote the stability of the consortium to a certain extent, while the subsidy represents a cost demanded by the developer. If the subsidy is too high, it will inhibit the developer's design change behavior, which is not conducive to their management of green projects, meaning there is a threshold applied to this subsidy.

## 6. Conclusions and Suggestions

### *6.1. Conclusions*

Based on the characteristics of bounded rationality and asymmetric information, this paper applies the evolutionary game theory to the research on the stability of the cooperative relationship of the design and construction consortium during the implementation of large-scale green building projects, establishes a tripartite evolutionary game model with the constructer, designer, and developer as the main body, and dynamically analyzes the influence of relevant parameters on the system evolution strategy. MATLAB2019b is used to simulate and analyze the data, which verifies the effectiveness of the evolutionary game and the influence of related factors on the stability of the consortium and obtains the stability conditions of the optimal strategy combination under different design changes. According to the influence relationship and stability conditions among various factors, corresponding countermeasures and suggestions are put forward for stakeholders. The main conclusions are as follows:

(1) A3 = (0, 1, 0), A4 = (1, 0, 0), A5 = (0, 1, 1), A6 = (1, 0, 1), A7 = (1, 1, 0), and A8 = (1, 1, 1) are evolutionary stability points, where A7 = (1, 1, 0) and A8 = (1, 1, 1) are ideal evolutionary stability points in two cases, respectively.

(2) If the absolute value of the difference between a and 1-a is smaller, its influence on the stability of the design and construction consortium will be reduced; if the opposite pertains, it will have greater influence, making the consortium more unstable.

(3) Increases in additional income $\Delta\pi_1$ will reduce the instability of the design and construction consortium at first. With continuous increases in $\Delta\pi_1$, the stability of the design and construction consortium will be strengthened.

(4) The payment of subsidies by the developer to the cooperative members will inhibit the non-cooperative behavior of the members and increase the stability of the design and construction consortium to a certain extent. In addition, the payment of such subsidies to the cooperative members will somewhat weaken the effects of extra income $\Delta\pi_1$ and default cost F on the stability of the consortium, but there is a threshold applied to the value of this subsidy.

(5) After adjusting the parameters in an attempt to optimize the game system, we found that values of A3 = (0, 1, 0), A4 = (1, 0, 0), A5 = (0, 1, 1), and A6 = (1, 0, 1) all contribute to the ideal state.

Therefore, this study not only investigates the role of the constructer, designer, and developer in the cooperation of the design and construction consortium from the perspective of design change management but also enriches the application of evolutionary game in the cooperation of the design and construction consortium, and also provides suggestions for the future development of green buildings in China. To summarize, this study provides a solution to the difficulties in the sustainable development of the cooperative relationship of China's Design and Construction Consortium. At the same time, the high-quality cooperation of the design and construction consortium is a method to solve the problem of design change, which is of great significance to the sustainable development of green buildings.

### *6.2. Suggestions*

The three-way evolutionary game model can effectively promote the dynamic analysis of multi-stakeholder behavior strategy evolution in the process of design and construction consortium cooperation. In this study, by changing the assignment of relevant parameters and combining them with stability analysis, the main factors affecting evolutionary balance are identified and the evolutionary path of participants' behavior strategies is simulated. According to the research results of this paper, in order to strengthen the stability of the design and construction consortium, avoid the occurrence of design changes or improve the quality of cooperation after the changes, promote the standard of green engineering projects, and ensure the vigorous development of green buildings in China, we put forward the following suggestions:

First, the consortium members should have clear rights and responsibilities, and fair income distribution should be ensured. All work should be carried out by the members within the scope of their duties, according to the contents of the general contract signed with the developer and the provisions of the consortium cooperation agreement. In the cooperation agreement, the responsibilities and obligations of the cooperative members should be clearly defined, and a responsible party should be established who shall bear the consequences if other members suffer losses due to a breach of contract. The setting of clear rights and responsibilities can help to effectively improve the subjective motivations of members, actively promote the development of the consortium model, and thus facilitate the smooth completion of green building projects. The design and construction consortium should establish a reasonable income distribution plan based on the principles of fairness and justice, as well as a maximum total income. Before this is set out, the division of powers and responsibilities between members should be fully considered, as should the degree of contribution of each to the final green product, the resource investments, the risk-sharing distribution, effort, and other factors. However, an income distribution system is not static and will help to maintain a stable cooperative relationship within the consortium for a long time if it is flexibly adjusted according to the actual operational conditions of green building projects. In a word, problems arising from design changes necessitated by substandard green buildings can be effectively avoided or actively dealt with so as to minimize their impact on the environment during project implementation, and to further promote the sustainable development of green buildings.

Second, the punishment mechanism for breaches of contract should be strengthened. The costs related to a breach of contract represent an important factor affecting the stability of the cooperative relationship within the consortium. Regardless of whether the developer initiates design changes, members of the consortium will usually take the default cost into consideration when making strategic choices. In order to prevent members from adopting non-cooperative strategies due to opportunism in the performance process, it is necessary to establish a reasonable punishment mechanism, increase the costs demanded of members as a result of breaches of contract, and prevent members of the consortium from defaulting [44]. This can help to ensure the active cooperation of all members, the successful completion of all the work according to the fair division of rights and obligations, improve the degree of cooperation between members, extract their best performances in service of the construction project, improve the quality of green buildings, maximize the benefits derived by stakeholders throughout the project, form a virtuous circle, and promote the sustainable development of green buildings.

Third, the internal management system within the consortium should be improved, which will improve the income generated by the consortium. In terms of technical management, the constructor should seek to learn about and develop their own technology, as well as update their construction equipment, so as to keep pace with the times and adapt to the requirements of green construction in relation to the given design scheme. The design unit should regularly organize technical training among designers and actively implement new national green design specifications or standards so as to meet the environmental protection requirements related to design [45]. An efficient consortium management team, comprising individuals of outstanding talent jointly selected by the constructer and designer, should seek to do the best job possible in terms of design management in the early stage of the project, find and solve problems that might arise in the design process over time, avoid design changes necessitated by irregular design, carry out the best job possible in terms of construction management in the later stage of the project, supervise the construction process, and avoid further design changes necessitated in the protection of the environment and its benefits. At the same time, if the developer does initiate design changes, they should ensure firm supervision of these changes and pay attention to their rationality and effectiveness [46]. Furthermore, an information communication platform should be established to ensure effective information management. If the units participating in the design and construction consortium share mutual trust and understanding, it will be easier to communicate the design changes, meaning the project can be more successfully completed, and this will encourage the cooperative members to establish a long-term and stable partnership.

Fourthly, when selecting the members of the design and construction consortium, the developer should start from the basis of the project itself, comprehensively and systematically consider the strengths of the individual enterprises, select those who show a high degree of technological suitability to the project being implemented, and reduce the need for design changes necessitated by substandard green design or green construction (or actively and effectively deal with such design changes caused by project optimization). After design changes are initiated, implementing an effective reward and punishment system will be beneficial to the management of the cooperative relationship within the consortium. The cooperative members should be encouraged to strengthen their cooperation, as a result of which, the total benefits created can reach a certain theoretical value, such that additional benefits can be created on the basis of the original income distribution. At the same time, it is necessary to set up a severe punishment system to help regulate the behavior of the cooperative members. For those members who fail to deal with design changes properly, that is, those who default when the developer initiates a design change, the developer should blacklist them and thus reduce their reputational value, which will help to effectively reduce the instability amongst the consortium's members, facilitate the smooth development of green building projects, improve the environmental benefits, and promote the sustainable development of green buildings.

### 6.3. Limitations

This study still has some limitations. First of all, each building project is unique, and it is impossible to completely list the influencing factors. Moreover, there are many complicated influencing factors in actual green projects, so the parameter setting needs to be further supplemented and improved. Secondly, the setting of parameter values is abstract, and the data may not be completely close to reality. Matlab2019b software is used to simulate and analyze the gradual stable equilibrium point, and the simulation results are consistent with the model results, which proves the theoretical and practical significance of this study. In order to better match reality, more data collection is needed, and empirical analysis is also the next research direction. Finally, the evolutionary game model of this paper focuses on the key participants of design change management, namely the constructer, designer, and developer, without considering the analysis of other interest groups, and should discuss the role of other stakeholders in the stability of design and construction consortium under design change more completely. Determining how to establish a more practical model considering more participants and more information according to the relationship between participants is also a promising research direction in the future.

**Author Contributions:** Conceptualization, Y.W. and L.L.; methodology, Y.W.; software, Y.W.; validation, Y.W. and L.L.; formal analysis, Y.W. and L.L.; investigation, L.R. and R.S.; resources, X.G.; data curation, X.G.; writing—original draft preparation, L.L.; writing—review and editing, Y.W. and X.G.; visualization, L.L.; supervision, Y.W. and X.G.; project administration, L.L.; funding acquisition, Y.W. All authors have read and agreed to the published version of the manuscript.

**Funding:** This research was funded by the State Key Program of National Social Science Foundation of China (grant no. 19AGL030), the National Social Science Foundation of China (grant no. 21BGL297), and the Major Program of Philosophy and Social Science Research in Jiangsu University (grant no. 2020SJZDA085).

**Data Availability Statement:** Not applicable.

**Acknowledgments:** The authors would like to thank the anonymous reviewers for their constructive comments and suggestions.

**Conflicts of Interest:** The authors declare no conflict of interest.

## References

1. Wang, X.; Huang, Y.; Xue, W. Research on collaborative management of prefabricated construction: Integration of design and construction. In Proceedings of the ICCREM, Guangzhou, China, 10–12 November 2017; pp. 1–9.
2. Cho, N.; El Asmar, M.; Underwood, S. Long-term performance benefits of the design-build delivery method applied to road pavement projects in the US. *J. Civ. Eng.* **2020**, *24*, 1049–1059.
3. Minchin, R.E., Jr.; Li, X.; Issa, R.R. Comparison of cost and time performance of design-build and design-bid-build delivery systems in Florida. *J. Constr. Eng. Manag.* **2013**, *139*, 04013007. [CrossRef]
4. Park, H.-S.; Lee, D.; Kim, S.; Kim, J.-L. Comparing project performance of design-build and design-bid-build methods for large-sized public apartment housing projects in Korea. *J. Asian Archit. Build. Eng.* **2015**, *14*, 323–330. [CrossRef]
5. Mohd Nawi, M.N.; Baluch, N.H.; Bahaudin, A.Y. Impact of fragmentation issue in construction industry: An overview. In Proceedings of the MATEC Web of Conferences, Langkawi, Malaysia, 31 May–2 June 2014; p. 01009.

6. Ko, O.Y.; Paek, J.H. Korea experience project management consortium on the US forces Korea relocation program. *J. Asian Archit. Build. Eng.* **2008**, *7*, 85–92. [CrossRef]
7. Chang, A.S.; Shen, F.Y.; Ibbs, W. Design and construction coordination problems and planning for design–build project new users. *Can. J. Civ. Eng.* **2010**, *37*, 1525–1534. [CrossRef]
8. Yeganeh, A.A.; Azizi, M.; Falsafi, R. Root causes of design-construction interface problems in iranian design-build projects. *J. Constr. Eng. Manag.* **2019**, *145*, 05019014. [CrossRef]
9. Lee, Z.P.; Rahman, R.A.; Doh, S.I. Key drivers for adopting design build: A comparative study between project stakeholders. *Phys. Chem. Earth* **2020**, *120*, 102945. [CrossRef]
10. Perkins, R.A. Sources of changes in design–build contracts for a governmental owner. *J. Constr. Eng. Manag.* **2009**, *135*, 588–593. [CrossRef]
11. Shen, W.; Tang, W.; Wang, S.; Duffield, C.F.; Hui, F.K.P.; You, R. Enhancing trust-based interface management in international engineering-procurement-construction projects. *J. Constr. Eng. Manag.* **2017**, *143*, 04017061. [CrossRef]
12. Sushil, S. Interpreting the interpretive structural model. *Glob. J. Flex. Syst. Manag.* **2012**, *13*, 87–106. [CrossRef]
13. Heravi, G.; Mohammadian, M. Investigating cost overruns and delay in urban construction projects in Iran. *Int. J. Constr. Manag.* **2019**, *21*, 958–968. [CrossRef]
14. Yap, J.B.H.; Skitmore, M. Investigating design changes in Malaysian building projects. *Arch. Eng. Des. Manag.* **2017**, *14*, 218–238. [CrossRef]
15. Park, H.; Han, S.H.; Rojas, E.M.; Son, J.; Jung, W. Social network analysis of collaborative ventures for overseas construction projects. *J. Constr. Eng. Manag.* **2011**, *137*, 344–355. [CrossRef]
16. Okunlola Ojo, S.; Aina, O.; Yakeen Adeyemi, A. A comparative analysis of the performance of traditional contracting and design-build procurements on client objectives in Nigeria. *J. Civ. Eng. Manag.* **2011**, *17*, 227–233.
17. Sedita, S.R.; Apa, R. The impact of inter-organizational relationships on contractors' success in winning public procurement projects: The case of the construction industry in the Veneto region. *Int. J. Proj. Manag.* **2015**, *33*, 1548–1562. [CrossRef]
18. Khalef, R.; El-adaway, I.H. Identifying Design-Build Decision-Making Factors and Providing Future Research Guidelines: Social Network and Association Rule Analysis. *J. Constr. Eng. Manag.* **2023**, *149*, 04022151. [CrossRef]
19. Dogan, S.Z.; Arditi, D.; Gunhan, S.; Erbasaranoglu, B. Assessing coordination performance based on centrality in an e-mail communication network. *J. Manag. Eng.* **2015**, *31*, 04014047. [CrossRef]
20. Gharaibeh, L.G.; Matarneh, S.T.; Arafeh, M.; Sweis, G. Factors leading to design changes in Jordanian construction projects. *Int. J. Product. Perform. Manag.* **2021**, *70*, 893–915. [CrossRef]
21. Yap, J.B.H.; Abdul-Rahman, H.; Wang, C.; Skitmore, M. Exploring the underlying factors inducing design changes during building production. *Prod. Plan. Control.* **2018**, *29*, 586–601. [CrossRef]
22. Shoar, S.; Chileshe, N. Exploring the Causes of Design Changes in Building Construction Projects: An Interpretive Structural Modeling Approach. *Sustainability* **2021**, *13*, 9578. [CrossRef]
23. Aslam, M.; Baffoe-Twum, E.E.; Saleem, F. Design Changes in Construction Projects Causes and Impact on the Cost. *Civ. Eng. J.* **2019**, *5*, 1647–1655. [CrossRef]
24. Bassa, M.; Reta, A.; Alyew, A.; Tora, M. Causes and Effects of Design Change in Building Construction Projects in Three Selected Southern Ethiopia Zones. *Int. J. Eng. Res.* **2020**, *8*, 757–761.
25. Abdul-rahman, H.; Wang, C.; Yap, J.B.H. Impacts of design changes on construction project performance: Insights from literature review. *Quant. Surv. Constr. Bus.* **2017**, *7*, 31–54.
26. Yana AA, G.A.; Rusdhi, H.A.; Wibowo, M.A. Analysis of factors affecting design changes in construction project with Partial Least Square (PLS). *Procedia Eng.* **2015**, *125*, 40–45. [CrossRef]
27. Saad, D.A.; Gharib, F.; El-Said, M. Simulation of design changes impact in healthcare construction projects using system dynamics. *Can. J. Civ. Eng.* **2021**, *48*, 554–569. [CrossRef]
28. Afsharghotli, A.; Yitmen, I. ANN model for assessment of design changes in gas–oil and petrochemical projects. *Arab. J. Sci. Eng.* **2020**, *45*, 4273–4284. [CrossRef]
29. Cohen, C.; Pearlmutter, D.; Schwartz, M. A game theory-based assessment of the implementation of green building in Israel. *Build. Environ.* **2017**, *125*, 122–128. [CrossRef]
30. Geng, X.; Lv, L.; Wang, Y.; Sun, R.; Wang, X. Evolutionary Game Research on Green Construction Considering Consumers' Preference under Government Supervision. *Environ. Res. Public Health* **2022**, *19*, 16743. [CrossRef]
31. Li, X.; Wang, C.; Kassem, M.A.; Liu, Y.; Ali, K.N. Study on Green Building Promotion Incentive Strategy Based on Evolutionary Game between Government and Construction Unit. *Sustainability* **2022**, *14*, 10155. [CrossRef]
32. Chen, Y.; Li, Z.; Xu, J.; Liu, Y.; Meng, Q. How Does the Government Policy Combination Prevents Greenwashing in Green Building Projects? An Evolutionary Game Perspective. *Buildings* **2023**, *13*, 917. [CrossRef]
33. Meng, Q.; Liu, Y.; Li, Z.; Wu, C. Dynamic reward and penalty strategies of green building construction incentive: An evolutionary game theory-based analysis. *Environ. Sci. Pollut. Res.* **2021**, *28*, 44902–44915. [CrossRef] [PubMed]
34. Chen, J.; Hua, C.; Liu, C. Considerations for better construction and demolition waste management: Identifying the decision behaviors of contractors and government departments through a game theory decision-making model. *J. Clean. Prod.* **2019**, *212*, 190–199. [CrossRef]

35. Su, Y.; Si, H.; Chen, J.; Wu, G. Promoting the sustainable development of the recycling market of construction and demolition waste: A stakeholder game perspective. *J. Clean. Prod.* **2020**, *277*, 122281. [CrossRef]
36. Ameh, O.J.; Soyingbe, A.A.; Odusami, K.T. Significant factors causing cost overruns in telecommunication projects in Nigeria. *J. Constr. Dev. Ctries.* **2010**, *15*, 49–67.
37. James, D.; Amusan Lekan, M.; Oloke, C.O.; Olusanya, O.; Tunji-Olayeni, P.F.; Dele, O.; Peter, N.J.; Omuh, I.O. Causes and effect of delay on project construction delivery time. *Int. J. Educ. Res.* **2014**, *2*, 197–208.
38. Gamil, Y.; Abdul Rahman, I. Assessment of critical factors contributing to construction failure in Yemen. *Int. J. Constr. Manag.* **2020**, *20*, 429–436. [CrossRef]
39. Maynard-Smith, J.; Price, G.R. The logic of animal conflict. *Nature* **1973**, *246*, 15–18. [CrossRef]
40. Becker, T.C.; Shane, J.S.; Jalselskis, E.J. Comparative Analysis of Lean Construction with Design-Build Using a Framework of Contractual Forms of Agreement. *J. Archit. Eng.* **2012**, *18*, 187–191. [CrossRef]
41. Liu, M. Behavior Analysis of Design and Construction Consortium Based on Game Theory. *West. Prospect. Proj.* **2008**, *145*, 179–182.
42. Smith, J.M. Game Theory and the Evolution of Fighting. In *On Evolution*; Edinburgh University Press: Edinburgh, UK, 1972.
43. Friedman, D. Evolutionary game in economics. *Econometrica* **1991**, *59*, 637–666. [CrossRef]
44. Chen, Z.; Jiang, Z. Implementing the spirit of Several Opinions to improve the ability and level of general contracting. *China Surv. Des.* **2016**, *7*, 23–27. [CrossRef]
45. Yang, J. Thoughts on the management of consortium cooperation mode led by designer in EPC project. *Archit. Econ.* **2022**, *43*, 43–48.
46. Huang, H. Key points of project management and control based on design enterprise leading consortium mode. *Water Resour. Hydropower Technol.* **2022**, *53*, 343–345.

buildings

MDPI

Article

# Evolutionary Mechanism of Government Green Development Behavior in Construction and Demolition Waste Recycling Projects: A Perspective of Ecological Compensation

**Weihong Chen, Wenjun Yin, Beiyu Yi, Shiqi Xu, Hao Zhang and Xingwei Li ***

College of Architecture and Urban-Rural Planning, Sichuan Agricultural University, Chengdu 611830, China; 202008146@stu.sicau.edu.cn (W.C.); 201808479@stu.sicau.edu.cn (W.Y.); 202108254@stu.sicau.edu.cn (B.Y.); 202008181@stu.sicau.edu.cn (S.X.); zhanghao2@stu.sicau.edu.cn (H.Z.)

* Correspondence: xwl@sicau.edu.cn

**Abstract:** Although construction and demolition waste (CDW) recycling projects have received increasing attention from national and regional governments, the mechanisms for the evolution of government green development behavior in such projects are not yet clear. From the perspective of ecological compensation for the cross-regional disposal of CDW, this study aims to reveal the evolutionary mechanism of government green development behavior through externality theory. The main findings are as follows. First, the initial probability of government adoption of green development behavior does not affect the final stable state of the system. Second, there is heterogeneity in the effects of the allocation coefficient of ecological benefits and the ecological compensation coefficient on different government green development behavior. Finally, ecological benefits can encourage the government to actively adopt green development behavior. This study introduces for the first time an ecological compensation perspective into the study of CDW project management, which not only enriches the knowledge system of the government green development behavior on project management but also provides a reference for the government to participate in the process of cross-regional disposal of CDW in favor of ecological compensation.

**Keywords:** construction and demolition waste recycling; ecological compensation; evolutionary game; externality; government green development behavior

## 1. Introduction

According to statistics, construction and demolition waste (CDW) accounts for approximately 36–40% of the world's total solid waste generation [1]. The large amount of CDW generation causes problems such as soil pollution, global warming, and deterioration of public health [2,3]. As an effective way to address CDW, CDW recycling has a positive impact on the sustainable development of the economy and environment [4,5]. However, the limited CDW treatment capacity still makes CDW management a great challenge [6], and CDW has become a global environmental issue [7].

Government green development behavior is an important initiative to address environmental management challenges, which mainly include incentive green development behavior and mandatory green development behavior [8]. Among them, incentives for green development behaviors include granting environmental subsidies and issuing tax breaks [9], promoting green consumption [10], etc. In contrast, strict regulation of the market and the issuance of mandatory and catalog regulations are mandatory green development behaviors [11]. Research shows that government green development behaviors are important for promoting regional environmental quality improvement [12]. By 2022, the Chinese government reduced domestic carbon emission intensity by 34.4% compared to a decade ago through green development behaviors such as environmental legislation and environmental supervision, reversing the undesirable trend of rapid growth of $CO_2$ emissions [13]. Therefore, as a concrete form of environmental management, the government green development behavior is also of great significance to CDW management.

How to effectively manage the large amount of CDW generated has become an important challenge for local governments [14]. Restricted by the limited regional disposal capacity, cross-regional disposal of CDW is considered a feasible strategy to alleviate the pressure of regional CDW management. In particular, CDW cross-regional disposal not only reduces resource consumption but also has significant ecological benefits [6]. Under the coordination of the higher government, the process of cross-regional CDW disposal mainly involves two subjects: The CDW-generating district government and the CDW disposal district government. The CDW disposal district government is responsible for accepting and distributing CDW to the corresponding disposal enterprises, while CDW disposal enterprises have two types of CDW disposal: direct landfill and resource utilization [15]. Compared with direct landfill disposal, CDW disposal by resource utilization not only saves land resources and transportation costs but also generates substantial environmental advantages [16]. However, most CDW is still disposed of directly into landfills [17]. It should be noted that the environmental pollution caused by the direct landfilling of CDW will inevitably generate negative externalities for neighboring areas due to the existence of environmental pollution spillover effects [18].

To improve CDW management and achieve sustainable green development in the region, CDW disposal district governments can adopt legislation [19], financial support [20], and other green development behaviors to urge enterprises to carry out CDW resource treatment. In addition, the environmental benefits generated by the governments of CDW disposal districts through development behaviors have significant positive externalities. Specifically, the CDW-generating district government can benefit from the environmental benefits generated by the green development behaviors of the CDW disposal district governments and promote their regional green development. It should be noted that to internalize positive environmental externalities, the beneficiary government should pay ecological compensation to the protector government. Ecological compensation can not only adjust the relationship between ecological protectors and ecological beneficiaries but also further promote the coordinated development of energy, the economy, and the environment [21]. It can be seen that there are green development behaviors between CDW-generating district governments and CDW-disposal district governments in the process of CDW disposal. Unfortunately, existing research lacks an investigation of government green development behaviors in the context of CDW recycling projects by using ecological compensation as a research perspective. Therefore, this paper aims to reveal the evolutionary mechanism of governmental green development behavior through externality theory under the perspective of ecological compensation in CDW cross-regional disposal.

This paper focuses on answering the following scientific question. How does ecological compensation affect the evolution of government green development behavior in the context of the cross-regional disposal of CDW? To answer this scientific question, this paper first constructs a game model of government green development behavior consisting of both governments of CDW cross-regional disposal using externality theory. Second, the stability of the equilibrium point is analyzed using the evolutionary game method, and the evolutionary stabilization strategies under different situations are determined. Finally, the effects of the initial probability, allocation coefficient of ecological benefits, ecological compensation coefficient, and ecological benefits on the evolutionary path of government green development behavior were analyzed by numerical simulation.

The main contributions of this paper are as follows: (1) studying the green development behavior of the government in the cross-regional collaborative management of CDW from the perspective of ecological compensation not only enriches the relevant studies on the green development behavior of the government but also provides a theoretical basis for the adoption of green development behavior by the government; (2) creatively introducing externality theory into the study of the green development behavior of the government, revealing the evolution mechanism of the green development behavior of the government considering the ecological compensation mechanism, which provides new evidence for the implementation of ecological compensation for CDW cross-regional disposal.

In summary, this paper shows important significance in theory and practice. On the one hand, this study not only enriches the relevant research in the field of CDW management from the perspective of government green development behavior but also provides new evidence from CDW cross-regional disposal for the study of ecological compensation mechanisms. On the other hand, in the context of CDW cross-regional disposal, this paper not only provides a reference for the government to formulate ecological compensation-related policies but also provides new ideas to promote the cross-regional collaborative management of CDW.

The rest of this paper is structured as follows. Section 2 provides a review of the relevant literature. Section 3 constructs an evolutionary game model of the government green development behavior under the perspective of ecological compensation for the cross-regional disposal of CDW. Section 4 analyzes the stability of each equilibrium point and discusses the evolutionary stabilization strategies under different situations. Section 5 conducts numerical simulations of key parameters and analyzes the influence of relevant parameters on the evolutionary path of the game subjects. Section 6 draws conclusions and management implications and summarizes the limitations of this paper.

## 2. Literature Review

This section reviews the relevant literature in three parts: externality theory, government green development behavior, and ecological compensation. Table 1 gives the studies related to government green development behavior and ecological compensation.

**Table 1.** Studies related to government green development behavior and ecological compensation.

| Research Topics | Dimensions | Sosurce Papers |
|---|---|---|
| Externality theory | Positive environmental externalities | [22] |
| | Negative environmental externalities | [23] |
| | Internalization of environmental externalities | [24–26] |
| Government green development behavior | The driving role of policies and regulations | [27,28] |
| | Environmental regulation mechanism | [29,30] |
| | The mechanism of government green development behavior | [8] |
| Ecological compensation | The role of ecological compensation | [21,31] |
| | Determination of ecological compensation standards | [32–35] |
| | Research background of ecological compensation | [36–38] |

### 2.1. Externality Theory

Externality theory explains the causes of externalities and how to solve them from an economic perspective [39]. Environmental externalities are the nonmarket impacts on the environment caused by the production and consumption of environmental agents [40]. Environmental externalities can be divided into positive environmental externalities and negative environmental externalities, depending on the impact generated. On the one hand, environmental protection behavior generates significant positive externalities. For example, in the process of environmental pollution management, prevention, and control, significant spillover benefits are created due to the improvement of ecosystem service functions [22]. On the other hand, transboundary pollution has negative externalities, resulting in a reduction in ecological quality in neighboring regions [23].

The problem of internalizing externalities is the core issue of externality theory. The optimal level of environmental protection can be achieved only by internalizing environmental externalities [24]. To this end, scholars have engaged in intense debate. Some scholars argue that the solution to the problem of negative externalities arising from environmental pollution focuses on the inclusion of environmental costs in prices and making the agent that generates the negative externalities bear the costs associated with waste disposal [25]. Other scholars, from the perspective of positive environmental externalities, propose internalizing positive environmental externalities by providing ecological compensation for the costs paid by ecological reserves to improve ecological service functions [26].

Unfortunately, most studies have focused on considering only unilateral environmental externalities. In fact, the behavior of the subject of environmental governance will continuously change and evolve under the influence of various factors. In particular, in the early stage of CDW governance, the government was unable to guide enterprises scientifically and effectively to govern CDW due to its insufficient resource awareness [41], which led to a large amount of CDW being directly landfilled and simply disposed of. However, the pollution of soil and water caused by direct landfills and simple disposal will generate negative externalities for neighboring areas. However, in the context of green development, the government may generate positive externalities by actively promoting CDW resource utilization [42]. Therefore, it is incomplete to consider only the unilateral environmental externalities generated by the behavior of environmental agents. However, existing research related to CDW management lacks consideration of both positive environmental externalities and negative environmental externalities internalized in depth.

### 2.2. Government Green Development Behavior

Green development behavior has received much attention from both enterprise and government researchers, especially government green development behavior [43,44]. Governmental green development behavior is an organizational behavior that uses the government as a vehicle to facilitate the achievement of governmental green development and economic growth goals [8]. The existing research on government green development behavior focuses on the following three aspects. First, there is research on policies and regulations. This type of research mainly focuses on the driving role of policies and regulations on the green development behavior of enterprises and the public [27,28]. The second is research related to environmental regulation mechanisms. This type of research explores the effects of government regulation and the role of environmental governance. The third is research related to the multisubject role mechanism of government green development behavior [29,30]. For example, studies have verified the mechanism of the role of government green development behavior in the multisubject interaction among government, enterprises, and the public through the meta-analysis method [8].

However, existing research on government green development behavior is still incomplete. On the one hand, some studies focus on considering the impact of government green development behaviors on enterprises, the public, and the ecological environment but lack research on the interactions among governments in adopting green development behaviors. On the other hand, as a finite rational subject, the government will continuously revise and improve green development behaviors according to the changing environment. Unfortunately, the dynamic evolutionary mechanism of governmental green development behavior has not been revealed in the field of CDW management.

### 2.3. Ecological Compensation

Ecological compensation is an institutional arrangement that adjusts the relationship between the interests of ecological beneficiaries and ecological protectors, mobilizes enthusiasm for ecological protection, and internalizes environmental externalities [45]. Ecological compensation was first applied in the fields of ecology and economics [46]. Through ecological compensation, the environmental costs and ecological benefits generated by environmental treatment can be integrated into the traditional economic cost-benefit analysis [21]. With the deepening of ecological compensation research, it continues to play an important role in biodiversity conservation [47], ecological restoration [48], and the coordination of regional environmental protection [49]. In addition, as an important tool for governments to coordinate cross-regional environmental governance [35], ecological compensation can effectively promote the regionalized cooperation of the government in environmental pollution management [31], thus forming a long-term mechanism of environmental governance with shared benefits and cooperation.

The ecological compensation standard is a decisive factor in the long-term and stable implementation of the ecological compensation mechanism. How to determine reasonable ecological compensation standards based on the beneficiary-pays and polluter-pays principles has become a difficult point in the establishment of ecological compensation mechanisms [50]. For this reason, many scholars have explored the feasibility of determining ecological compensation standards through the ecological footprint method, the ecosystem service value method, and the opportunity cost method [32–35].

Although an increasing number of ecological compensation studies focus on watershed ecological compensation [36], cultivated land ecological compensation [37], and grassland ecological compensation [38], they have not yet been able to explain the ecological compensation mechanism for the cross-regional disposal of CDW.

In summary, existing studies confirm that ecological compensation can effectively internalize environmental externalities and facilitate the green development behavior of the government to collaboratively manage cross-regional pollution. However, previous related studies on governmental green development behavior have not revealed the evolutionary mechanism of government green development behavior from the perspective of ecological compensation, and the existing related studies on ecological compensation have not considered the context of CDW management.

## 3. Problem Description and Model Assumption

Consider an evolutionary game model consisting of a G1 government and a G2 government. Among them, the G1 government is the CDW generating district government, while the G2 government is the CDW disposal district government. The total amount of CDW generated in the administrative area of the G1 government exceeds its local disposal capacity and needs to be disposed of off-site across districts, and the higher-level government coordinates the G2 government with disposal capacity to dispose off-site. The G2 government can promote the resource utilization of CDW by taking incentives to absorb and dispose of excess CDW and improve the green development level of the region. At the same time, the G2 government can also dispose of CDW in landfills by enterprises without incentives. However, the G2 government needs to supervise enterprises without incentives to avoid illegal dumping. The G2 government's environmental protection behavior of promoting CDW resource utilization has positive externalities, and the G1 government that benefits from environmental protection should pay ecological compensation to the G2 government. In addition, the G1 government has already paid the costs associated with CDW disposal, and the G1 government can also choose not to pay ecological compensation. The game model formed is shown in Figure 1.

In this paper, the G1 government and the G2 government are selected as game subjects with finite rationality, and ecological compensation is used as the research perspective to reveal the evolutionary mechanism of government green development behavior through externality theory. The parameter settings and descriptions are shown in Table 2.

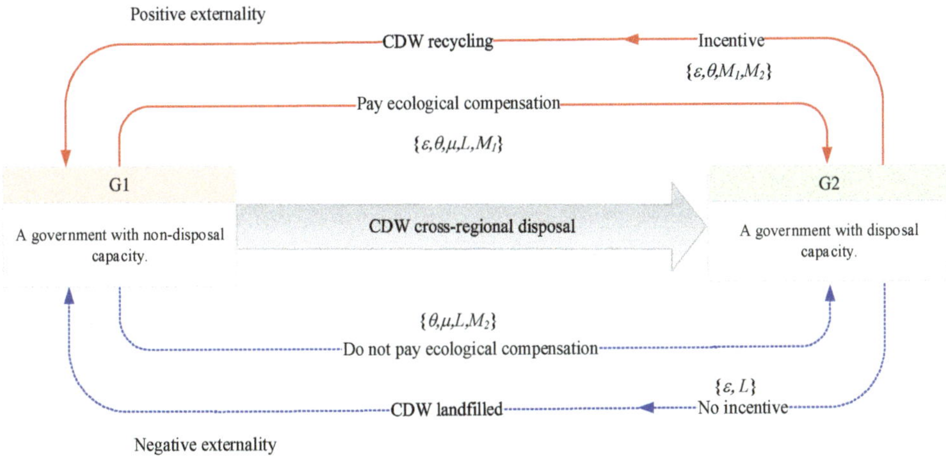


**Figure 1.** Construction and demolition waste (CDW) cross-regional disposal model. Note: G1 and G2 represent two different local governments within the same country. $\varepsilon$ is the ecological compensation coefficient; $\theta$ is the allocation coefficient of ecological benefits of G2 government; $\mu$ is the environmental loss spillover coefficient; $L$ is the environmental loss when G2 government does not adopt the incentive strategy; $M_1$ is the overall ecological benefits when G1 government compensates and G2 government adopts the incentive strategy; $M_2$ is the overall ecological benefits when G1 government does not compensate and G2 government adopts the incentive strategy.

**Table 2.** Parameter settings and descriptions.

| Parameter | Parameter Description |
|---|---|
| $C_0$ | Base cost of G1 government cross-regional disposal of CDW. |
| $C_1$ | Additional costs for G1 government adoption of ecological compensation. |
| $C_2$ | Additional costs for G2 government adoption of ecological compensation. |
| $C_3$ | Cost when G2 government adopts incentive strategy. |
| $C_4$ | Cost when G2 government does not adopt incentive strategy. |
| $S$ | Reputation gained when G2 government adopts incentive strategy. |
| $L$ | Environmental losses when G2 government does not adopt incentive strategy. |
| $M_1$ | The overall ecological benefits when G1 government compensates and G2 government adopts the incentive strategy. |
| $M_2$ | The overall ecological benefits when G1 government does not compensate and G2 government adopts the incentive strategy. |
| $\alpha$ | The weight coefficient of ecological environment in local government performance appraisal. |
| $\varepsilon$ | Ecological compensation coefficient $(0 < \varepsilon < 1)$. |
| $\theta$ | The allocation coefficient of ecological benefits of G2 government. $(0 < \theta < 1)$. |
| $R$ | The amount of incentives given by the central government when the G1 government pays ecological compensation. |
| $J$ | Green subsidies given by the central government when the G2 government adopts incentive strategy. |
| $\mu$ | Environmental loss spillover coefficient. |

Note: The proportion of overall ecological benefits received by the G1 government is $1 - \theta$.

The basic assumptions are as follows:

**Assumption 1.** *The higher-level government coordinates the G2 government with disposal capacity to handle the excess CDW generated within the administrative area of the G1 government; the G1 government bears the costs related to the cross-regional and off-site disposal of CDW; and the basic cost of cross-regional and off-site disposal of CDW by the G1 government is* $C_0$.

**Assumption 2.** *When the G1 government and the G2 government establish horizontal ecological compensation mechanisms, they need to pay transaction costs $C_1$ and $C_2$, respectively. The regulation of the central government is an important factor in ensuring the implementation of ecological compensation. To encourage local governments to establish horizontal ecological compensation mechanisms, the central government subsidizes G1 government that establish horizontal ecological compensation mechanisms R to ensure the long-term effectiveness of ecological compensation mechanisms [51].*

**Assumption 3.** *To improve the G2 government's motivation to promote CDW resource utilization, the G1 government can choose to pay ecological compensation to the G2 government to reward its eco-environmental protection behavior. The ecological compensation paid by the G1 government is determined according to the input cost of the G2 government to manage CDW [52].*

**Assumption 4.** *The government plays an important role in promoting the resource utilization of CDW [53]. The G2 government can promote CDW resource utilization by adopting incentive measures to increase the willingness of enterprises and social capital to participate in CDW resource utilization. The cost of the G2 government to adopt incentive measures is $C_3$, including the cost of subsidizing CDW recycling enterprises to provide financial security, accelerating the promotion and application of CDW recycled products, etc. [54]. The G2 government actively promotes CDW resource utilization and receives green subsidies from the central government for environmental governance J [55]. At the same time, the G2 government will also gain a good government reputation S for effective CDW management [56].*

**Assumption 5.** *The G2 government's promotion of CDW resource utilization will generate overall ecological benefits. When the G1 government pays ecological compensation, the overall ecological benefits generated by the G2 government adopting the incentive strategy is $M_1$ [57]; the overall ecological benefit generated by the G2 government adopting the incentive strategy when the G2 government does not pay ecological compensation is $M_2$ [58], where $M_1 > M_2$. The external spatial spillover effect of ecology allows the G1 government to also receive part of the ecological benefits, $\theta$ is the ecological benefit allocation coefficient of the G2 government, and the proportion of the overall ecological benefit received by the G1 government is $1-\theta$ [35]. In addition, the choice of environmental strategies by local governments is influenced by the weight of environmental indicators $\alpha$ in the performance appraisal system [59].*

**Assumption 6.** *The higher cost of CDW resource treatment, longer payback period, and low profitability of recycled construction materials [60] lead to the reluctance of enterprises to participate in CDW resource treatment. In the event that the G2 government chooses not to take incentives, enterprises tend to choose the direct landfill method to dispose of CDW. In this case, the G2 government needs to strengthen the regulation of illegal dumping activities in the cross-regional disposal of CDW, and the cost of regulation is $C_4$. Environmental pollution has strong externalities and can flow across borders. The direct landfill approach can contaminate soil and water sources and cause environmental losses L [61]. The spillover effect of environmental pollution $\mu$ will affect the ecological environment within the administrative region of the G1 government [62], and the negative externality effect caused within the administrative region of the G1 government is $\mu L$.*

**Assumption 7.** *The probability that government G1 chooses to pay ecological compensation is $x$, while the probability that G1 chooses not to pay ecological compensation is $1-x(0 \le x \le 1)$; the probability that G2 chooses to adopt an incentive strategy is $y$, while the probability that G2 chooses not to adopt an incentive strategy is $1-y(0 \le y \le 1)$. $x$ and $y$ are functions of time.*

Based on the above seven assumptions, the payoff matrix between the G1 government and the G2 government under different strategies is established in this paper, as shown in Table 3.

**Table 3.** The payoff matrix of the evolutionary game model.

| | | G1 Government | |
|---|---|---|---|
| | | **Payment of Ecological Compensation** | **No Payment of Ecological Compensation** |
| G2 government | Incentive | $R+\alpha(1-\theta)M_1-\varepsilon C_3-C_0-C_1$<br>$J+\alpha\theta M_1-(1-\varepsilon)C_3+S+C_0-C_2$ | $\alpha(1-\theta)M_2-C_0$<br>$J+\alpha\theta M_2+S+C_0-C_3$ |
| | No incentive | $R-\varepsilon C_4-C_0-C_1-\mu L$<br>$C_0-(1-\varepsilon)C_4-C_2-L$ | $-C_0-\mu L$<br>$C_0-L-C_4$ |

## 4. Evolutionary Game Model Analysis

### 4.1. Stability Point Calculation

According to the benefit matrix, the expected benefits of the G1 government paying ecological compensation and not paying ecological compensation strategies are shown in Equations (1) and (2), respectively, and the average expected benefit is shown in Equation (3):

$$U_{A_0}=R-C_0-C_1+y(\alpha(1-\theta)M_1-\varepsilon C_3)-(1-y)(\varepsilon C_4+\mu L) \tag{1}$$

$$U_{A_1}=y\alpha(1-\theta)M_2-C_0-(1-y)\mu L \tag{2}$$

$$\overline{U_A}=xU_{A_0}+(1-x)U_{A_1} \tag{3}$$

The resulting replication dynamic equation is shown in Equation (4):

$$F(x)=x(1-x)(R-C_1+y\alpha(1-\theta)(M_1-M_2)+y\varepsilon(C_4-C_3)-\varepsilon C_4) \tag{4}$$

According to the benefit matrix, the expected benefits of the G2 government with and without the incentive strategy are shown in Equations (5) and (6), respectively, and the average expected benefit is shown in Equation (7):

$$U_{B_0}=J+S+C_0+\alpha\theta M_2-C_3+x(\alpha\theta(M_1-M_2)+\varepsilon C_3-C_2) \tag{5}$$

$$U_{B_1}=C_0-L-x((1-\varepsilon)C_4+C_2)-(1-x)C_4 \tag{6}$$

$$\overline{U_B}=yU_{B_0}+(1-y)U_{B_1} \tag{7}$$

The resulting replication dynamic equation is shown in Equation (8):

$$F(y)=y(1-y)(J+S+\alpha\theta(x(M_1-M_2)+M_2)+(1-x\varepsilon)(C_4-C_3)+L) \tag{8}$$

Let $F(x)=0$ and $F(y)=0$, we obtain the 5 equilibrium points of the evolutionary game: (0,0), (0,1), (1,0), (1,1), and $(x^*,y^*)$. Among them, $x^*=\frac{C_3-\alpha\theta M_2-J-S-L-C_4}{\alpha\theta(M_1-M_2)+\varepsilon(C_3-C_4)}$, $y^*=\frac{\varepsilon C_4+C_1-R}{\alpha(1-\theta)(M_1-M_2)+\varepsilon(C_4-C_3)}$.

### 4.2. Evolutionary Equilibrium Stability Analysis

In this paper, the Jacobi matrix is constructed by referring to Jiang et al. [63], as shown in Equation (9). The stability of the equilibrium point of the replicated dynamic system can be obtained by analyzing the local stability of the Jacobi matrix.

$$J=\begin{bmatrix}\frac{\partial F(x)}{\partial x} & \frac{\partial F(x)}{\partial y}\\ \frac{\partial F(y)}{\partial x} & \frac{\partial F(y)}{\partial y}\end{bmatrix}=\begin{bmatrix}(1-2x)(y(\alpha(1-\theta)(M_1-M_2)-\varepsilon C_3)-(1-y)\varepsilon C_4+R-C_1) & x(1-x)(\alpha(1-\theta)(M_1-M_2)-\varepsilon C_3+\varepsilon C_4)\\ y(1-y)(\alpha\theta(M_1-M_2)+\varepsilon(C_3-C_4)) & (1-2y)(x(\alpha\theta(M_1-M_2)+\varepsilon(C_3-C_4))+J+S+L+\alpha\theta M_2+C_4-C_3)\end{bmatrix} \tag{9}$$

The determinant of the Jacobi matrix is shown in Equation (10):

$$det(J) = \frac{\partial F(x)}{\partial x} \cdot \frac{\partial F(y)}{\partial y} - \frac{\partial F(x)}{\partial y} \cdot \frac{\partial F(y)}{\partial x} \tag{10}$$

The trace of the Jacobi matrix is shown in Equation (11):

$$tr(J) = \frac{\partial F(x)}{\partial x} + \frac{\partial F(y)}{\partial y} \tag{11}$$

The determinant and trace of the system at each equilibrium point are shown in Tables 4 and 5.

**Table 4.** The determinant at the equilibrium point of the system.

| Point | *Det*(*J*) |
|---|---|
| (0,0) | $(R - C_1 - \varepsilon C_4)(J + S + L + \alpha\theta M_2 - C_3 + C_4)$ |
| (0,1) | $(R - C_1 - \varepsilon C_3 + \alpha(1-\theta)(M_1 - M_2))(-(J + S + L + \alpha\theta M_2 - C_3 + C_4))$ |
| (1,0) | $(-(R - C_1 - \varepsilon C_4))(J + S + L + \alpha\theta M_1 + (1-\varepsilon)(C_4 - C_3))$ |
| (1,1) | $(R - C_1 - \varepsilon C_3 + \alpha(1-\theta)(M_1 - M_2))(J + S + L + \alpha\theta M_1 + (1-\varepsilon)(C_4 - C_3))$ |
| $(x^*, y^*)$ | $\frac{-(\varepsilon C_4 + C_1 - R)(\alpha(1-\theta)(M_1 - M_2) + R - C_1 - \varepsilon C_3)(-J - S - L - \alpha\theta M_2 + C_3 - C_4)(J + S + L + \alpha\theta M_1 + (1-\varepsilon)(C_4 - C_3))}{(\alpha\theta(M_1 - M_2) + \varepsilon(C_3 - C_4))(\alpha(1-\theta)(M_1 - M_2) + \varepsilon(C_4 - C_3))}$ |

**Table 5.** The trace at the equilibrium point of the system.

| Point | *Tr*(*J*) |
|---|---|
| (0,0) | $(R - C_1 - \varepsilon C_4) + (J + S + L + \alpha\theta M_2 - C_3 + C_4)$ |
| (0,1) | $(R - C_1 - \varepsilon C_3 + \alpha(1-\theta)(M_1 - M_2)) + (-(J + S + L + \alpha\theta M_2 - C_3 + C_4))$ |
| (1,0) | $(-(R - C_1 - \varepsilon C_4)) + (J + S + L + \alpha\theta M_1 + (1-\varepsilon)(C_4 - C_3))$ |
| (1,1) | $-((R - C_1 - \varepsilon C_3 + \alpha(1-\theta)(M_1 - M_2)) + (J + S + L + \alpha\theta M_1 + (1-\varepsilon)(C_4 - C_3)))$ |
| $(x^*, y^*)$ | 0 |

When the equilibrium point satisfies both conditions $det(J) > 0$ and $tr(J) < 0$, the equilibrium point that replicates the dynamic equations is the system evolutionary stability point (ESS). The *tr(J)* at the point $(x^*, y^*)$ is equal to 0 and does not satisfy the condition of $tr(J) < 0$. Therefore, this point is not an evolutionary stability point. Therefore, only the stability of the remaining four points $(0,0)$, $(0,1)$, $(1,0)$ and $(1,1)$ need to be analyzed. Six scenarios of equilibrium stability were obtained based on the equilibrium points of the evolutionary model and their stability conditions, as shown in Tables 6–11. Among them, $A = R - C_1 - \varepsilon C_4$, $B = J + S + L + \alpha\theta M_2 - C_3 + C_4$, $D = R - C_1 - \varepsilon C_3 + \alpha(1-\theta)(M_1 - M_2)$, $E = J + S + L + \alpha\theta M_1 + (1-\varepsilon)(C_4 - C_3)$.

**Table 6.** Stability analysis of the points in case (1).

| Point | $A < 0, B < 0, D < 0$ | | | $A < 0, B < 0, D > 0, E < 0$ | | |
|---|---|---|---|---|---|---|
| | *Det*(*J*) | *Tr*(*J*) | Stability | *Det*(*J*) | *Tr*(*J*) | Stability |
| (0,0) | + | - | ESS | + | - | ESS |
| (0,1) | - | ? | Saddle point | + | + | Unstable point |
| (1,0) | + | + | Unstable point | - | ? | Saddle point |
| | - | ? | Saddle point | | | |
| (1,1) | - | ? | Saddle point | - | ? | Saddle point |
| | + | + | Unstable point | | | |

Note: "+" indicates that the calculation result is greater than 0, "-" indicates that the calculation result is less than 0, and "?" indicates uncertainty; ESS system evolution is stable.

**Table 7.** Stability analysis of the points in case (2).

| Point | $A<0, B>0, D<0, E>0$ | | | $A>0, B>0, D<0, E>0$ | | |
|---|---|---|---|---|---|---|
| | $Det(J)$ | $Tr(J)$ | Stability | $Det(J)$ | $Tr(J)$ | Stability |
| (0,0) | - | ? | Saddle point | + | + | Unstable point |
| (0,1) | + | - | ESS | + | - | ESS |
| (1,0) | + | + | Unstable point | - | ? | Saddle point |
| (1,1) | - | ? | Saddle point | - | ? | Saddle point |

Note: "+" indicates that the calculation result is greater than 0, "-" indicates that the calculation result is less than 0, and "?" indicates uncertainty; ESS system evolution is stable.

**Table 8.** Stability analysis of the points in case (3).

| Point | $A>0, B<0, D<0, E<0$ | | | $A>0, B<0, D>0, E<0$ | | |
|---|---|---|---|---|---|---|
| | $Det(J)$ | $Tr(J)$ | Stability | $Det(J)$ | $Tr(J)$ | Stability |
| (0,0) | - | ? | Saddle point | - | ? | Saddle point |
| (0,1) | - | ? | Saddle point | + | + | Unstable point |
| (1,0) | + | - | ESS | + | - | ESS |
| (1,1) | + | + | Unstable point | - | ? | Saddle point |

Note: "+" indicates that the calculation result is greater than 0, "-" indicates that the calculation result is less than 0, and "?" indicates uncertainty; ESS system evolution is stable.

**Table 9.** Stability analysis of the points in case (4).

| Point | $B>0, D>0, E>0$ | | | $A>0, B<0, D>0, E>0$ | | |
|---|---|---|---|---|---|---|
| | $Det(J)$ | $Tr(J)$ | Stability | $Det(J)$ | $Tr(J)$ | Stability |
| (0,0) | - | ? | Saddle point | - | ? | Saddle point |
| | + | + | Unstable point | | | |
| (0,1) | - | ? | Saddle point | + | + | Unstable point |
| (1,0) | + | + | Unstable point | - | ? | Saddle point |
| | - | ? | Saddle point | | | |
| (1,1) | + | - | ESS | + | - | ESS |

Note: "+" indicates that the calculation result is greater than 0, "-" indicates that the calculation result is less than 0, and "?" indicates uncertainty; ESS system evolution is stable.

**Table 10.** Stability analysis of the points in case (5).

| Point | $A<0, B<0, D>0, E>0$ | | |
|---|---|---|---|
| | $Det(J)$ | $Tr(J)$ | Stability |
| (0,0) | + | - | ESS |
| (0,1) | + | + | Unstable point |
| (1,0) | + | + | Unstable point |
| (1,1) | + | - | ESS |

Note: "+" indicates that the calculation result is greater than 0, "-" indicates that the calculation result is less than 0, and ESS system evolution is stable.

**Table 11.** Stability analysis of the points in case (6).

| Point | $A > 0, B < 0, D < 0, E > 0$ | | |
|---|---|---|---|
| | ***Det*(*J*)** | ***Tr*(*J*)** | **Stability** |
| (0,0) | - | ? | Saddle point |
| (0,1) | - | ? | Saddle point |
| (1,0) | - | ? | Saddle point |
| (1,1) | - | ? | Saddle point |

Note: "-" indicates that the calculation result is less than 0, and "?" indicates uncertainty.

The results of judging the stability of the equilibrium point under case (1) are given in Table 6.

Table 6 shows that in case (1), the only stable equilibrium point for the evolution of the G1 government and G2 government is $(0,0)$ when $A < 0, B < 0, D < 0$ or $A < 0$, $B < 0, D > 0, E < 0$. In this case, the cost of paying ecological compensation by the G1 government is higher than the benefit, regardless of which strategy the G2 government chooses. Therefore, the G1 government chooses not to pay ecological compensation. For the G2 government, regardless of which strategy the G1 government chooses, the cost of the G2 government's incentive for enterprises to resource CDW is higher than the benefit. Therefore, the G2 government tends to choose not to adopt the incentive strategy.

The results of judging the stability of the equilibrium point under case (2) are given in Table 7.

Table 7 shows that in case (2), the only stable equilibrium point for the evolution of the G1 government and G2 government is $(0,1)$ when $A < 0, B > 0, D < 0, E > 0$ or $A > 0, B > 0, D < 0, E > 0$. In this case, the benefits of the G2 government incentivizing firms to resource CDW are improved, and the benefits are higher than the costs. Therefore, the G2 government chooses to adopt an incentive strategy. For the G1 government, the benefits of not paying ecological compensation are higher when the G2 government chooses to adopt an incentive strategy. Therefore, the G1 government chooses not to pay ecological compensation.

The results of judging the stability of the equilibrium point under case (3) are given in Table 8.

Table 8 shows that in case (3), the only stable equilibrium point for the evolution of the G1 government and G2 government is $(1,0)$ when $A > 0, B < 0, D < 0, E < 0$ or $A > 0$, $B < 0, D > 0, E < 0$. In this case, the G1 government pays ecological compensation with improved benefits. Therefore, the G1 government chooses to pay ecological compensation. For the G2 government, the benefits of the G2 government's incentive for firms to resource CDW treatment are lower than the benefits from not adopting the incentive strategy. Therefore, the G2 government chooses not to adopt the incentive strategy.

The results of judging the stability of the equilibrium point under case (4) are given in Table 9.

Table 9 shows that in case (4), the only stable equilibrium point for the evolution of the G1 government and G2 government is $(1,1)$ when $B > 0, D > 0, E > 0$ or $A > 0$, $B < 0, D > 0, E > 0$. In this case, the choice to pay ecological compensation brings large benefits to the G1 government. At the same time, the G2 government incentivizes firms to carry out CDW resource treatment, which brings large benefits. Therefore, the G1 government and the G2 government will actively participate in ecological compensation for CDW cross-regional disposal. That is, the G1 government chooses to pay ecological compensation, and the G2 government chooses to adopt an incentive strategy to promote CDW recycling disposal.

The results of judging the stability of the equilibrium point under case (5) are given in Table 10.

Table 10 shows that in case (5), when $A < 0, B < 0, D > 0, E > 0$, the equilibrium points of the G1 government and G2 government evolution are $(0,0)$ or $(1,1)$. In this case, based on the characteristics of cost sharing and benefit sharing of the ecological compensation mechanism, the G1 government and the G2 government form a close game relationship. When the G1 government chooses not to pay ecological compensation, the G2 government chooses not to adopt an incentive strategy to obtain higher benefits; when the G1 government chooses to pay ecological compensation, the G2 government has more incentive to promote CDW resource utilization, thus improving social and environmental benefits. Therefore, the G2 government chooses to adopt the incentive strategy.

The results of judging the stability of the equilibrium point under case (6) are given in Table 11.

Table 11 shows that in case (6), when $A > 0, B < 0, D < 0, E > 0$, the G1 government and G2 government evolve without a stable point. At this time, the G1 government and the G2 government form a close game relationship; both parties pursue their own interest maximization, and it is difficult to form a stable system. Therefore, the strategy choice of the G1 government and the G2 government has uncertainty.

## 5. Numerical Simulation and Discussion

According to the analysis in Section 4, the most ideal steady state is that the G1 government chooses to pay ecological compensation and the G2 government chooses to adopt an incentive strategy, i.e., strategy $(1,1)$. To analyze the effects of initial probability, the allocation coefficient of ecological benefits, the ecological compensation coefficient, and ecological benefits on the evolutionary process of green development behavior of the G1 government and the G2 government, numerical simulations and analytical discussions of key variables are conducted in this paper using MATLAB R2020b. The values of the following parameters were determined according to the literature: $C_1$ [35] = 3, $C_3$ [4] = 4, $C_4$ [64] = 8, $S$ [56] = 4, $L$ [61] = 4, $\alpha$ [59] = 0.4, $J$ [55] = 7, $M_1$ [58] = 35, $M_2$ [58] = 15. The values of the following parameters were determined based on expert interviews: $R$ = 5, $\varepsilon$ = 0.5, $\theta$ = 0.5.

### *5.1. Effect of Initial Probability on Game Equilibrium*

Figure 2 shows the effect of the initial probability on the evolutionary process and final state of the green development behavior of the G1 government and the G2 government.

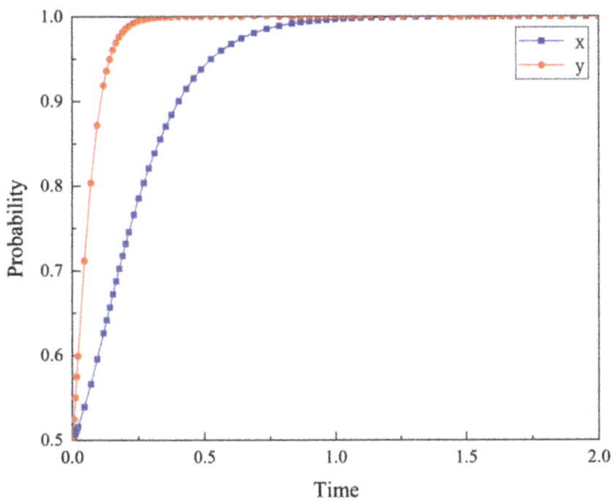


**Figure 2.** Effect of initial probability on the evolutionary strategy of green development behavior of G1 government and G2 government.

Figure 2 shows that with all initial probabilities of 0.5, the ESS of the game model is $(1, 1)$. In this case, the G1 government chooses to pay ecological compensation, and the G2 government chooses to adopt an incentive strategy to promote CDW resource utilization. Both governments take active measures to realize the collaborative management of CDW across regions and improve the utilization rate of CDW resources. The results of this study are contrary to those of Wang et al. [65]. The study by Wang et al. shows that the initial probability has a strong influence on the final decision of both upstream and central governments. When the initial willingness to participate is low, upstream and central governments tend not to participate in ecological compensation. Unlike Wang et al.'s study, the initial probability in this paper is taken to be 0.5. Thus, the results of the two studies are inconsistent in the case of inconsistent initial values. In addition, G2 government reaches the steady state of green development behavior faster than G1 government. This shows that the G2 government is more sensitive to the initial probability than the G1 government.

### 5.2. Effect of the Allocation Coefficient of Ecological Benefits on Game Equilibrium

The allocation coefficient of ecological benefits on the evolutionary process and final states of green development behavior of the G1 government and the G2 government are given in Figure 3.

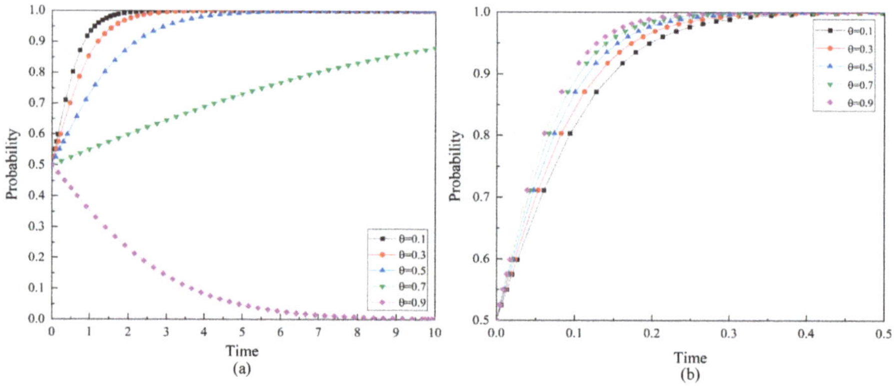


**Figure 3.** Impact of the allocation coefficient of ecological benefits on the evolutionary strategy of government green development behavior. Note: Figure (**a**) shows the G1 government, and figure (**b**) shows the G2 government. For the G1 government, 0.1, 0.3, 0.5, 0.7 and 0.9 represent the highest, high, normal, low, and lowest allocation coefficients of ecological benefits, respectively. For the G2 government, 0.1, 0.3, 0.5, 0.7 and 0.9 represent the proportions of lowest, low, normal, high, and highest allocated ecological benefits, respectively.

Figure 3a shows that as the value of $\theta$ increases, the strategy of the G1 government changes from choosing to pay ecological compensation to not paying ecological compensation. This is because an increase in the value of $\theta$ means that the G1 government has fewer ecological benefits to allocate, and the G1 government has less incentive to pay ecological compensation. Therefore, the G1 government will change its strategic choice when it can allocate fewer ecological benefits. This result is contrary to the findings of Gao et al. The reason for the difference may be that this paper considers both the allocation coefficient of ecological benefits and the weighting coefficient of the ecological environment in the local government performance appraisal system. Gao et al. [35] showed that when the allocation coefficient of ecological benefits reaches 0.8, the downstream government's strategy changes from not paying ecological compensation to paying ecological compensation. In the context set in this paper, the size of the allocation coefficient of ecological benefits affects the amount of ecological benefits received by the G1 government, which in turn affects the performance appraisal and strategy choice of the G1 government.

Figure 3b shows that as the size of the allocation coefficient of ecological benefits $\theta$ varies from 0.1 to 0.9, the evolution of the G2 government always results in the adoption of incentives, but the system converges to a steady state at a faster rate. Therefore, an increase in the value of $\theta$ only accelerates the rate of system stabilization and has little effect on the G2 government's strategy choice. Thus, it can be seen that the G2 government tends to adopt incentives to achieve CDW resource utilization and thus improve regional green development, regardless of the amount of ecological benefits they can obtain.

### *5.3. Effect of the Ecological Compensation Coefficient on Game Equilibrium*

Figure 4 shows the effects of the ecological compensation coefficient on the evolutionary process and final state of the green development behavior of the G1 government and the G2 government.

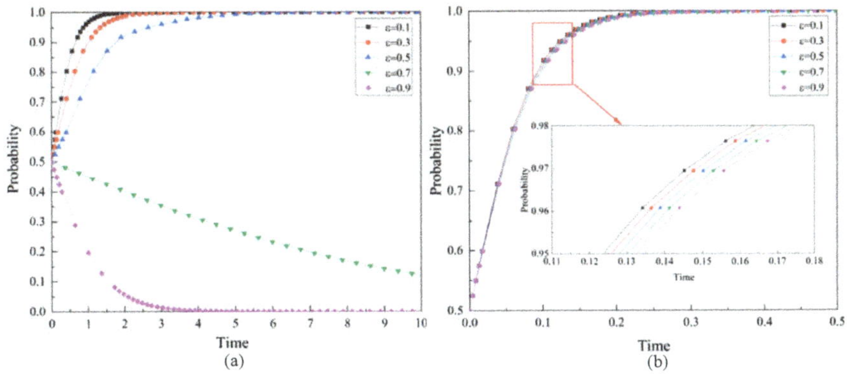

**Figure 4.** Effect of the ecological compensation coefficient on the evolutionary strategy of government green development behavior. Note: Figure (**a**) shows the G1 government, and figure (**b**) shows the G2 government. Values of 0.1, 0.3, 0.5, 0.7, and 0.9 represent the lowest, low, normal, high and highest ecological compensation levels, respectively.

Figure 4a shows that when $\varepsilon$ is not greater than 0.5, the convergence time for the G1 government to reach the steady state eventually stabilizes at 1, although it increases with the increase of $\varepsilon$. However, when $\varepsilon$ is greater than 0.5, the G1 government eventually stabilizes at 0. In particular, when $\varepsilon = 0.7$, the G1 government pays a large amount of ecological compensation. Having already borne the cost of CDW disposal across the area, the G1 government considers it its right to obtain a good ecological environment and thus gradually tends not to pay ecological compensation. This result is consistent with the findings of Wang et al. [66]. Therefore, the size of the $\varepsilon$ value affects the G1 government's strategy choice. In addition, as the value of $\varepsilon$ increases, the G1 government converges to the steady state at a faster rate. This is because an oversized $\varepsilon$ will increase the financial burden of the G1 government and thus reduce the G1 government's willingness to pay ecological compensation [67].

Figure 4b shows that the G2 government strategy is always to take incentives to promote CDW resource utilization, regardless of how much $\varepsilon$ increases. In addition, the G2 government converges very quickly and takes a very short time to reach the steady state. This indicates that the G2 government is sensitive to $\varepsilon$ and that the payment of ecological compensation by the G1 government to the G2 government can effectively encourage the G2 government to adopt an incentive strategy. Therefore, it is important to determine a reasonable ecological compensation coefficient to promote the establishment of a CDW cross-regional synergistic management mechanism between the G1 government and the G2 government. This finding is contrary to that of Wang [66] et al. The reason for this difference may be due to the different research contexts between Wang et al. and this paper. Wang et al. found that upstream governments are not sensitive to ecological compensation paid by downstream governments in watershed ecological compensation, while the research context of this paper is based on ecological compensation for CDW cross-regional disposal.

### 5.4. Effect of Ecological Benefits $M_1$ on Game Equilibrium

Figure 5 shows the effects of ecological benefits $M_1$ on the evolutionary process and end state of the green development behavior of the G1 government and the G2 government, respectively.

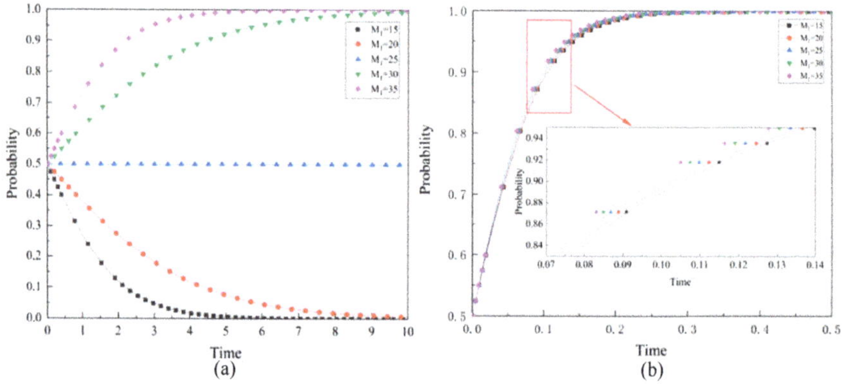


**Figure 5.** Impact of ecological benefits $M_1$ on the evolutionary strategy of government green development behavior. Note: Figure (**a**) shows the G1 government, and figure (**b**) shows the G2 government. 15, 20, 25, 30, and 35 represent the least, less, normal, more, and most ecological benefits $M_1$, respectively.

Figure 5a shows that the G1 government's strategy choice changes from not paying ecological compensation to paying ecological compensation with the increase in $M_1$. Comparing the two curves that converge to 1 in Figure 5a, the larger the value of $M_1$, the faster the convergence to a steady state where the G1 government pays ecological compensation. Therefore, the larger the value of $M_1$, the greater the G1 government tends to pay ecological compensation.

Figure 5b shows that as the ecological benefits $M_1$ increase, the ESS remains constant, but the system converges to the steady state of taking incentives at an accelerated rate. Unlike the effect of $M_1$ on the G1 government, the increase in $M_2$ only accelerates the convergence of the system without affecting the strategy choice of the G2 government. The results of this study are consistent with the findings of Lu [58] et al.

### 5.5. Effect of Ecological Benefits $M_2$ on the Game Equilibrium

Figure 6 shows the effects of ecological benefits $M_2$ on the evolutionary process and end state of the green development behavior of the G1 government and the G2 government, respectively.

Figure 6a shows that when the value of $M_2$ changes from 10 to 30, the G1 government's strategy changes from paying ecological compensation to not paying ecological compensation. When the value of $M_2$ is less than 25, to encourage the G2 government to actively promote CDW recycling to generate more ecological benefits, the G1 government will choose to pay ecological compensation. As the value of $M_2$ increases, the overall ecological benefits increase. However, this will lead to free-rider behavior of the G1 government [68], and the G1 government will stop paying ecological compensation to reduce fiscal expenditure. However, the free-rider behavior of the G1 government will undoubtedly affect the enthusiasm of the G2 government for environmental management and negatively affect the promotion of CDW resource utilization. Therefore, specific policies and regulations should be used to strengthen the regulation of the G1 government to avoid the free-rider phenomenon as much as possible.

Figure 6b shows that the G2 government's strategy is always to take incentive measures regardless of the increase in the value of $M_2$. However, as the value of $M_2$ increases, the rate of convergence of the system to the steady state accelerates. In addition, the convergence time of the G2 government is much shorter than that of the G1 government compared to the convergence time of the G1 government to the steady state at the same $M_2$ value. Therefore, the G2 government is more sensitive to $M_2$, and $M_2$ plays a positive role in promoting the G2 government to take incentive measures for CDW resource utilization.

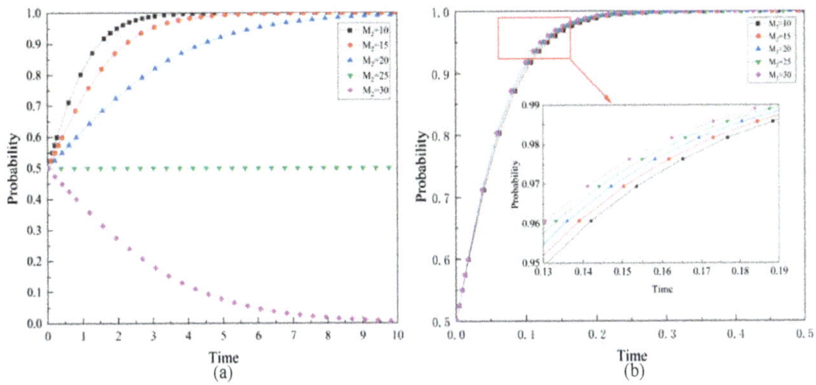


**Figure 6.** Ecological benefits $M_2$ impact of the evolutionary strategy of government green development behavior. Note: Figure (**a**) shows the G1 government, and figure (**b**) shows the G2 government. 10, 15, 20, 25, and 30 represent the least, less, normal, more, and most ecological benefits $M_2$, respectively.

## 6. Conclusions and Implications

### 6.1. Conclusions

Under the perspective of ecological compensation of CDW cross-regional disposal, this paper constructs a game model of the evolution of government green development behavior based on externality theory, which reveals the mechanism of the evolution of government green development behavior. The main findings are as follows:

(1) In the process of CDW cross-regional disposal, the initial probability of different governments adopting green development behavior does not affect the final steady state of the system. Among them, the steady state of the G1 government's green development behavior is paying ecological compensation, and the steady state of the G2 government's green development behavior is taking incentive measures to promote CDW resource utilization. However, with the same initial probability of government green development behavior, the time required for the G1 government to reach the steady state is longer than that required for the G2 government to reach the steady state. This may affect the cooperation between the G1 government and the G2 government to manage cross-regional CDW collaboratively.

(2) In the cross-regional treatment of CDW, there is heterogeneity in the influence of the allocation coefficient of ecological benefits and the ecological compensation coefficient on different government green development behaviors. On the one hand, the increase in the allocation coefficient of ecological benefits implies that the ecological benefits gained by the G1 government adopting green development behavior will be reduced, resulting in a lower willingness of the G1 government to participate in ecological compensation mechanisms. However, the larger the allocation coefficient of ecological benefits is, the greater the G2 government tends to realize CDW resource utilization through incentive measures. On the other hand, although the ecological compensation

coefficient is not conducive to the adoption of green development behavior by the G1 government, it can effectively promote the adoption of an incentive strategy by the G2 government. That is, an increase in the ecological compensation coefficient will increase the expenditure cost of the G1 government and lead to a change in the G1 government's strategic choice from paying ecological compensation to not paying ecological compensation. However, the G2 government will eventually reach the same steady state regardless of the size of the ecological compensation coefficient. In addition, the G2 government reaches a steady state in a shorter period of time compared to the G1 government. Therefore, the payment of ecological compensation by the G1 government can effectively promote the green development behavior of the G2 government to adopt incentive measures.

(3) In the process of CDW cross-regional disposal, if the ecological benefits brought by the ecological compensation system are high enough, different governments will actively participate in CDW cross-regional disposal cooperation through green development behavior. However, when only G2 government adopts green development behavior, a free-rider effect may occur, and the ecological benefits will negatively affect G1 government's green development behaviors.

### 6.2. Implications

Based on the above conclusions, the following management implications are obtained from this paper:

(1) To enhance the effectiveness of the cross-regional treatment of CDW, it is recommended that the different local governments all actively adopt green development behavior. For example, on the one hand, it is suggested that the central government strengthen policy propaganda to raise local governments' awareness of horizontal ecological compensation. In addition, the central government can also provide incentives for government green development behavior through vertical financial transfer payments and the establishment of special funds to increase the participation of local governments. On the other hand, it is suggested that local governments strengthen communication and coordination between departments and build a platform for information sharing and exchange to increase the willingness of each government to participate in ecological compensation.

(2) To allocate more ecological benefits to both governments that adopt green development behavior, it is suggested that different governments actively realize cross-regional resource utilization of CDW through cross-regional collaboration. Cross-regional collaboration by governments not only improves environmental governance capacity but also helps to produce more environmental benefits and improve regional green development. For example, eight prefecture-level cities in China, including Guangzhou, Shenzhen, and Zhuhai, have successfully achieved balanced cross-regional disposal of CDW [69]. In addition, the government's ecological compensation efforts should match its financial capacity. On the one hand, it is suggested that the beneficiary government take into account the environmental treatment cost and treatment effectiveness of the ecological reserve and provide moderate compensation to the protection government. On the other hand, to reduce the financial pressure caused by the direct payment of ecological compensation fees, the government should adopt market-oriented and diversified ecological compensation methods. For example, the government can broaden the financing channels of ecological compensation by attracting social capital to enter and purchase ecological products such as CDW recycled materials to promote the development of green industries in protected areas.

(3) It is suggested that the government actively promote CDW resource utilization through green development behavior, such as subsidizing costs, formulating policies, and promoting guidance, to improve ecological and environmental quality and generate more ecological benefits. What needs special attention is that to reduce or avoid the phenomenon of government free riders in the process of cross-regional disposal

of CDW, the central government should strengthen the incentives and constraints for local governments. For example, the central government should implement the ecological compensation responsibilities of beneficiary governments by improving the ecological compensation appraisal and evaluation mechanism and increasing rewards and punishments.

This paper takes ecological compensation as the perspective and takes into account the allocation coefficient of ecological benefits, ecological compensation coefficient, ecological benefits, and other influencing factors in the model to reveal the evolution mechanism of the government green development behavior in the ecological compensation of CDW cross-regional disposal. However, there are still some limitations. (1) Ecological compensation for CDW cross-regional disposal also involves other stakeholders, such as enterprises responsible for transporting and disposing of CDW and the public near the CDW disposal site. However, the consideration of these stakeholders is missing in this paper. (2) Different forms of ecological compensation are not explored in this paper. Therefore, researchers could construct a game model considering the government, enterprises, and the public together in the research framework, and the effects of different ecological compensation methods on the behavior of compensation subjects and objects can be considered comprehensively.

**Author Contributions:** Methodology, validation, formal analysis, investigation, resources, data curation, writing—original draft, writing—review and editing, visualization, W.C.; conceptualization, methodology, writing—original draft, supervision, project administration, X.L.; writing—review and editing, W.Y., B.Y., S.X. and H.Z. All authors have read and agreed to the published version of the manuscript.

**Funding:** This work was supported by the National Natural Science Foundation of China (grant number 72204178), the Sichuan Science and Technology Program, Natural Science Foundation of Sichuan, China (grant number 2023NSFSC1053), and the National College Students Innovation and Entrepreneurship Training Plan (grant number 202310626041).

**Institutional Review Board Statement:** Not applicable.

**Informed Consent Statement:** Not applicable.

**Data Availability Statement:** Not applicable.

**Conflicts of Interest:** The authors declare no conflict of interest.

## References

1. Islam, S.; Islam, J.; Hoque, N.M.R. Improvement of consolidation properties of clay soil using fine-grained construction and demolition waste. *Heliyon* **2022**, *8*, e11029. [CrossRef] [PubMed]
2. Wu, H.; Duan, H.; Wang, J.; Wang, T.; Wang, X. Quantification of carbon emission of construction waste by using streamlined LCA: A case study of Shenzhen, China. *J. Mater. Cycles Waste Manag.* **2015**, *17*, 637–645. [CrossRef]
3. Prakash, R.; Thenmozhi, R.; Raman, S.N.; Subramanian, C. Characterization of eco-friendly steel fiber-reinforced concrete containing waste coconut shell as coarse aggregates and fly ash as partial cement replacement. *Struct. Concr.* **2020**, *21*, 437–447. [CrossRef]
4. Ma, L.; Zhang, L. Evolutionary game analysis of construction waste recycling management in China. *Resour. Conserv. Recycl.* **2020**, *161*, 104863. [CrossRef]
5. Liu, Y.; Hao, J.; Li, C.; Li, Y.; Zhou, C.; Zheng, H.; Xu, S.; Chen, W.; Li, X. How Can Construction and Demolition Waste Recycling Public-Private Partnership Projects Performance Compensate during the Operation Period? A Two-Stage Perspective of Recycling and Remanufacturing. *Systems* **2023**, *11*, 170. [CrossRef]
6. He, L.; Yuan, H.; Wu, H. Collaborative mechanism for promoting the cross-regional management of construction and demolition waste. *J. Clean. Prod.* **2022**, *372*, 133706. [CrossRef]
7. Zheng, H.; Li, X.; Zhu, X.; Huang, Y.; Liu, Z.; Liu, Y.; Liu, J.; Li, X.; Li, Y.; Li, C. Impact of Recycler Information Sharing on Supply Chain Performance of Construction and Demolition Waste Resource Utilization. *Int. J. Environ. Res. Public Health* **2022**, *19*, 3878. [CrossRef]
8. Li, X.; Dai, J.; Zhu, X.; He, J.; Li, J.; Liu, X.; Huang, Y.; Shen, Q. What Is the Mechanism of Government Green Development Behavior Considering Multi-Agent Interaction? A Meta-Analysis. *Int. J. Environ. Res. Public Health* **2022**, *19*, 8263. [CrossRef]

9. Chen, G.; Wei, B.; Zhu, R. The impact of environmental subsidy on the performance of corporate environmental responsibility: Evidence from China. *Front. Environ. Sci.* **2022**, *10*, 972328. [CrossRef]
10. Akhtar, R.; Sultana, S.; Masud, M.M.; Jafrin, N.; Al-Mamun, A. Consumers' environmental ethics, willingness, and green consumerism between lower and higher income groups. *Resour. Conserv. Recycl.* **2021**, *168*, 105274. [CrossRef]
11. Fu, Y.; Dong, N.; Ge, Q.; Xiong, F.; Gong, C. Driving-paths of green buildings industry (GBI) from stakeholders' green behavior based on the network analysis. *J. Clean. Prod.* **2020**, *273*, 122883. [CrossRef]
12. Wu, L.; Ma, T.; Bian, Y.; Li, S.; Yi, Z. Improvement of regional environmental quality: Government environmental governance and public participation. *Sci. Total Environ.* **2020**, *717*, 137265. [CrossRef] [PubMed]
13. Promoting Green Development and Building a Beautiful China—A Series of Press Conferences on the Theme of "China in This Decade" Focus on the Achievements of Ecological Civilization in the New Era. Available online: https://www.gov.cn/xinwen/2022-09/16/content_5710157.htm (accessed on 30 May 2023).
14. Zhou, C.; He, J.; Li, Y.; Chen, W.; Zhang, Y.; Zhang, H.; Xu, S.; Li, X. Green Independent Innovation or Green Imitation Innovation? Supply Chain Decision-Making in the Operation Stage of Construction and Demolition Waste Recycling Public-Private Partnership Projects. *Systems* **2023**, *11*, 94. [CrossRef]
15. Ding, Z.; Zhu, M.; Wu, Z.; Fu, Y.; Liu, X. Combining AHP-Entropy Approach with GIS for Construction Waste Landfill Selection-A Case Study of Shenzhen. *Int. J. Environ. Res. Public Health* **2018**, *15*, 2254. [CrossRef] [PubMed]
16. Barbudo, A.; Ayuso, J.; Lozano, A.; Cabrera, M.; Lopez-Uceda, A. Recommendations for the management of construction and demolition waste in treatment plants. *Environ. Sci. Pollut. Res.* **2020**, *27*, 125–132. [CrossRef] [PubMed]
17. Han, Y.; Zheng, H.; Huang, Y.; Li, X. Considering Consumers' Green Preferences and Government Subsidies in the Decision Making of the Construction and Demolition Waste Recycling Supply Chain: A Stackelberg Game Approach. *Buildings* **2022**, *12*, 832. [CrossRef]
18. Ortiz, C.; Alvarado, R.; Mendez, P.; Flores-Chamba, J. Environmental impact of the shadow economy, globalisation, and human capital: Capturing spillovers effects using spatial panel data approach. *J. Environ. Manag.* **2022**, *308*, 114663. [CrossRef]
19. Nguyen, H.G.; Nguyen, D.T.; Nghiem, H.T.; Tran, V.C.; Kato, A.; Matsuno, A.; Isobe, Y.; Kawasaki, M.; Kawamoto, K. Current Management Condition and Waste Composition Characteristics of Construction and Demolition Waste Landfills in Hanoi of Vietnam. *Sustainability* **2021**, *13*, 10148. [CrossRef]
20. Tan, R.; Jin, H.; Yu, M.; Yang, J.; Zhang, J. Research on Construction Waste Recycling Subsidy Model Considering Contractor's Environmental Awareness. *Sustainability* **2023**, *15*, 2333. [CrossRef]
21. Yu, B.; Xu, L. Review of ecological compensation in hydropower development. *Renew. Sust. Energ. Rev.* **2016**, *55*, 729–738. [CrossRef]
22. Chen, W.Y. Environmental externalities of urban river pollution and restoration: A hedonic analysis in Guangzhou (China). *Landsc. Urban Plan.* **2017**, *157*, 170–179. [CrossRef]
23. Du, Z.; Zhu, X.; Shan, W. External Diseconomy Theory and Environmental Management: Optimal Taxation Model in China Marine Management. *J. Coast. Res.* **2020**, *109*, 174–180. [CrossRef]
24. Bithas, K. Sustainability and externalities: Is the internalization of externalities a sufficient condition for sustainability? *Ecol. Econ.* **2011**, *70*, 1703–1706. [CrossRef]
25. Eidelwein, F.; Collatto, D.C.; Rodrigues, L.H.; Lacerda, D.P.; Piran, F.S. Internalization of environmental externalities: Development of a method for elaborating the statement of economic and environmental results. *J. Clean. Prod.* **2018**, *170*, 1316–1327. [CrossRef]
26. Deng, H.; Zheng, P.; Liu, T.; Liu, X. Forest Ecosystem Services and Eco-Compensation Mechanisms in China. *Environ. Manag.* **2011**, *48*, 1079–1085. [CrossRef]
27. Luo, J.; Huang, M.; Bai, Y. Promoting green development of agriculture based on low-carbon policies and green preferences: An evolutionary game analysis. *Environ. Dev. Sustain.* **2023**, *Published online*. [CrossRef]
28. Li, X.; Li, J.; Huang, Y.; He, J.; Liu, X.; Dai, J.; Shen, Q. Construction enterprises' adoption of green development behaviors: An agent-based modeling approach. *Hum. Soc. Sci. Commun.* **2022**, *9*, 244. [CrossRef]
29. Pillay, Y.P.; Buschke, F.T. Misaligned environmental governance indicators and the mismatch between government actions and positive environmental outcomes. *Environ. Sci. Policy* **2020**, *112*, 374–380. [CrossRef]
30. Li, X.; Dai, J.; Li, J.; He, J.; Liu, X.; Huang, Y.; Shen, Q. Research on the Impact of Enterprise Green Development Behavior: A Meta-Analytic Approach. *Behav. Sci.* **2022**, *12*, 35. [CrossRef]
31. Farley, J.; Costanza, R. Payments for ecosystem services: From local to global. *Ecol. Econ.* **2010**, *69*, 2060–2068. [CrossRef]
32. Luo, W.; Bai, H.; Jing, Q.; Liu, T.; Xu, H. Urbanization-induced ecological degradation in Midwestern China: An analysis based on an improved ecological footprint model. *Resour. Conserv. Recycl.* **2018**, *137*, 113–125. [CrossRef]
33. Jiang, Y.; Guan, D.; He, X.; Yin, B.; Zhou, L.; Sun, L.; Huang, D.; Li, Z.; Zhang, Y. Quantification of the coupling relationship between ecological compensation and ecosystem services in the Yangtze River Economic Belt, China. *Land Use Policy* **2022**, *114*, 105995. [CrossRef]
34. Yi, Y.; Wei, Z.; Fu, C. A Differential Game of Transboundary Pollution Control and Ecological Compensation in a River Basin. *Complexity* **2020**, *2020*, 6750805. [CrossRef]
35. Gao, X.; Shen, J.; He, W.; Sun, F.; Zhang, Z.; Guo, W.; Zhang, X.; Kong, Y. An evolutionary game analysis of governments' decision-making behaviors and factors influencing watershed ecological compensation in China. *J. Environ. Manag.* **2019**, *251*, 109592. [CrossRef] [PubMed]

36. Lv, C.; Xu, X.; Guo, X.; Feng, J.; Yan, D. Basin water ecological compensation interval accounting based on dual perspectives of supply and consumption: Taking Qingyi River Basin as an example. *J. Clean. Prod.* **2023**, *385*, 135610. [CrossRef]
37. Wang, K.; Ou, M.; Wolde, Z. Regional Differences in Ecological Compensation for Cultivated Land Protection: An Analysis of Chengdu, Sichuan Province, China. *Int. J. Environ. Res. Public Health* **2020**, *17*, 8242. [CrossRef] [PubMed]
38. Hou, L.; Xia, F.; Chen, Q.; Huang, J.; He, Y.; Rose, N.; Rozelle, S. Grassland ecological compensation policy in China improves grassland quality and increases herders' income. *Nat. Commun.* **2021**, *12*, 4683. [CrossRef]
39. Sun, J.; Ruze, N.; Zhang, J.; Shi, J.; Shen, B. Capacity planning and optimization for integrated energy system in industrial park considering environmental externalities. *Renew. Energy* **2021**, *167*, 56–65. [CrossRef]
40. Pajewski, T.; Malak-Rawlikowska, A.; Golebiewska, B. Measuring regional diversification of environmental externalities in agriculture and the effectiveness of their reduction by EU agri-environmental programs in Poland. *J. Clean. Prod.* **2020**, *276*, 123013. [CrossRef]
41. Shao, Z.; Li, M.; Yu, D. Bibliometric analysis of construction and demolition waste recycling and utilization: Review and prospect. *Proc. Inst. Civ. Eng.-Eng. Sustain.* **2022**, *175*, 283–292. [CrossRef]
42. Sun, Y.; Gu, Z. Implementation of Construction Waste Recycling under Construction Sustainability Incentives: A Multi-Agent Stochastic Evolutionary Game Approach. *Sustainability* **2022**, *14*, 3702. [CrossRef]
43. Li, X.; He, J.; Huang, Y.; Li, J.; Liu, X.; Dai, J. Predicting the factors influencing construction enterprises' adoption of green development behaviors using artificial neural network. *Hum. Soc. Sci. Commun.* **2022**, *9*, 238. [CrossRef]
44. Li, X.; Dai, J.; Zhu, X.; Li, J.; He, J.; Huang, Y.; Liu, X.; Shen, Q. Mechanism of attitude, subjective norms, and perceived behavioral control influence the green development behavior of construction enterprises. *Hum. Soc. Sci. Commun.* **2023**, *10*, 266. [CrossRef]
45. Busch, J.; Ring, I.E.; Akullo, M.; Amarjargal, O.; Borie, M.; Cassola, R.S.; Cruz-Trinidad, A.; Droste, N.; Haryanto, J.T.; Kasymov, U.; et al. A global review of ecological fiscal transfers. *Nat. Sustain.* **2021**, *4*, 756–765. [CrossRef]
46. Qu, Q.; Tsai, S.B.; Tang, M.; Xu, C.; Dong, W. Marine Ecological Environment Management Based on Ecological Compensation Mechanisms. *Sustainability* **2016**, *8*, 1267. [CrossRef]
47. Vaissiere, A.C.; Meinard, Y. A policy framework to accommodate both the analytical and normative aspects of biodiversity in ecological compensation. *Biol. Conserv.* **2021**, *253*, 108897. [CrossRef]
48. Cai, Z.; Yang, X. Research on Restoration of Heavy Metal Contaminated Farmland Based on Restoration Ecological Compensation Mechanism. *Sustainability* **2023**, *15*, 5210. [CrossRef]
49. Zhai, T.; Wu, L.; Chen, Y.; Nazir, M.; Chang, M.; Ma, Y.; Cai, E.; Ding, G.; Zhao, C.; Li, L.; et al. Ecological Compensation in the Context of Carbon Neutrality: A Case Involving Service Production-Transmission and Distribution-Service Consumption. *Land* **2022**, *11*, 2321. [CrossRef]
50. Zhao, Y.; Wu, F.; Li, F.; Chen, X.; Xu, X.; Shao, Z. Ecological Compensation Standard of Trans-Boundary River Basin Based on Ecological Spillover Value: A Case Study for the Lancang-Mekong River Basin. *Int. J. Environ. Res. Public Health* **2021**, *18*, 1251. [CrossRef]
51. Li, M.; Lu, S.; Li, W. Stakeholders' ecological-economic compensation of river basin: A multi-stage dynamic game analysis. *Resour. Policy* **2022**, *79*, 103083. [CrossRef]
52. Jiang, K.; You, D.; Li, Z.; Shi, S. A differential game approach to dynamic optimal control strategies for watershed pollution across regional boundaries under eco-compensation criterion. *Ecol. Indic.* **2019**, *105*, 229–241. [CrossRef]
53. Wu, Z.; Yu, A.T.W.; Shen, L. Investigating the determinants of contractor's construction and demolition waste management behavior in Mainland China. *Waste Manag.* **2017**, *60*, 290–300. [CrossRef] [PubMed]
54. Liu, J.; Teng, Y. Evolution game analysis on behavioral strategies of multiple stakeholders in construction waste resource industry chain. *Environ. Sci. Pollut. Res.* **2023**, *30*, 19030–19046. [CrossRef]
55. Chu, Z.; Bian, C.; Yang, J. How can public participation improve environmental governance in China? A policy simulation approach with multi-player evolutionary game. *Environ. Impact Assess. Rev.* **2022**, *95*, 106782. [CrossRef]
56. Wang, W.; You, X.; Liu, K.; Wu, Y.; You, D. Implementation of a Multi-Agent Carbon Emission Reduction Strategy under the Chinese Dual Governance System: An Evolutionary Game Theoretical Approach. *Int. J. Environ. Res. Public Health* **2020**, *17*, 8463. [CrossRef]
57. Zheng, S.; Yu, L. The government's subsidy strategy of carbon-sink fishery based on evolutionary game. *Energy* **2022**, *254*, 124282. [CrossRef]
58. Lu, Z.; Wu, X.; Zhang, S.; Li, J.; Cai, F.; Xu, R. A study of ecological compensation in watersheds based on the three-way decisions theory. *J. Clean. Prod.* **2022**, *368*, 133166. [CrossRef]
59. Li, J.; Jiang, S. How can governance strategies be developed for marine ecological environment pollution caused by sea-using enterprises?-A study based on evolutionary game theory. *Ocean Coast. Manag.* **2023**, *232*, 106447. [CrossRef]
60. Ding, Z.; Yi, G.; Tam, V.W.Y.; Huang, T. A system dynamics-based environmental performance simulation of construction waste reduction management in China. *Waste Manag.* **2016**, *51*, 130–141. [CrossRef]
61. Wang, L.; Pan, F.; Li, Y. Evolutionary Game Strategies Analysis of Economic Development and Environmental Protection between Local Governments under Central Supervision Mechanism in China. *Sustainability* **2022**, *14*, 12498. [CrossRef]
62. Chu, Z.; Bian, C.; Yang, J. Joint prevention and control mechanism for air pollution regulations in China: A policy simulation approach with evolutionary game. *Environ. Impact Assess. Rev.* **2021**, *91*, 106668. [CrossRef]

63. Jiang, K.; Zhang, X.; Wang, Y. Stability and influencing factors when designing incentive-compatible payments for watershed services: Insights from the Xin'an River Basin, China. *Mar. Policy* **2021**, *134*, 104824. [CrossRef]
64. Chen, J.; Hua, C.; Liu, C. Considerations for better construction and demolition waste management: Identifying the decision behaviors of contractors and government departments through a game theory decision-making model. *J. Clean. Prod.* **2019**, *212*, 190–199. [CrossRef]
65. Wang, Q.; Wang, N.; Wang, H.; Xiu, Y. Study on Influencing Factors and Simulation of Watershed Ecological Compensation Based on Evolutionary Game. *Sustainability* **2022**, *14*, 3374. [CrossRef]
66. Wang, Y.; Wu, X.; Shen, J.; Chi, C.; Gao, X. Analysis on Decision-Making Changes of Multilevel Governments and Influencing Factors in Watershed Ecological Compensation. *Complexity* **2021**, *2021*, 6860754. [CrossRef]
67. Shen, J.; Gao, X.; He, W.; Sun, F.; Zhang, Z.; Kong, Y.; Wan, Z.; Zhang, X.; Li, Z.; Wang, J.; et al. Prospect theory in an evolutionary game: Construction of watershed ecological compensation system in Taihu Lake Basin. *J. Clean. Prod.* **2021**, *291*, 125929. [CrossRef]
68. Zhang, M.; Li, H.; Song, Y.; Li, C. Study on the heterogeneous government synergistic governance game of haze in China. *J. Environ. Manag.* **2019**, *248*, 109318. [CrossRef]
69. Provincial Housing and Urban-Rural Development Department Held a Symposium on Cross-Regional Balanced Disposal of Construction and Demolition Waste. Available online: http://zfcxjst.gd.gov.cn/xwzx/zxdt/content/post_3462043.html#:~:text=%E6%8E%92%E6%94%BE%E5%9C%B0%E8%A6%81%E5%88%87%E5%AE%9E%E6%8B%85%E8%B4%9F,%E7%9B%91%E7%AE%A1%E5%92%8C%E5%A6%A5%E5%96%84%E5%A4%84%E7%BD%AE%E3%80%82 (accessed on 4 July 2023).

buildings

MDPI

Article

# Integrated Platform-Based Tool to Improve Life Cycle Management and Circularity of Building Envelope Components

**Luca Morganti [1,2], Marco Demutti [3], Ioakeim Fotoglou [4,5], Eva Alessandra Coscia [3], Paolo Perillo [4] and Alessandro Pracucci [1,*]**

1 Focchi Spa, Via Cornacchiara 805, 47824 Rimini, Italy; l.morganti@focchi.it
2 Department of Architecture, University of Ferrara, Via Quartieri 8, 44121 Ferrara, Italy
3 R2M Solution Srl, Via Fratelli Cuzio 92, 27100 Pavia, Italy; marco.demutti@r2msolution.com (M.D.); eva.coscia@r2msolution.com (E.A.C.)
4 HOLONIX Srl, Corso Italia 8, 20821 Meda, Italy; ioakeim.fotoglou@holonix.it (I.F.); paolo.perillo@holonix.it (P.P.)
5 Department of Applied Informatics, University of Macedonia, Egnatias 156, 54636 Thessaloniki, Greece
* Correspondence: a.pracucci@focchi.it

**Abstract:** The construction industry impacts resource consumption and waste production and thus necessitates solutions that promote sustainable practices. In particular, it must address challenges surrounding data collection and digital management services that often hinder the recycling or reuse of materials. The purpose of this study is to develop an integrated platform-based digital tool that seamlessly incorporates circular economy principles into construction processes, with a particular focus on circular resource utilization in the building envelope components' life cycle. This tool aims to address the challenges in the construction industry related to resource consumption, waste production, and sustainability by promoting sustainable practices and improving material management. This paper reports industrial joint research to develop an integrated platform-based digital tool that seamlessly incorporates circular economy principles into construction processes, focusing on circular resource utilization in the building envelope components life cycle. By combining different platforms and services from the previous research, this study's results enhance and integrate comprehensive thinking approaches and services. These include promoting raw material substitution, reducing supply chain arrangement time, minimizing waste throughout the building component lifecycle, enhancing disassembly processes, and improving life cycle environmental assessments. The tool design was informed by user-driven narratives and requirements, and demo case testing procedures, ensuring usability and industry relevance. Furthermore, integration with product lifecycle management software and data management platforms enhanced data sharing and accuracy. This study highlights the potential of integrated tools to revolutionize supply chains and promote circular economy, transforming construction paradigms. They can contribute to the reshaping of material management towards environmental consciousness, fostering resource-efficient green building solutions, and enabling more circularity in the construction industry.

**Keywords:** supply chain management; material management; construction waste recycling; eco-design tool; data management platform; augmented reality; production management and innovation

## 1. Introduction

The construction sector extensively utilizes resources and energy during its life stages. Still, there is a lack of a comprehensive life cycle perspective when implementing building components and systems, with consequential resource and energy loss occurring in end-of-life scenarios. This sector is estimated to account for 50% of all raw material usage and 36% of global energy consumption [1,2]. In the EU, around 924 million tons of waste were generated from the disposal of buildings in 2016, more than 35% of the total. Moreover, in China, a total of 2.36 billion tons of waste were generated in 2018. This production of construction waste is anticipated to rise further in the upcoming years [3]. The recycling and recovery of building materials varies greatly across the EU, with rates ranging from less than 10% to over 90%. Despite the high recycling rates of some countries, secondary materials are rarely utilized in construction. For instance, while The Netherlands recycles more than 95% of its construction waste stream, only 3–4% of secondary materials are used in the construction sector, compared to over 50% in civil engineering [4,5]. Nonetheless, building stages offer multiple opportunities to reduce waste and promote circularity, and numerous approaches and concepts have been devised by the construction industry to facilitate the shift towards a circular economy. Nonetheless, these methods often fall short in terms of reusability and are restricted to design-centric approaches. Hence, there is a pressing requirement to establish the circularity of constructions and establish corresponding evaluation mechanisms [6]. Surely, the designing stage is crucial to allow life cycle optimization, and is capable to address for making modular and easily disassembled components, aiding manufacturers to prioritize separating materials, reducing complexity, and minimizing energy use during manufacturing [7]. Yet it is essential to consider several factors when these components and materials reach their end-of-life stage, such as overseeing demolition and renovation work to ensure the quality of materials for recycling and reuse [8,9]. The actions to be taken in this last stage consist of performing material audits before demolition, devising waste management plans, segregating high-quality materials for recycling, executing precise and selective disassembly operations, preparing materials for re-utilization and recycling, and enhancing traceability, quality assessment, and certification of construction and demolition waste streams [10].

These actions occur during different stages of the life cycle of building components and require tools to facilitate their implementation. However, the construction sector faces several barriers that must be overcome to achieve these goals and ensure successful completion:

- Poor quality data and difficulty in digitizing information about the origin of waste and recycled aggregate materials;
- Lack of digital services for creating new value chains and organizing information on raw materials;
- Uncompetitive pricing of secondary materials, and lack of trust in their quality;
- A long delay between implementing circular actions and their effect on waste management;
- A need for guidance in calculating sustainability parameters for new recycled materials [11,12].

To address these barriers, the utilization of data-driven digital tools and circularity plans have emerged as a key driver for promoting not only circularity but also a low-carbon economy and social development in the construction sector. This trend has gained significant momentum globally (with different emphases depending on regional characteristics), as it challenges the factors that impede sustainable growth and progress in this industry [13–16]. In this scientific scenario, this article will present the digital composing, methodologies, development, and results of industrial joint research aimed at developing a digital tool to improve and integrate circular economy thinking and services in the building sector, focusing on building envelope components (BECs). This research was undertaken under the framework of the CE4Con (Circular Economy for Construction Industries) project. CE4Con's objective is to develop a one-stop Integrated Platform-Based Tool (IPBT) to aid manufacturers of BECs that seek to create building envelope products tracking LCA and circularity Key Performance Indicators (KPIs) from the designing stage to the dismantling phase. Particularly, the developed digital tool aims to act in two life cycle stages of the BECs: (1) the design stage, assisting in selecting building materials for the processes and products under development, considering recycled composite, near-distance market, and low-carbon aspects; and (2) the construction and dismantling stage, supporting material auditing, product disassembly, and material pre-treatment to facilitate recycling. Finally, to validate the tool, prefabricated facade modules were chosen as real demo testers for their advantages in circular and sustainable development [17]. Indeed, the lean management and practices of that type of facade industry further contextualize CE4Con in their main research dimensions, which include data-driven intelligent decision making and activity process optimization [18].

Adopting an IPBT for circularity solutions can contribute to implementing the best practices indicated above, as opposed to the barriers that need to be overcome. The solution also provides comprehensive services to establish connections and collaborations among various construction and circular economy stakeholders.

In the following sections, this paper presents the materials and methods (Section 2) employed in this study, providing an overview of the materials used, including the cloud-based multi-sided platform for circularity data management (Section 2.1.1), the product lifecycle management software (Section 2.1.2), the Life Cycle Assessment Simulation Engine (Section 2.1.3), the augmented reality snag list and dismantling support tool (Section 2.1.4), the blockchain platform (Section 2.1.5), as well as the use cases for the demo tester (Section 2.1.6). The subsections detailing the methods encompass the explanation of the stages behind the platform's development, including the analysis and investigation (Section 2.2.1), the adoption of an end-user-centric approach (Section 2.2.2), the definition of the platform's functionalities and development (Section 2.2.3), and the final tool validation (Section 2.2.4). Building upon the materials identified and the methodology set up, this article presents the results (Section 3), commencing with the structure definition and data collection for circular entities (Section 3.1) and subsequently the definition of user stories and requirements (Section 3.2) to identify the most promising approach for the operational flowcharts and tool development (Section 3.3) and finally the overall tool validation for early stage tests (Section 3.4). The paper proceeds to offer insights into the discussion (Section 4), providing an analysis of the correspondence between the development's goals and the achieved results for the integrated platform. In the conclusions (Section 5), the article summarizes the outcomes achieved and highlights open issues that point to opportunities for further investigation.

## 2. Materials and Methods

The following sections present the materials and methods used for developing the one-stop Integrated Platform-Based Tool (IPBT).

Software and tools integrated at the starting point of the development are considered materials for this research as well as the use case demo testers chosen to validate the IPBT. On the other hand, the methodology applied focused on an end-user-centered approach for developing the tool, which is centered on eight specific usage scenarios (e.g., Scenario #3: management of the supply chain).

### 2.1. Materials

Figure 1 illustrates the software used for the IPBT implementation and typical users through the life cycle of the building components. These platforms and tools, which are considered materials of the research, include the following:

- A cloud-based multi-sided platform for circularity data management.
- A product lifecycle management software.
- A Life Cycle Assessment Simulation Engine.
- An augmented reality snag list and dismantling support tool.
- A blockchain platform.

#### 2.1.1. Cloud-Based Multi-Sided Platform for Circularity Data Management

To promote circularity in construction, the IPBT proposal relies on a data management cloud-based multi-sided platform (DMP). This platform allows users, such as designers, managers, quality departments, and procurement teams, to manage and obtain pertinent information, thanks to a user-friendly solution. The DMP used in this study is designed to handle digital polymorphic entities subject to changes throughout their life cycle (including rework, reuse, or recycling), and collects all the available environmental and circularity data about supplier options and storing data about their life cycle. Namely, in CE4Con, the employed DMP is REUse, which enables the recycling of composite materials through access to digitalized information and services [19]. REUse facilitates the entire value chain by matching demand for new materials with the availability of recycled products. Additionally, it gathers information on the latest materials and waste treatment processes to ensure the ongoing availability of recycled materials while also considering their technical properties. REUse was developed for the needs of the FiberEUse H2020 project [20], which conducted studies and campaigns for the same purpose [21].

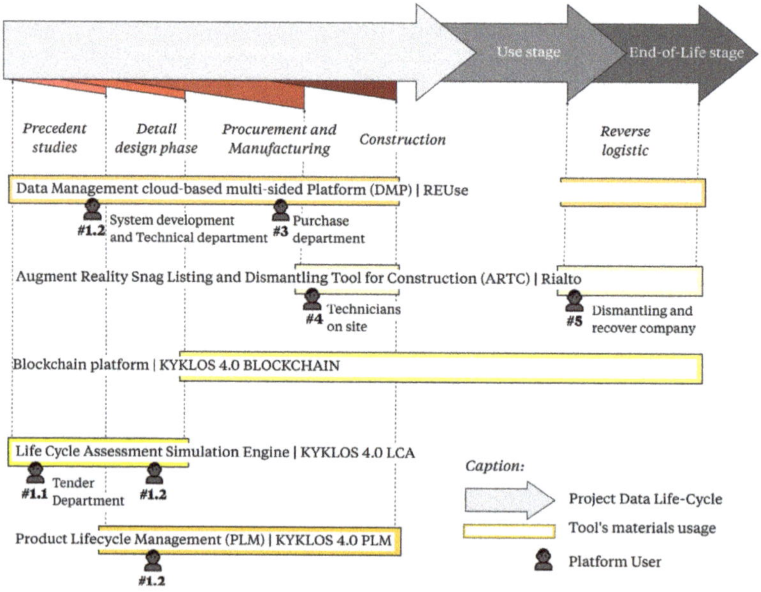


**Figure 1.** Project data cycle and materials timeline.

### 2.1.2. Product Lifecycle Management Software

In addition to the so-called environmental data related to the supply chain, which are collected in the DMP, it is essential to integrate the IBPT with computer-aided technologies used by building component manufacturing companies, particularly for retrieving project data related to building materials and components. Linking the IPBT to a product lifecycle management (PLM) is essential since it acts as a central repository, enabling seamless data sharing and coordination across different project phases and users, e.g., designers and manufacturers. Through it, the IPBT would gain access to crucial data and information about materials, processes, and stakeholders. This integration enhances the accuracy of material data, improves supply chain transparency, and streamlines decision-making [22,23]. Lastly, a PLM ensures a holistic approach to sustainable construction, enabling better-informed choices throughout the entire project life cycle [24].

Specifically, the KYKLOS 4.0 PLM was integrated into the CE4Con tool. KYKLOS 4.0 is an H2020 project whose primary goal is to establish a circular manufacturing ecosystem that generates customized goods with prolonged lifespans utilizing advanced technologies [25]. By using international standard formats for data exchange and storage, the PLM is a vital contributor to CE4Con. The PLM module uses ISO 10303 standards for data exchange, sharing, and archiving, including file exchange, APIs, and web services [26].

### 2.1.3. Life Cycle Assessment Simulation Engine

To handle the possibility that the suppliers cannot provide all the required data for an accurate environmental assessment of the analyzed products through IPBT, it was decided to incorporate an ecological impact quantification tool. This tool can perform a Life Cycle Assessment (LCA) [27] of the product and gather secondary environmental data from external databases to fill any data gaps left by the suppliers. This can help various departments, e.g., design teams and quality departments.

Linking the IPBT to a LCA tool is crucial in achieving comprehensive sustainability goals in the construction sector [28,29]. It allows to draw a comparison of different project options based not only on economic aspects, technical factors, and end-of-life possibilities but also on relative greenhouse gas emissions and other environmental impacts [30]. The tool can detect hotspots in all aspects of the supply chain, from raw materials to manufacturing, distribution, use, and disposal. The LCA tool used for this purpose in CE4Con is also part of the ecosystem of KYKLOS 4.0 [25].

#### 2.1.4. Augmented Reality Snag List and Dismantling Support Tool

Within the specific IPBT developed in this project, an Augment Reality Snag Listing and Dismantling Tool for Construction (ARTC), based on a mobile app, was integrated. A snag list is an auditing report performed by on-site technicians and used in construction projects. It includes a list of all the detected faults and defects in construction items [31]. In this way, the contribution of the tool to the platform can be two-fold. Firstly, the "snagging" function allows the user to take note of damages in on-site installed components so they can be more easily cataloged and remain visible during dismantling in the future. Through the use of a mobile application, it makes issue management more efficient and less time-consuming. Secondly, the dismantling section provides real-time information on the availability and recyclability of building materials and components.

The AR-based tool developed as ARTC used in CE4Con is Rialto, which, compared to current snag list applications, adds other advanced features to the IPBT, such as the automatic location and measurements through mobile cameras and sensors, automatic displaying information over objects, overlaying and comparing design models to as-built models and voice entry capabilities. This allows for the entering of snag list entities, quickly and rapidly. Moreover, integrating Rialto with the digital platform allows bidirectional communication between devices and workers on site and the backend systems on cloud (Figure 2).

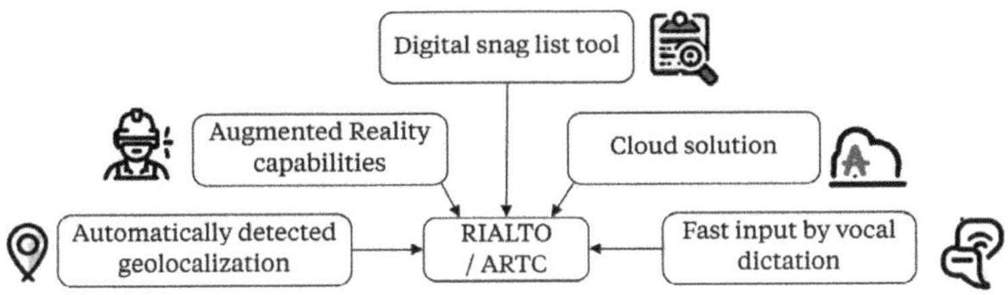


**Figure 2.** Main features diagram of Rialto.

#### 2.1.5. Blockchain Platform

Blockchain technology provides transparent, immutable, and traceable data, which makes it suitable for applications that require trust and accountability, fitting for cases such as circular economy. CE4Con can leverage blockchain to guarantee that data related to recycled and reused materials and certifications are tamper-proof and easily accessible to all stakeholders. This technology fosters trust between participants, verifies the authenticity of materials, and facilitates the development of a sustainable circular economy by providing a secure and auditable data infrastructure.

In the KYKLOS 4.0 ecosystem, the blockchain platform used in CE4CON is another essential component. It has been studied as a potential solution to the problem of securely storing designs in an immutable way that can also be verified externally.

2.1.6. The Use Case Demo Tester

While integrating these IT materials and validating the tool, a real demo tester use case was deemed necessary. In choosing the appropriate type of building envelope component (BEC) for this purpose, prefabricated and custom facade modules were selected. In the construction sector, the prefabricated facade technology brings some advantages thanks to the off-site construction in lean manufacturing, which leads to the optimization of the use of materials, enhancing the assembly and dismantling phases, and reducing waste generation, as well as saving time and money [32–34]. Prefabricated facades are complex and require detailed information on energy use, carbon footprint, and material sources. This makes it challenging to evaluate alternatives and disclose the data on the end-of-life stages. For this reason, the CE4Con solution has been validated on Prefabricated Modules for Curtain Walls (PMCW).

The PMCWs selected as use cases are as follows:

- Two facade modules designed in the H2020 Project RenoZeb to develop Plug and Play Facade integrating insulation and energy generation systems to address building renovation of existing stocks [35,36].
- One spandrel facade module with external aluminum fins designed for the 1 Broadgate building, a major fourteen-stores office-led development located in London [37].

Pictures of these modules are in Figure 3. These facade types offered a sufficiently broad scenario of the possibilities of materials, components, and processes that a typical PMCW can contain and a perspective in both the research scenario and the actual market.

(a) (b)

**Figure 3.** (**a**) Installed on-site RenoZeb facade modules; (**b**) 1 Broadgate project's external aluminum fins modules mockup.

*2.2. Methods*

The CE4Con methodological development is divided into four stages:

- Preventive analysis and investigation of required information;
- Definition through an end-user-centric approach of the requirements of the tool;
- Definition and development of the platform functionalities;
- Tool validation.

2.2.1. Preventive Analysis and Investigation

The first step in this methodology involves carefully selecting data that meet the requirements of the IPBT's integrated services for circular and ecological construction components. To achieve this, specific rules and prerequisites defined by rating systems, environmental assessment methods (such as LCA), scientific literature, and construction companies' targets are considered. This process requires a comprehensive analysis of the life cycle of BECs, gathering relevant information to accurately describe and track them, and evaluating any difficulties or practices associated with the circular economy.

#### 2.2.2. End-User-Centric Approach

The main goal of the second phase is to define the technical requirements of the IPBT. This is completed by analyzing the typical commission job actors that could affect the life cycle of a typical PMCW and who typically manage the data highlighted in the first stage. To achieve this goal, a depicted platform-based framework that connects stakeholders involved across the construction life cycle is necessary. The data gathered in phase one, together with this framework, highlight different types of interaction with the platform-based tool and describe the actors involved and their related data inputs and outputs. This leads to the definition of roles, scenarios, and system requirements that shape the architecture of CE4Con.

#### 2.2.3. Definition of the Platform's Functionalities and Development

In the third phase, the design of the software architecture of the IPBT occurs. This involves establishing a relationship between the various materials that make up the one-stop tool to ensure that all the necessary data are available for the system to work effectively. In practice, this consists of defining the relationships among the various materials that will be part of CE4Con and the flowcharts useful for planning, communicating, and developing the complex processes of the IPBT. The different components are able to obtain the data from others via their APIs, which relate to the relay of product or process information.

#### 2.2.4. Tool Validation

Lastly, functional testing, which involves systematic testing of each feature and UI evaluation, and user acceptance testing, involving feedback from end users as well as test scenarios to evaluate real-world usage, were performed to validate CE4Con functionalities.

## 3. Results

This section presents the development and related outcomes of t study, following the methodological stages.

### *3.1. Circular Entities Structure Definition and Data Collection*

REUse, the DMP integrated into CE4Con, focuses on collecting, connecting, and enriching data to associate with digital object entities. The atomic unit of the data model is the circular entity (CE) and its attributes, which were already defined in the platform and adapted for CE4Con.

CEs are defined as representations of categories of objects, or sets of objects, with similar characteristics. In fact, in REUse, the concept of CE replaces the concept of a static product entity. It is designed as a CE, i.e., an entity that can transform itself during its life, interact with processes, re-manufacturing, and logistics, and cope with non-standardized and dynamically changing parameters. From an IT standpoint, these kinds of CEs are designed to satisfy specific architectural requirements to deal with the needs of changing parameters, such as those often encountered in the circular economy. As shown in the class diagram in Figure 4, attributes can be as numerous as required. This means that the entity data model is not predefined but can be created according to user needs. The attributes describe parameters that to help define the entity. As it is possible to add many attributes, they have been designated as expandable attributes.

The class diagram reflects the concept of polymorphic data. Polymorphism of data defines thedata as extensible and able to serve different product histories through high-level object-oriented design [38]. These functionalities are helpful for the first step of the methodology, which involves identifying required data and their organization in a standardized system for the adaptation to PMCW.

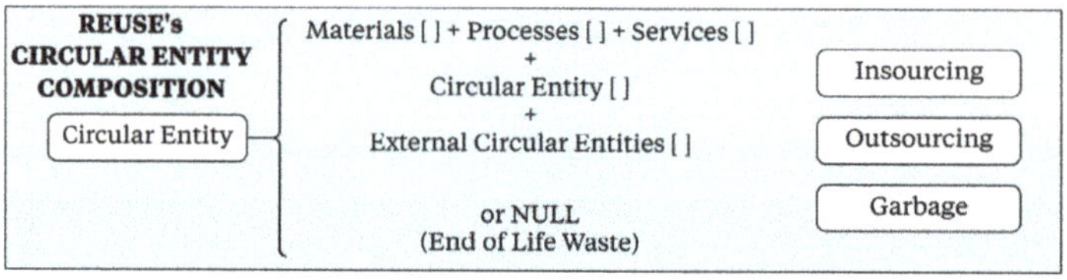


**Figure 4.** Definition of circular entity for company recursion in REUse.

Although there is a lack of specific policies regarding construction components, there are proposed solutions that could be implemented to improve circular design [39]. These solutions include performance indicators based on legal frameworks, product circularity documents, and LCA methodology [40,41]. Regarding facade-related static and dynamic data, these encompass administrative documents, such as manuals for assembly and maintenance, detailed drawings of technical systems, components, and materials, performance statistics, and connections to building ratings and certificates. Establishing the necessary criteria must consider LCA impact categories, specific rules for labeled scopes, and other certifications [42]. Moreover, when evaluating intricate building components such as PMCW, it is crucial to categorize them by their various parts and materials. This categorization should distinguish materials with different usage cycles and lifespans [43]. Therefore, the first step consists of identifying the CEs and their attributes that need to be stored in the DMP, as previously described. To that end, the ideal life cycle of a PMCW was analyzed, with attention being paid to the components and materials that could ideally compose it. There are two classifications of CEs: those related to the PMCW, called "Façade Module CE", and those associated with the individual materials or components that make up the module, called "Component Supplier CE". Their definition according to the REUse logic and their association with the real component of a typical PMCW are depicted in Figure 5.

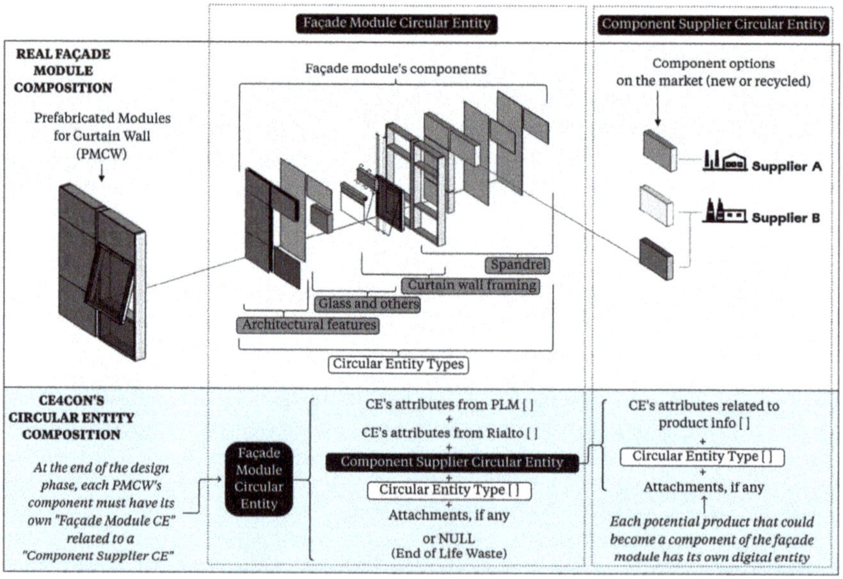


**Figure 5.** Definition of the two circular entities in CE4Con.

In addition, it is essential to identify "CE Types" capable of assigning a specific field to each CE, to facilitate the association between the "Façade Module CE" and "Component Supplier CE". These "CE Types" can be further subdivided into several "Material Types" that can encapsulate within them the categories of all the elements of the facade module (e.g., extruded profile, gaskets, glass sheet). Based on this list of keys, "Material Types", a list of attributes describing the CEs has been developed, which is able to describe CEs during their life cycle. In Figure 6, the list of "Material Type" and each of the CEs' attributes are listed, specifying which category they belong to and in what form this data will be collected and transmitted. It is important to note that each "Façade Module CE" is linked to a "Component Supplier CE" representing the material or component that characterizes it. Finally, Figure 6 specifies which information collected in the DMP can typically be retrieved from the PLM system and which can be uploaded or displayed on the ARTC.

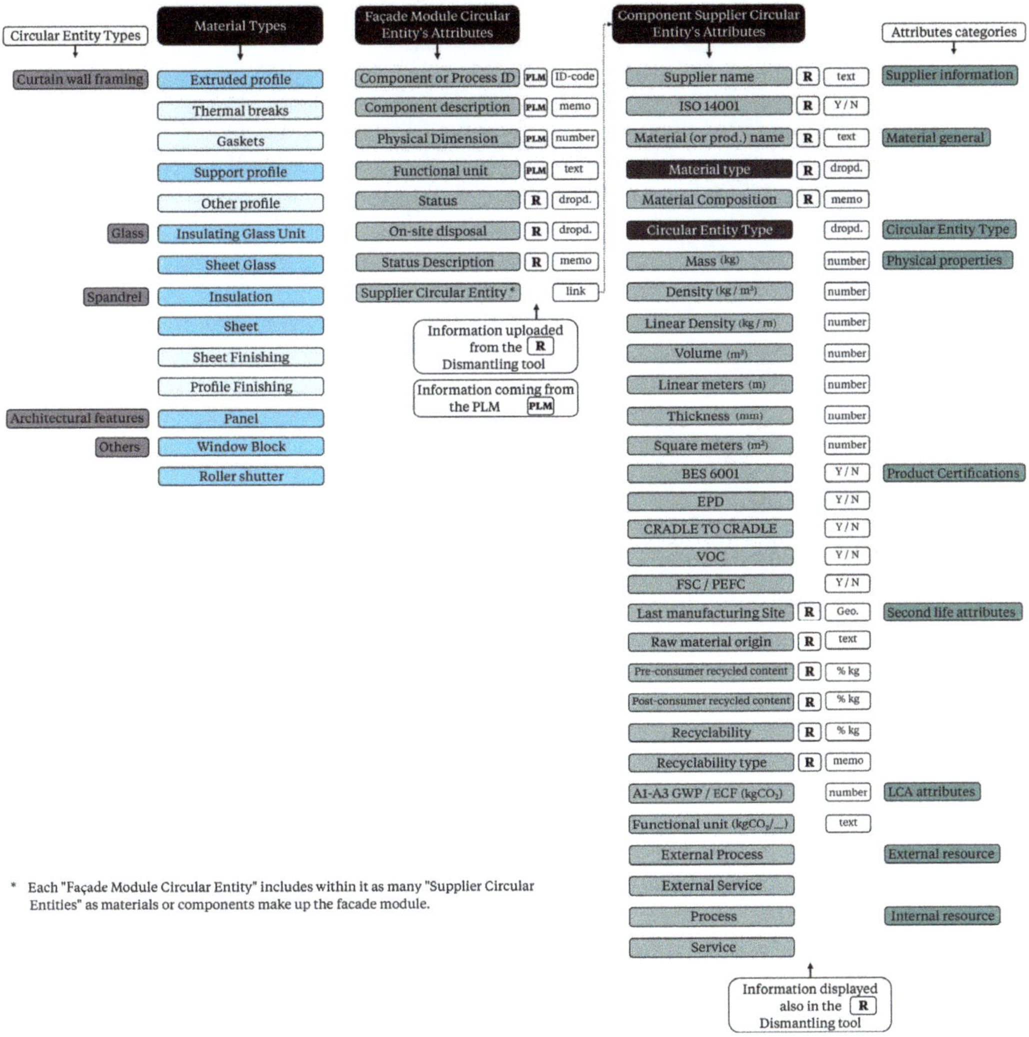


**Figure 6.** CE4Con's circular entities and attributes diagram.

Considering the components of the demo cases, appropriate CEs have been gathered in REUse, focusing on authentic supplier products that could be used for the PMCW's constitution. Additionally, the facade manufacturing company that proceeded with the demo cases conducted an internal data retrieval operation to collect the project data that would later be stored in the PLM. The collected data were then analyzed and prepared for upload to both REUse and PLM for further processing. The data collected was chosen based on industry standards and best practices found in the existing literature, ensuring it was relevant and aligned with the industry's standards.

### 3.2. *Definition of User Stories and Requirements*

To achieve the purpose of the one-stop Integrated Platform-Based Tool (IPBT), it is necessary to create a comprehensive framework that connects stakeholders across the entire construction life cycle. Based on the data highlighted by the CE's attribute selection, various types of interactions with the IPBT have been identified throughout the data life cycle.

All the steps involved in the data life cycle are illustrated in Figure 7, which represents the complete life cycle processes, the roles of the users involved, and the type of platform through which they exchange data during a commission job [44]. This includes defining the types of interaction between the platform and the user (i.e., CEs and attributes upload, CEs and attributes consultation, documentation) and the types of exchanged data (i.e., project data, use phase data, end-of-life data). A list of roles and scenarios has been identified from each interaction analyzed and then described in the figure.

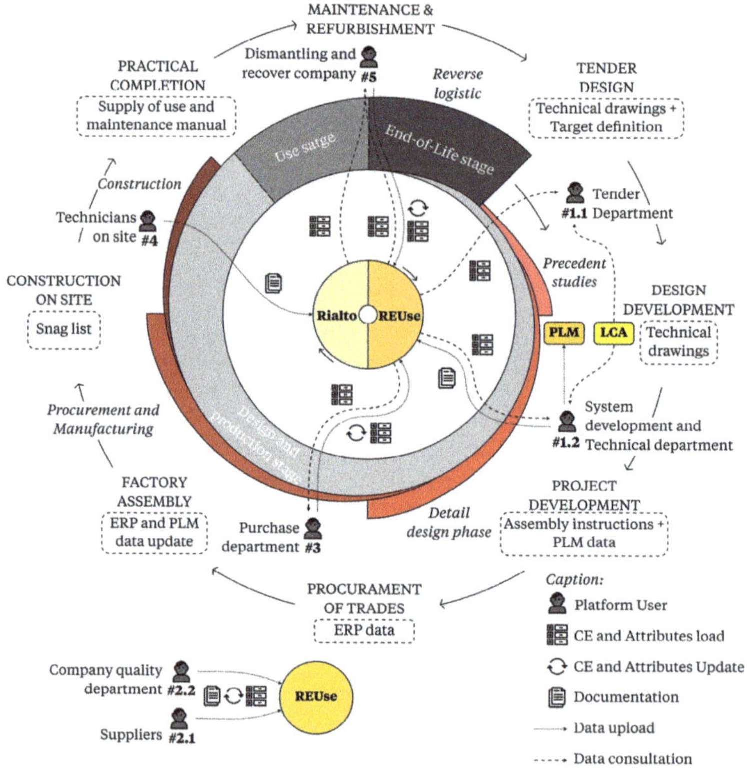


**Figure 7.** Circular entities' data life-cycle diagram; process and interactions with the platforms and users involved.

The platform caters to multiple stakeholders with distinct roles. These roles are identified based on their involvement in different phases of the construction process, such as design, procurement, installation, and end-of-life considerations. By mapping out these roles, the platform design team gained insights into the range of interactions and needs to be addressed in different scenarios. A scenario accurately predicts how actors will interact with a technical system. In Appendix A, Tables A1–A7 reporting these are presented. The tables include information about actors involved in the scenarios, their goals, starting points [45], and user stories. These stories describe a connected sequence of actions that should be taken by the operators [46]. By using an end-user-centric approach, scenarios are used to establish the requirements of the platform. These technical requirements form the foundation of the platform's abilities and features, and they connect the user stories to the actual implementation of the platform. The requirements include various aspects such as user interfaces, data storage, integration with other systems, and performance metrics. Here is a list of the eight main requirements that have resulted from the analysis.

#### 3.2.1. R.1. User Web Interface for CE and Attribute Management (REUse)

The platform necessitates an intuitive web interface administered by a backend, enabling stakeholders along the supply chain (e.g., suppliers, quality department) to update CE data and generate new entries. This interface should facilitate effortless uploading and retrieval of documentation such as Environmental Product Declarations (EPDs) for CEs.

A streamlined interface enhances user experience, facilitating data navigation and action execution. The platform acts as a collaborative hub, organizing a responsive supply chain that caters to dynamic manufacturing needs while optimizing the synergy between processing facilities and logistics services.

#### 3.2.2. R.2. Dynamic Attributes for Comprehensive CE Descriptions (REUse)

In addition to fixed attributes, the solution must empower users to provide supplementary attributes pertinent to CE recycling, reuse, or future disposal. These dynamic attributes enrich the description of CEs, conforming to international standards and physical properties essential for end-of-life considerations.

#### 3.2.3. R.3. Search Template for CE Identification (CE4Con's Frontend Tool)

The platform needs a specialized search tool that can filter and identify environmentally friendly CEs based on specific parameters, exploiting key "CE Type" and "Material Types". This search feature is vital whenever manufacturers introduce new facades to the platform, as it helps align components with CEs that meet the necessary requirements.

#### 3.2.4. R.4. Integration with LCA Component (KYKLOS 4.0)v

Environmental data from the IPBT integrated with the LCA component align with the project data to facilitate environmental assessments, enabling sustainability evaluation and circular envelope design. An ideal platform-LCA integration empowers environmental assessment for sustainable design, aligning data sources for comprehensive LCA calculations.

#### 3.2.5. R.5. Integration of a Conversion Factor (CE4Con's Frontend Tool)

The LCA data can come from different suppliers, regions, or sources, each using their functional units. These units might need to be converted to a standard unit to enable meaningful comparisons and analyses. A conversion factor allows systems to transform the data into a standard unit, ensuring that LCA results are accurate and can be compared across products.

#### 3.2.6. R.6. Integration with PLM Component (KYKLOS 4.0)

Integrating IPBT and PLM components speeds up environmental assessments, correlating project data (e.g., dimensions, profiles, glass stratigraphy) with CE attributes for efficient module assessment and end-of-life scenarios.

#### 3.2.7. R.7. ARTC Integration for Snag List Facilitator (Rialto)

The tool must have the capability to integrate the ARTC mobile app with the DMP backend, facilitating on-site examination of mounted facades. This integration should enable technicians to utilize augmented reality (AR) to identify and monitor defects.

The inclusion of AR features with the ARTC within the tool should be emphasized, as they provide a distinct advantage for defect identification and monitoring. These AR capabilities should significantly improve accuracy and efficiency compared to currently available solutions.

#### 3.2.8. R.8. ARTC Integration for Dismantling (Rialto)

ARTC integration with the DMP backend is a mandatory feature, enabling the retrieval of dismantling and pre-treatment data. This integration is crucial for optimizing disassembly processes and ensuring the accurate updating of end-of-life information for disassembled CEs. Operators in dismantling and recovery companies can benefit greatly from ARTC's integration. This integration streamlines data retrieval and ensures timely updates of CE end-of-life information.

These functional requirements collectively shape a comprehensive, user-centric platform catering to the circular economy's demands within the construction sector.

### 3.3. *Definition of the Operational Flowcharts and Tool Development*

Moving forward with the application of the methodology, it focuses on the detailed planning of CE4Con as the IPBT developed in the project. As graphically summarized in Figure 8, the main integration tool, which is taking advantage of KYKLOS 4.0 components and the other materials, is a frontend tool developed for designers and the related backend logic, which is highlighted as necessary by a previous analysis. However, the DMP is the central core of the system.

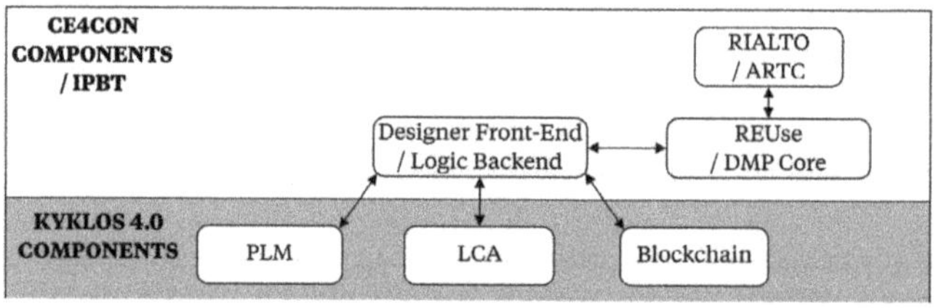


**Figure 8.** CE4Con's materials integration diagram.

#### 3.3.1. Multi-Sided Platform for Data Management Set Up: REUse

The DMP represents the core component for both major stages of use in CE4Con: (1) assisting in selecting building materials for the processes and products under development and (2) supporting end-of-life stage decisions. This is the first fundamental part of the platform's operation is the "Component Supplier CE" upload. To do that, the suppliers or the manufacturing company's quality department connect to REUse to create a new CE, and log in into the web interface of REUse where they can also modify or delete old CEs. These constitute actions available on REUse are represented in the flowchart in Figure 9. Having registered the suitable components and kept them updated, this essentially creates the base of thedata necessary for further design steps.

The objective related to the DMP core within the project was to create a sufficient number of CEs on REUse to provide diverse options for the various PMCW components within the selected demo cases. In total, 38 CEs were created, reflecting the diversity of the data required for different components and use cases. These data were adequate for comparing various options and attributes within the project's scope, ensuring that only few attributes were left empty or non-applicable.

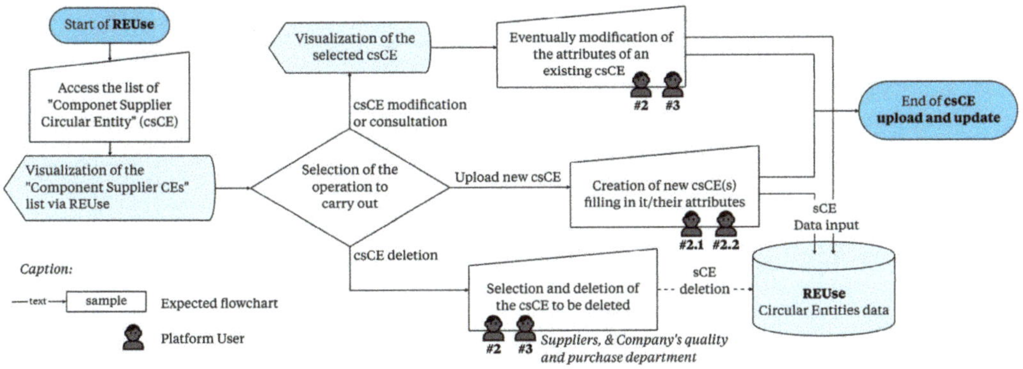


**Figure 9.** REUse flowchart.

3.3.2. Frontend Tool Setup: The "Designer Interface"

The frontend is the tool conceived to move from preliminary studies to the detailed design phase in the early stages of the product life cycle. A flowchart related to the frontend tool showing integration with REUse and Kyklos 4.0 components during the design support phase is presented in Figure 10. User icons show the scenarios involved in the different parts of the process.

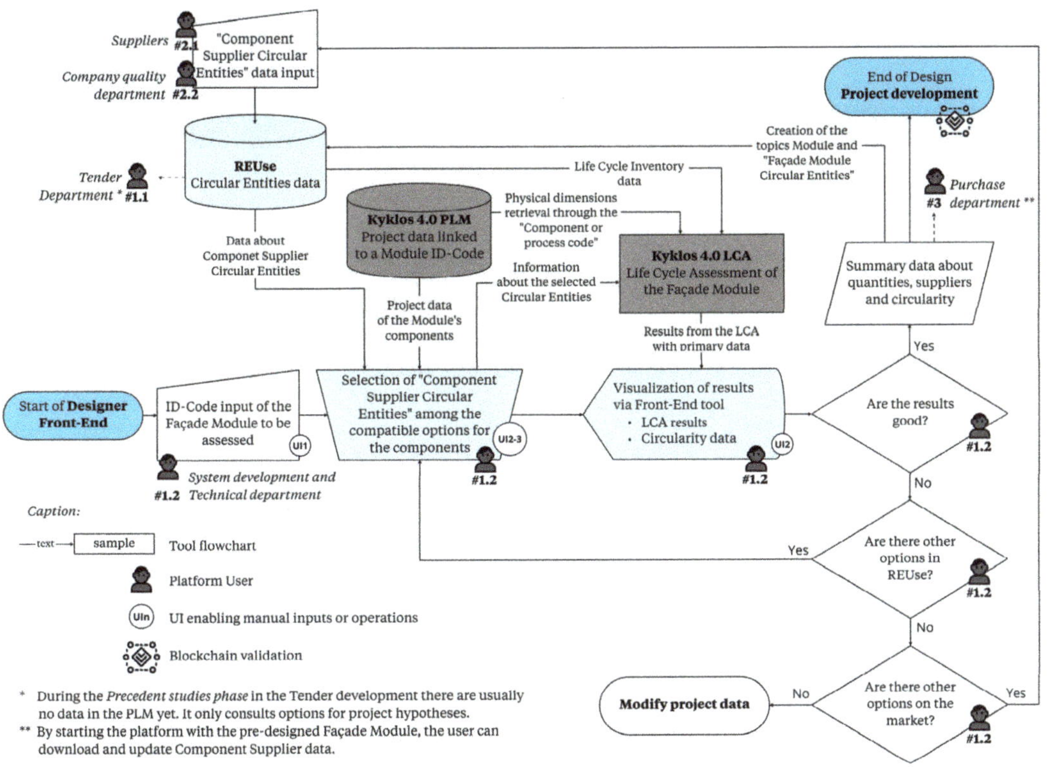


**Figure 10.** Designer frontend tool flowchart.

As seen in Section 3.3.1, the suppliers or manufacturing company's quality department, having registered the possible components, created and developed the DMP necessary for further design steps.

The next stage, which during the study has defined as the "Design stage", is where integration with other materials practically happens. By accessing the frontend tool, designers can identify suppliers for each BEC with the lowest environmental impact and better circularity characteristics (Scenario #1.2).

Initially, when accessed, User Interface 1 (UI1) prompts the designer to input an ID code of the PMCW (Figure 11). By entering this, the backend connects to the Kyklos 4.0 PLM service to load all the project data linked to the particular PMCW code, taking the designer to the UI2 screen, represented in Figure 12, and displaying all the information related to the PMCW being designed.

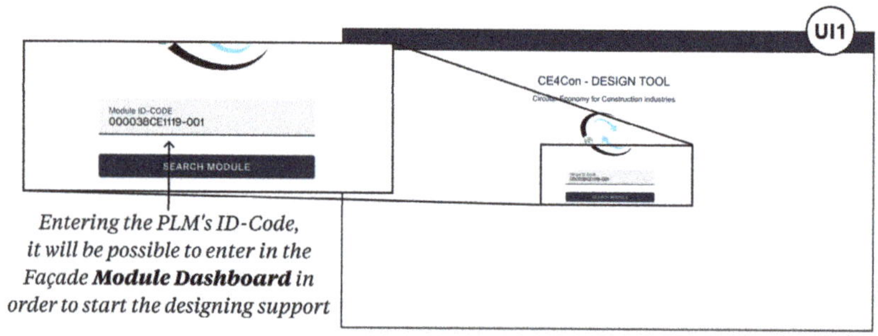


**Figure 11.** Frontend tool for design support—UI1 graphic.

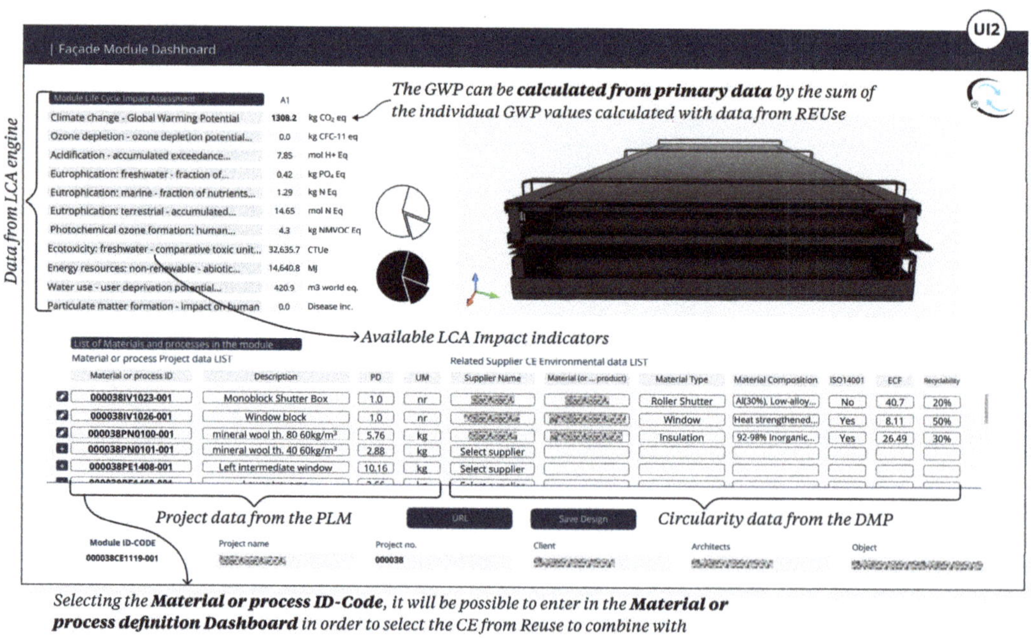


**Figure 12.** Frontend tool for design support—UI2 graphic.

The UI2 concept presents a universal interface that allows designers to compose a list of materials from the REUse components. They can select and add materials to this interface, which consolidates the information in one place.

The "List of Materials and processes" in the facade module is automatically filled in with the data from PLM following the entry of the PMCW identification code. At the beginning of the "Design stage", no CE from REUse was initially selected as an option for the various facade module components. By clicking on each empty entry in the list, the designer is taken to another section of the interface (UI3, Figure 13), which is the "Façade Material and Component definition Dashboard". Here, two of the areas displayed are linked with Kyklos 4.0, and one area is linked to REUse. The designer can select the materials in the "Material type" and "Circular Entity Type" fields and enter keys for searching for possible circular entities in REUse, as identified in requirement R.3. Lastly, if necessary, they can apply a correction to the LCA attributes by inserting a conversion factor that is multiplied with the attributes depending on their functional units, and is identified in requirement R.5.

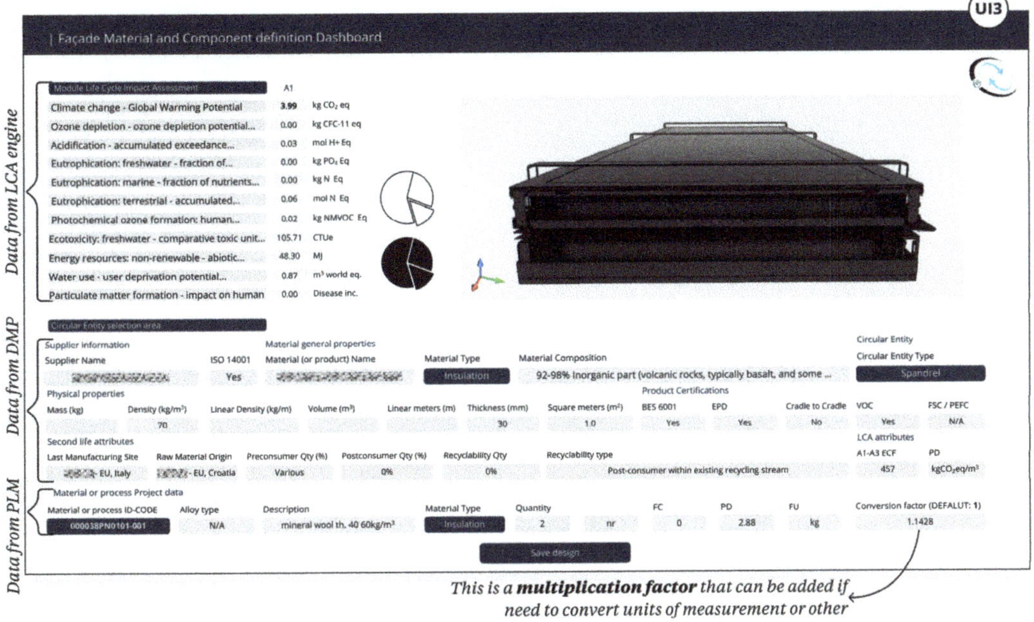


**Figure 13.** Frontend tool for design support—UI3 graphic.

By selecting these two necessary data sources from the PLM service and REUse, the backend has enough data to communicate and make a REST call to the Kyklos LCA to obtain the necessary impact assessment information. Thus, it populates the "Module Life Cycle Impact Assessment" fields.

At this point, the designer has a complete overview of all the material data, process data, impact assessment data, and circular data available to them to make a suitable choice. They can confirm or modify their design until they are satisfied with the choices that have been made, after which they click on the save option and are taken back to the UI2 screen.

Moving forward, the designer clicks on the next "Material or process ID" field and proceeds to follow the steps again until the list is a complete representation of the facade design.

When the design stage is completed, UI2 will now represent a complete picture that comprises the information from all the PMCW. At this stage, all the information the designer wants to know to finalize the design is available to them in a unified interface, and the module design is complete.

As an additional feature, the designer can save their work, which creates all the "Façade Module CEs" in REUse containing the PLM information along with a link to the chosen "Component Supplier CEs". These "Façade Module CEs" serve as the foundation for the next stages of the CE4Con process, which encompasses the construction and dismantling phases. Additionally, it is possible to make a request to the blockchain module to store these design data to ensure its immutability. This marks the end of the design phase and satisfies all the requirements of the design scenarios.

#### 3.3.3. Rialto Platform Set up and Integration

While the frontend is the CE4Con's implementation tool in the early stages of the process, the ARTC is the main tool for its implementation in the construction and end-of-life stages. The role of Rialto is to support the core functionalities of the REUse platform in the following two ways:

- By examining the mounted PMCW to obtain or update a snag list;
- By allowing the retrieval of dismantling and pre-treatment information to facilitate the disassembly and the update of end-of-life information of the PMCW.

The diagram shown in Figure 14 details the interaction of the users with Rialto, as well as its integration with the REUse DMP. For each functionality, the figure also reports the targeted scenarios.

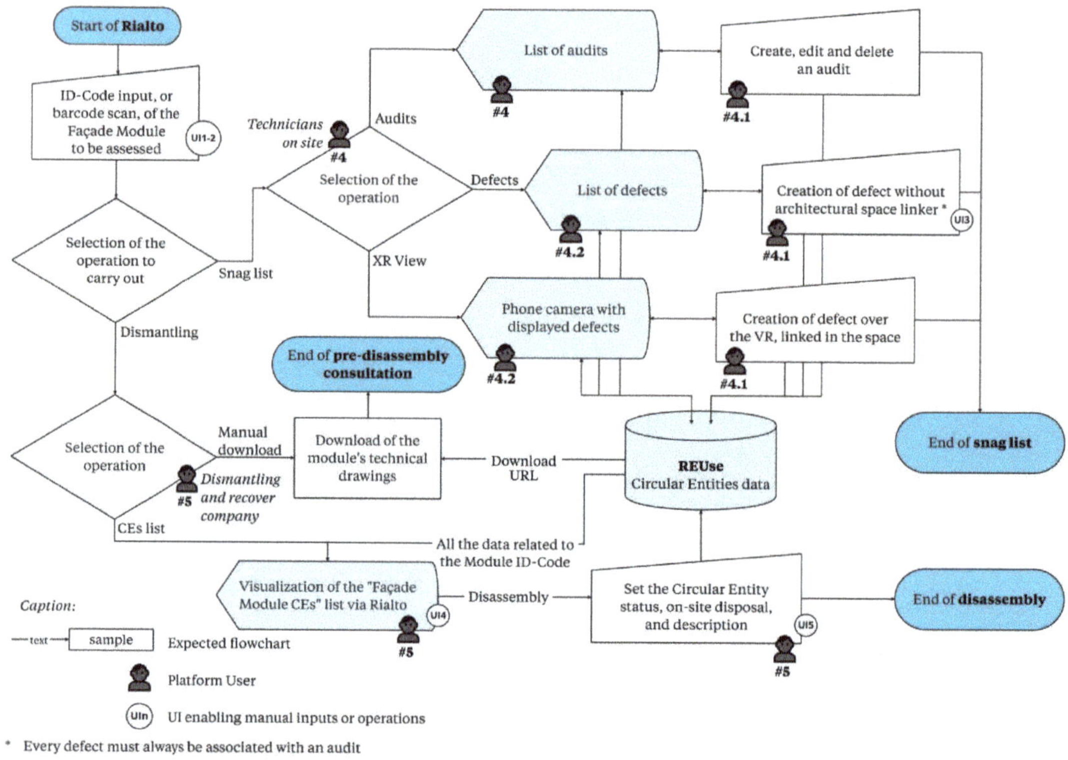


**Figure 14.** Rialto app flowchart.

On the Rialto home page, the user is asked to enter the unique ID code of the PMCW to be assessed. The app also supports barcode or QR code scans, which is useful because those who take care of the operations only have access to the facade and do not know the production codes that are used in the PLM. Once the facade module data have been identified, the user can navigate through the bottom navigation bar and choose between the snag listing of a given module (Scenarios #4.1 and #4.2: identification and monitoring of defects in the facade) or its dismantling (Scenario #5: end-of-life scenario). In the snag list section of the app, the user can see and edit existing audits, as well as create new ones. The same holds for defects: once an audit has been activated, it can be associated with one or multiple defects. The defect can be placed in the 3D space using the AR or just by providing a text description (written or converted from voice).

In the dismantling section, the user can download the module's technical drawings to guide and ease the dismantling of the module. Then, through the "CEs list", it is possible to see the list of components (i.e., circular entities) within the module. For each of them, Rialto allows the user to see the relevant properties, notify the dismantling by setting the corresponding status (dismantled, not dismantled, unknown), and provide a description of the dismantling process. Finally, textual on-site disposal describes the possible end-of-life scenario of the component.

Audits, defects, and CEs are stored both locally and in REUse through the use of REST APIs. This allows making the data available in subsequent audits as well as to other team members in real time.

Figure 15 shows some of the main screenshots of the application, namely the home page, the barcode scanning functionality, the list of circular entities in a facade module, and the dismantling of a circular entity.

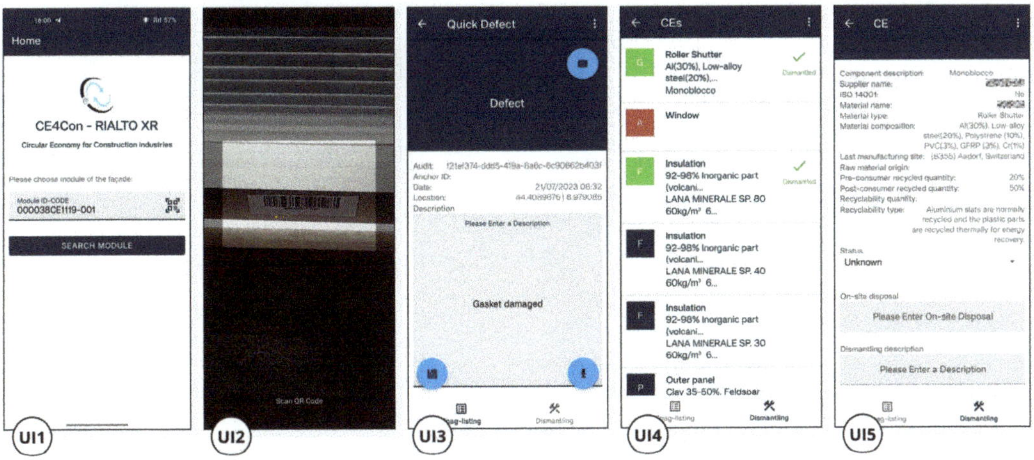


**Figure 15.** Rialto interface screenshots.

Rialto introduces three main innovations in the construction site: a virtual environment for team working, the adoption of on-site AR applications, and the bidirectional communication between on-site devices and workers and backend systems on remote servers or the Cloud. This allows for a virtual cooperation place where team members can meet from different locations and support/monitor people in the construction site. In this regard, compared to Rialto's main competitors, which are often part of ecosystems with proprietary platforms, the possibility of having bidirectional communication with any CDE through APIs allows Rialto to achieve better interoperability with a variety of applications. This allowed to easily integrate Rialto with the REUse platform.

It is claimed that digital collaboration solutions, mobile and AR devices, and advanced information management platforms play a vital role in the construction sector's digitalization [47]. AR technology merges physical surroundings with the computer-generated data, driving efficiency, accuracy, and confidence in construction projects. It reduces errors, saves manhours, and cuts costs [48].

### 3.4. Tool Validation

In order to validate the functionalities of CE4Con, the digital tool underwent both functional testing and user acceptance testing [49]. Efficiency evaluation is an ongoing process that requires periodic review as technology advances, user needs evolve, and business objectives shift. It is crucial to involve stakeholders, such as users and IT personnel, in the evaluation process to gain a comprehensive understanding of the tool's efficiency from different perspectives. This holistic approach will ensure that CE4Con remains effective in meeting its intended purposes. The validation phase has been completed up to the current state of development, which is prototyping (TRL 6–7). Recognizing the importance of third-party market evaluations is critical as they provide a real-world operational and commercial environment for the tool. These external assessments are invaluable in confirming the tool's actual functionality, mitigating cognitive biases, exploring potential business models, and delivering empirical evaluations that can significantly enhance the tool's performance and relevance in the broader market landscape. Table 1 provides an overview of the stakeholders who were involved in validating the various development phases of the tool and those expected to participate in future market testing and evaluation.

**Table 1.** Tool development process of stakeholder outline (X: actor involved in the phase; X *: the actor involved was assumed by the facade manufacturer company).

| | Prototyping (TRL 6-7) | | | On Market (TRL 8-9) |
|---|---|---|---|---|
| | **User Requirements Definition** | **Tool Development** | **Validation in a Relevant Environment** | **Customer Validation on the Market** |
| **Facade designer** ∣ Scenario #1 | X | | X * | X |
| **Supplier** ∣ Scenario #2.1 | X * | | | X |
| **Quality department** ∣ Scenario #2.2 | X | | X * | X |
| **Purchase department** ∣ Scenario #3 | X * | | | X |
| **Technician on site** ∣ Scenario #4 and #5 | X * | | X * | X |
| **Front- and backend developers** | | X | X | |
| **SW technology providers** | | X | X | |

Functional testing involves a thorough and systematic examination of each function and feature of the tool, ensuring that it performed as expected. Additionally, the user interface was evaluated to ensure that it was intuitive and user-friendly. Automated testing frameworks and tools were also utilized in order to provide comprehensive coverage and demonstrate the tool's efficiency. These tests revealed the strengths of CE4Con, as well as areas that could be further optimized. One of the notable strengths of CE4Con is its efficiency in completing tasks related to design and on-site stages. The use of "Material Types" and "Circular Entity Types" proved to be a functional approach for the agile selection of design and supply options. However, for particularly complex facade modules, assigning each element a CE can be time consuming, although still less so than traditional methods.

The error rate assessment considered both the frequency and severity of errors. Still, no high-severity errors were found during development. While a few marginal errors did occur infrequently, which were resolved before the validation phase. Finally, workflow integration was also evaluated in order to minimize disruption to existing processes. Integration between CE4Con and Kyklos 4.0 components was achieved through APIs. The adaptability of the tool to different workflows and systems was assessed, considering the data formats, protocols, and compatibility. The validation stage of the digital tool's efficiency assessment provided confirmation of its performance. Overall, the tool demonstrated a high level of efficiency in tasks completion, resources utilization, and error management.

On the other hand, the user acceptance testing (UAT) plays a vital role in evaluating user satisfaction with a given tool. This process involves obtaining feedback from end users or stakeholders to ensure that the tool meets their needs [50]. The UAT applied in the validation of CE4Con is an integral part of the adoption process and methodology, and is based on the Concerns-Based Adoption Model [51]. That is why adoption of new technologies and innovations is a complex process that involves multifaceted factors, including social, cognitive, and developmental aspects. Therefore, any attempts to facilitate technology adoption must address a broad range of concerns, from emotional to factual to contextual. The primary objective of the UAT is to obtain validation and acceptance of the CE4Con IT platform by users and pilots. Acceptability is evaluated through a survey questionnaire that covers various aspects such as the environment, setup and deployment, usability, training, handling of new business requirements, collaboration channels, and tester knowledge; the survey asked for an evaluation from 1 (strongly negative) to 5 (strongly positive). Participants reported being either satisfied or very satisfied with the functionalities they were concerned with and found the environment adequate or very suitable for development. Table 2 reports average evaluations of the UAT carried out in CE4Con.

**Table 2.** Summary of CE4Con UAT results (1: strongly negative; 5: strongly positive).

| | Environment, Setup and Deployment Satisfaction | Usability Satisfaction | Basic Training Functionality | Perception of the Possible Impact on Business |
|---|---|---|---|---|
| **Facade manufacturing company** | 4.0 | 3.7 | 5.0 | 4.5 |
| **Front- and backend developers** | 4.5 | 4.7 | 4.7 | 4.5 |
| **SW Technology providers** | 4.5 | 3.7 | 4.0 | 3.5 |

## 4. Discussion

This chapter provides a comprehensive analysis of the outcomes obtained during CE4Con, with a discussion on how the achievements have contributed to addressing the barriers that hinder circular transition in the construction sector.

### 4.1. *Quality and Digitization of Circularity Data*

The REUse platform within CE4Con provides a comprehensive and centralized repository to collect the technical data on various materials and products available on the market. Besides collecting data about leftover and dismantled materials from BECs, this can enable stakeholders to make informed decisions about raw material substitutions. Moreover, the integration of REUse with Kyklos 4.0 components ensures seamless data sharing and enhances the availability of information functional for decision making during the entire life cycle process. The project has made progress in researching working strategies to reduce the setup time needed for supply chain organization. This platform integration allows for streamlined communication and matchmaking between manufacturing companies and suppliers, promoting efficient reuse and recycling of materials, which is also enabled due to the tracking of materials and components throughout their life cycle.

Furthermore, blockchain integration introduces a layer of transparency, traceability, and immutability to the circular economy ecosystem. Besides standardizing the data formats and protocols, reinforcing accountability [52], and elevating the level of certification and transparency of materials and components within the construction industry, this integration is expected to enhance the tracking of reusable materials according to predefined scenarios. However, the construction industry appears unwilling to adopt blockchain technologies. Many stakeholders lack the necessary knowledge and expertise to implement and use blockchain effectively in their operations. Education and training are essential but time-consuming processes.

### *4.2. Digital Services for Circular Management Support*

The CE4Con interface for building materials suppliers and manufacturing companies envisions a future where construction waste and second-hand materials can be reintroduced into their businesses, further optimizing resource utilization. The projected achievement of a reduction in setup time signifies the potential of the platform to revolutionize supply chain processes. In addition, a streamlined interface enhances user experience, facilitating data navigation and action execution. The platform acts as a collaborative hub, organizing a responsive supply chain that caters to dynamic manufacturing needs while optimizing the synergy between processing facilities and logistics services. For all those reasons, it helps to overcome the barrier of limited existing digital services for creating new value chains and organizing information on raw materials.

As anticipated, the design phase is the most crucial step to reduce the environmental impact of building envelope components. In this sense, CE4Con's design support interface was developed, and became a vital point of the value proposition.

### *4.3. Secondary Materials Warranty and Cost Analysis Considerations*

CE4Con can reduce waste by matching supplier production waste with potential buyers or recycling facilities. However, businesses' lack of confidence in products from reverse logistics may be a detriment to their use. To address this lack of trust in secondary material quality, CE4Con has implemented some valuable features. First, the transparent and traceable information system can contribute to this goal. Blockchain technology ensures accountability, maintaining detailed records and documentation for each batch of component suppliers. Moreover, the inclusion of reputable and established suppliers of recycled materials among the platform users is crucial. Suppliers with specific certifications and a track record of providing high-quality materials are more likely to instill trust. However, while CE4Con offers the conditions for using virtuous materials, it is necessary to acknowledge that the ultimate decision lies with the designer using the platform.

CE4Con faced challenges when it came to integrating economic evaluation aspect within its functions. Gathering the precise economic data from the circular entity providers proved to be difficult, which made it challenging to incorporate a feature that would allow for a direct component cost comparison. Unlike the certified data on environmental impact or product composition, component costs can vary and are dependent on numerous factors. This makes it inaccurate to rely on stored prices in enterprise resource planning software, as previous purchase costs may have differed, and component suppliers are often hesitant to publish their prices, which would also have to be continually updated on REUse. As of this point, it has not been possible to integrate effective economic evaluation features while considering the materials embedded in CE4Con. However, as the time required for material identification and organizing reverse logistics can be reduced, significant cost reductions for recycling and new product manufacturing can be achieved.

### *4.4. Accelerating Circular Actions Implementation and Waste Management*

In addition to the design phase, the second stage of the BEC's life cycle in which CE4Con focuses is the disassembly and disposal part. In this stage, the challenge of long delays between implementing circular actions is addressed with Rialto functions, which offer a significant contribution. Rialto integrates Kyklos 4.0 PLM and REUse to collect comprehensive material and component information for PMCWs, which is essential for monitoring, reporting, and efficiently dismantling operations. In this way, CE4Con can enable a real-time data update system for tracking circular initiatives. This allows stakeholders to input and access data as soon as changes occur, documenting and tracking recycling or reusing materials. To expedite the process further, CE4Con can integrate with existing waste management systems used by construction companies and municipalities, thanks to standardized languages. This integration would allow for efficient communication between circular actions taken on the platform and waste collection and processing activities.

Additionally, the implementation of a snag listing support function in Rialto enhances the ability to catalog damages and defects during the facade's lifecycle which aids in keeping a log of performed maintenance operations up to date. The project aims to reduce the dismantling time, and Rialto can play a crucial role in achieving this goal by expediting the retrieval of essential information, thereby enhancing disassembly processes and contributing to project timelines.

### 4.5. Supporting Guidance about Sustainable and Circular Parameters

The CE4Con project aims to promote holistic sustainability evaluation in the construction industry by integrating the methodology for environmental assessment and encouraging the use of primary data, which significantly improves the accuracy of analysis [53]. To achieve this, the platform combines the primary project data from the Kyklos PLM with primary environmental data from suppliers' information uploaded on REUse. The goal is to bring the adoption of LCA practices to 100% for all products analyzed in the platform, while also using secondary data provided by the Kyklos LCA when the primary data are unavailable. This achievement signifies the contribution of CE4Con-like tools to promoting sustainability in construction.

By using CE4Con, stakeholders can define Key Performance Indicators (KPIs) to assess circularity based on their scope and targets. These KPIs could include metrics such as waste reduction, supply chain setup time, successful material matches, and increased components reuse. By quantifying these outcomes, CE4Con users can demonstrate their effectiveness in promoting circular economy practices, track progress toward sustainability goals, and make data-driven decisions for future projects. Likewise, Rialto functionalities enable users to set KPIs related to waste reduction on site and receive regular updates on their achievement (e.g., percentage of materials sent to reverse logistics instead of end-of-life processes). This transparency can motivate stakeholders to take more immediate action. However, it is important to note that sustainability standards and best practices evolve over time. Therefore, CE4Con should commit to keeping its standards up to date with the latest developments in sustainability assessment methodologies.

The CE4Con solution presents a robust framework for promoting circular economy practices in the construction sector. The achievements of the project highlight the potential of the IPBT to reshape traditional construction processes, optimize resource utilization, and foster sustainable building practices for a more environmentally conscious industry.

## 5. Conclusions

In response to the need for sustainable practices within the construction industry, this study has developed an Integrated Platform-Based Tool (IPBT) with a high-potential solution for the design, construction, and disassembly management and innovation of BECs. It combines a Data Management Platform (DMP), an Augmented Reality Snag Listing and Dismantling Tool (ARTC), and integrates the LCA simulation engine, PLM software, and blockchain technology. This IPBT, named CE4Con, explored a wide range of services, including environmental and circularity assessments, supply chain optimization, and circular material management. CE4Con's achievement of information integration and interoperability empowers stakeholders with accessible and trustworthy data, driving efficiency and collaboration within the construction industries across different phases of the BECs' life cycle. User stories played a key role in shaping the tool's functionalities. They were taken into consideration, ensuring a user-centric design approach focused on industrial and label needs. The platform's interface was carefully customized for different user categories, optimizing usability and ensuring the integration of investigated requirements. Demo cases provided evidence of the platform's effectiveness and successful integration with its components. The data collections in DMP were executed with proficiency, resulting in a robust and diverse dataset. The integration of its primary environmental data into the Kyklos LCA through the DMP database significantly enhances the accuracy and reliability of environmental assessments. However, further developments and accurate economic data from suppliers would be necessary for implementing cost analysis considerations of potential materials and components. Active engagement of manufacturing suppliers can enable robust economic assessments and a clearer understanding of the cost implications of recycling and using recycled materials. As these platforms evolve and more suppliers join them, the potential for achieving economic benefits through enhanced cost comparisons becomes increasingly viable.

The project outlines other potential avenues for future advancements. An area of notable potential lies in the possible integration of IPBTs with Digital Product Passports (DPPs) and Digital Building logbook frameworks. By collecting and organizing data regarding products' end-of-life dispositions, they can contribute to standardized DPPs, which streamline future disassembly processes and reverse logistics [54]. Furthermore, the establishment of Digital Building logbooks offers comprehensive and accessible repositories of building information from inception to completion. Acknowledged by the European Commission, these logbooks hold promise in promoting sustainability. To maximize their adoption, a systematic and enhanced approach to data capture, processing, exchange, and storage is essential [55].

A further and influential possible development is the integration of the ecosystem logic within the supply chain, marking a pivotal development in fostering virtuous networks in the construction and deconstruction industry. IPBTs have the potential to connect waste providers, producers, constructors, manufacturers, designers, and various stakeholders with waste recyclers and reverse logistics providers. This incentivizes sustainable practices and drives demand for eco-friendly materials. IPBTs have become a catalyst for market transformation, encouraging innovation in material design and waste management. This aligns with the global shift towards sustainability and circular economy practices and allows for the exploration of diverse alternatives [56].

In conclusion, CE4Con serves as an example of how integrated solutions can bring a transformation in traditional construction practices. By fostering resource efficiency and sustainability, these solutions have the capacity to make a significant impact on the construction industry's environmental impact and waste output. By adopting a circular economic outlook and a resolute commitment to responsible material management, the platform offers a direction towards a sustainable future for the construction industry.

**Author Contributions:** Conceptualization, L.M., E.A.C. and A.P.; methodology, L.M., M.D. and A.P.; software, M.D., I.F. and P.P.; validation, L.M. and A.P.; formal analysis, L.M., M.D. and I.F.; investigation, L.M. and M.D.; resources, E.A.C., P.P. and A.P.; data curation, L.M.; writing—original draft preparation, L.M. and M.D.; writing—review and editing, L.M., M.D., I.F., E.A.C., P.P. and A.P.; visualization, L.M.; supervision, E.A.C., P.P. and A.P.; project administration, E.A.C.; funding acquisition, E.A.C., P.P. and A.P. All authors have read and agreed to the published version of the manuscript.

**Funding:** This work was supported by the European Project H2020 "KYKLOS 4.0" under grant agreement no. 872570.

**Data Availability Statement:** The data presented in this study are available upon request from the corresponding author, with the permission of third parties. The data are not publicly available due to the third parties' privacy regulations.

**Acknowledgments:** The results in this study as well as this study are part of the results obtained in the CE4Con project: "Circular Economy for Construction Industries" (2023). This information only reflects the author's views and neither the Agency nor the Commission are responsible for any use that may be made of the information contained herein.

**Conflicts of Interest:** The authors declare that they have no financial interest or personal relationships that could have influenced the work presented in this article.

## Appendix A

This section summarizes scenarios by describing the types, goals, and operations of different CE4Con users.

Table A1. Scenario #1 summary table.

| Scenario #1: Curtain Wall Module Design (Attribution of Circular Entities) | |
|---|---|
| **Purpose**<br>Identify and select Component Supplier Circular Entities (CEs) for the facade module materials and components designed. | |
| **User**<br>Unit system developers, tender, and system developer department. | **Equipment**<br>Laptop (CE4Con's frontend tool). |
| **Input**<br>Project data regarding unit components (from PLM). Those have to match with existing Component Supplier CEs. The system department provides the "Disassembly instruction" document useful for end-of-life information. | **Output**<br>Facade module CEs. |
| **Starting Point**<br>A new facade module from the manufacturer has to be added to the platform through a frontend tool in order to evaluate and select Component Supplier CEs depending on their environmental impact and circularity value. | |
| **CE4CON solution**<br>With a frontend tool, the platform will permit searching for the Component Supplier CEs that satisfy the parameters required and to choose the feasible option with the lowest CO2eq impact and with the highest recycled content. This can encourage adoption of more virtuous materials in the design of new products. | |
| **Scenario**<br>1. User#1 identifies the facade module that they would like to assess;<br>2. User#1 connects to the platform through the frontend as a manufacturer;<br>3. User#1 adds the facade module and its components to the platform by entering a module identification code;<br>4. User#1 verifies the requirements that each material and component, and the whole module has to satisfy;<br>5. User#1 selects the components of the unit that has to be linked with a Component Supplier CE;<br>6. User#1 selects the search template available for the selected component;<br>7. User#1 chooses one of the available Component Supplier CEs that fulfilled the requirement;<br>8. Repeat 5 to 7 until all the components of the unit have been linked with a Component Supplier CE;<br>9. User#1 considers the results of the whole module assessment and eventually can modify some Component Supplier CEs. | |

Table A2. Scenario #2.1 summary table.

| Scenario #2.1: Addition of Component Supplier Circular Entities in the Database (the Supplier) | |
|---|---|
| **Purpose**<br>Provide new Component Supplier Circular Entities (CEs) available for the market, or update ones already in the platform. Eventually, upload documentation (e.g., EPD, etc.). | |
| **User**<br>Materials, components, or treatments from the supplier. | **Equipment**<br>Laptop (REUse). |
| **Input**<br>Component Supplier CEs and their attributes of their new products, update of their Component Supplier CEs. | **Output**<br>Component Supplier CEs information. |
| **Starting Point**<br>New materials, components, or treatments from the supplier have to be added to the platform as CE available for the customers. | |
| **CE4CON solution**<br>Upload on REUse information about the products in order to make it easier for the clients to access the needed data. This can help in organizing new supply chain, reducing setup time. Moreover, REUse services could be used by suppliers to optimize recycling of production waste. | |
| **Scenario**<br>1. User#2.1 has a new CE (or some CEs) to offer to the market that has finished its tests;<br>2. User#2.1 connects to the platform through REUse as a supplier (login, etc.);<br>3. User#2.1 uploads (or updated) the CEs and their attributes;<br>4. Eventually, User#2.1 uploads the documentation regarding the CE. | |

**Table A3.** Scenario #2.2 summary table.

| **Scenario #2.2: Addition of Component Supplier Circular Entities in the Database (Quality Department)** | |
|---|---|
| **Purpose**<br>Provide new Component Supplier Circular Entities (CEs) available for the market, or update ones already in the platform. Eventually, upload documentation (e.g., EPD, etc.). | |
| **User**<br>Quality department. | **Equipment**<br>Laptop (REUse). |
| **Input**<br>Component Supplier CEs and their attributes of their new products, update of their Component Supplier CEs, and eventually update the attributes of the facade module CEs. | **Output**<br>Old Component Supplier CEs. |
| **Starting Point**<br>New element materials, components, or treatment singled out from the quality department have to be added to the platform as the available Component Supplier CE for the manufacturing company. | |
| **CE4CON solution**<br>Upload on REUse information about the products in order to make it easier for the clients to access the needed data. This can help in organizing new supply chain, reducing setup time. Moreover, REUse services could be used by suppliers to optimize recycling of production waste. | |
| **Scenario**<br>1. User#2.2 has a new CE (or some CEs) to upload on the platform;<br>2. User#2.2 connects to the platform through REUse as a quality specialist (login, etc.);<br>3. User#2.2 uploads (or updates) the CEs and their attributes;<br>4. Eventually, User#2.2 uploads the documentation regarding the CE. | |

**Table A4.** Scenario #3 summary table.

| **Scenario #3: Management of the Effective Supply Chain** | |
|---|---|
| **Purpose**<br>Support the procurement of facade modules' components. | |
| **User**<br>Purchase department. | **Equipment**<br>Laptop (CE4Con's frontend tool). |
| **Input**<br>Update the attributes of the facade module circular entities (CEs). | **Output**<br>Component Supplier CEs selected during the design development for the units. |
| **Starting Point**<br>The materials and components that will be part of the assembled module have to be purchased and supplied. Eventually, the Component Supplier CEs' attributes selected during the design have to be updated. | |
| **CE4CON solution**<br>The purchase department has the ability to preview which suppliers are able to meet project requirements, including from a sustainability standpoint, that have already been selected by system designers. This speeds up and facilitates the process of retrieving materials and components. | |
| **Scenario**<br>1. User#3 identifies the elements' requirements of the unit;<br>2. User#3 connects to the platform through the frontend as a purchasing department (login, etc.);<br>3. User#3 verifies that the requirements that each component and the whole module must satisfy are still satisfied by the selected Component Supplier CEs;<br>4. Check the available suppliers. | |

**Table A5.** Scenario #4.1 summary table.

<table>
<tr><th colspan="2">Scenario #4.1: Snag List: Identification of Defects in the Installed Facade</th></tr>
<tr><td colspan="2">Purpose<br>Performing the examination of the installed facade modules to obtain a snag list (faults and defects description).</td></tr>
<tr><td>User<br>Technician on site.</td><td>Equipment<br>Android mobile device (Rialto).</td></tr>
<tr><td>Input<br>Mounted facade.</td><td>Output<br>Snag list on the facade thanks to XR.</td></tr>
<tr><td colspan="2">Starting Point<br>User#4.1 arrives and needs to add a defect after auditing.</td></tr>
<tr><td colspan="2">CE4CON solution<br>Through Rialto, the possibility is offered to group all building snag lists related to the facade in one place. These indications will then also be visible during disassembly because they are saved in REUse.</td></tr>
<tr><td colspan="2">Scenario<br>1. User#4.1 identifies the facade;<br>2. User#4.1 starts auditing the installed facade<br>3. User#4.1 identify a snag in the facade;<br>4. User#4.1 uses Rialto on their mobile phone to record the defect, possibly attaching images and descriptions, and describing the issue;<br>5. Rialto gets the location in which User#4.1 wants to put a tag (anchor);<br>6. Information is stored in the Rialto Cloud backend system.</td></tr>
</table>

**Table A6.** Scenario #4.2 summary table.

<table>
<tr><th colspan="2">Scenario #4.2: Snag List: Monitoring of Defects in the Installed Facade</th></tr>
<tr><td colspan="2">Purpose<br>Performing the examination of defects in the installed facade module that were detected in previous audits.</td></tr>
<tr><td>User<br>Technician on site.</td><td>Equipment<br>Android mobile device (Rialto).</td></tr>
<tr><td>Input<br>Mounted facade, AR snag list.</td><td>Output<br>Snag list on the facade thanks to XR.</td></tr>
<tr><td colspan="2">Starting Point<br>User#4.2 arrives and has to audit the facade.</td></tr>
<tr><td colspan="2">CE4CON solution<br>Through Rialto it is possible to view and retrieve both in a summary list and through XR the facade defects.</td></tr>
<tr><td colspan="2">Scenario<br>1. User#4.2 identifies the facade;<br>2. Thanks to the Cloud system, User#4.2 sees in the AR the tags where the issues were detected;<br>3. User#4.2 retrieves the description, pictures, and other information about the issues from the application;<br>4. User 4 decides if the issue is solved, eventually including some pictures about the current status and some textual information.</td></tr>
</table>

**Table A7.** Scenario #5 summary table.

<table>
<tr><th colspan="2">Scenario #5: End-of-Life Scenario</th></tr>
<tr><td colspan="2">Purpose<br>Retrieve dismantling and pre-treatment information and use it to plan the dismantling of the facade modules. Update end-of-life information on the dismantled circular entities in the platform.</td></tr>
<tr><td>User<br>Operator of the dismantling and recovery company.</td><td>Equipment<br>Android mobile device (Rialto).</td></tr>
</table>

**Table A7.** *Cont.*

| **Scenario #5: End-of-Life Scenario** | |
|---|---|
| **Input**<br>Existing facade module CEs and related component supplier CEs' attributes. | **Output**<br>End-of-life information about the CEs in the facade. |
| **Starting Point**<br>Information about the facade to dismantle and its circular entities already in the platform. | |
| **CE4CON solution**<br>Facilitate the disassembly and the update of end-of-life information of the facade, time reduction, and related reduction in cost. | |
| **Scenario**<br>1. User#5 identifies the facade;<br>2. User#5 retrieves dismantling and pre-treatment information on the facade;<br>3. User#5 starts the dismantling;<br>4. User#5 confirms dismantling and adds entities to the platform. | |

## References


1. Norouzi, M.; Chàfer, M.; Cabeza, L.F.; Jiménez, L.; Boer, D. Circular Economy in the Building and Construction Sector: A Scientific Evolution Analysis. *J. Build. Eng.* **2021**, *44*, 102704. [CrossRef]
2. Osobajo, O.A.; Oke, A.; Omotayo, T.; Obi, L.I. A Systematic Review of Circular Economy Research in the Construction Industry. *Smart Sustain. Built Environ.* **2020**, *11*, 39–64. [CrossRef]
3. Ginga, C.P.; Ongpeng, J.M.C.; Daly, M.K.M. Circular Economy on Construction and Demolition Waste: A Literature Review on Material Recovery and Production. *Materials* **2020**, *13*, 2970. [CrossRef]
4. Migliore, M.; Talamo, C.; Paganin, G. Construction and Demolition Waste. In *Strategies for Circular Economy and Cross-Sectoral Exchanges for Sustainable Building Products: Preventing and Recycling Waste*; Migliore, M., Talamo, C., Paganin, G., Eds.; Springer Tracts in Civil Engineering; Springer International Publishing: Cham, Switzerland, 2020; pp. 45–76. [CrossRef]
5. European Commission: Energy, Climate Change, Environment. Construction and Demolition Waste. Available online: https://environment.ec.europa.eu/topics/waste-and-recycling/construction-and-demolition-waste_en (accessed on 7 August 2023).
6. Charef, R.; Lu, W.; Hall, D. The Transition to the Circular Economy of the Construction Industry: Insights into Sustainable Approaches to Improve the Understanding. *J. Clean. Prod.* **2022**, *364*, 132421. [CrossRef]
7. Zeng, R.; Chini, A.; Ries, R. Innovative Design for Sustainability: Integrating Embodied Impacts and Costs during the Early Design Phase. *Eng. Constr. Archit. Manag.* **2020**, *28*, 747–764. [CrossRef]
8. Sparrevik, M.; de Boer, L.; Michelsen, O.; Skaar, C.; Knudson, H.; Fet, A.M. Circular Economy in the Construction Sector: Advancing Environmental Performance through Systemic and Holistic Thinking. *Environ. Syst. Decis.* **2021**, *41*, 392–400. [CrossRef]
9. Charef, R. Supporting Construction Stakeholders with the Circular Economy: A Trans-Scaler Framework to Understand the Holistic Approach. *Clean. Eng. Technol.* **2022**, *8*, 100454. [CrossRef]
10. Díaz-López, C.; Bonoli, A.; Martín-Morales, M.; Zamorano, M. Analysis of the Scientific Evolution of the Circular Economy Applied to Construction and Demolition Waste. *Sustainability* **2021**, *13*, 9416. [CrossRef]
11. Charef, R.; Morel, J.-C.; Rakhshan, K. Barriers to Implementing the Circular Economy in the Construction Industry: A Critical Review. *Sustainability* **2021**, *13*, 12989. [CrossRef]
12. Kirchherr, J.; Hekkert, M.; Bour, R.; Kostense-Smit, E.; Muller, J. *Breaking the Barriers to the Circular Economy*; Deloitte: London, UK, 2017.
13. Oluleye, B.I.; Chan, D.W.M.; Antwi-Afari, P.; Olawumi, T.O. Modeling the Principal Success Factors for Attaining Systemic Circularity in the Building Construction Industry: An International Survey of Circular Economy Experts. *Sustain. Prod. Consum.* **2023**, *37*, 268–283. [CrossRef]
14. Shooshtarian, S.; Hosseini, M.R.; Kocaturk, T.; Arnel, T.; Garofano, N.T. Circular Economy in the Australian AEC Industry: Investigation of Barriers and Enablers. *Build. Res. Inf.* **2023**, *51*, 56–68. [CrossRef]
15. Xing, Z.; Huang, J.; Wang, J. Unleashing the Potential: Exploring the Nexus between Low-Carbon Digital Economy and Regional Economic-Social Development in China. *J. Clean. Prod.* **2023**, *413*, 137552. [CrossRef]
16. Jemal, K.M.; Kabzhassarova, M.; Shaimkhanov, R.; Dikhanbayeva, D.; Turkyilmaz, A.; Durdyev, S.; Karaca, F. Facilitating Circular Economy Strategies Using Digital Construction Tools: Framework Development. *Sustainability* **2023**, *15*, 877. [CrossRef]
17. Machado, N.; Morioka, S.N. Contributions of Modularity to the Circular Economy: A Systematic Review of Literature. *J. Build. Eng.* **2021**, *44*, 103322. [CrossRef]
18. Du, J.; Zhang, J.; Castro-Lacouture, D.; Hu, Y. Lean Manufacturing Applications in Prefabricated Construction Projects. *Autom. Constr.* **2023**, *150*, 104790. [CrossRef]

19. European Commission: Funding & Tender Opportunities. rEUse Platform. Available online: https://ec.europa.eu/info/funding-tenders/opportunities/portal/screen/opportunities/horizon-results-platform/36709;resultId=36709;isExactMatch=false (accessed on 7 August 2023).
20. Large Scale Demonstration of New Circular Economy Value-Chains Based on the Reuse of End-of-Life Fiber Reinforced Composites | FiberEUse Project | Fact Sheet | H2020. CORDIS | European Commission. Available online: https://cordis.europa.eu/project/id/730323/it (accessed on 7 August 2023).
21. Arabsolgar, D.; Musumeci, A. FiberEUse: Large-Scale Demonstration of New Circular Economy Value Chains Based on the Reuse of End-of-Life Fiber-Reinforced Composites—A Circular It Platform to Manage Innovative Design and Circular Entities. *Proceedings* **2021**, *65*, 23. [CrossRef]
22. Avvaru, V.S.; Bruno, G.; Chiabert, P.; Traini, E. Integration of PLM, MES and ERP Systems to Optimize the Engineering, Production and Business. In *Product Lifecycle Management Enabling Smart X*; Nyffenegger, F., Ríos, J., Rivest, L., Bouras, A., Eds.; IFIP Advances in Information and Communication Technology; Springer International Publishing: Cham, Switzerland, 2020; pp. 70–82. [CrossRef]
23. Lennartsson, M.; André, S.; Elgh, F. PLM Support for Design Platforms in Industrialized House-Building. *Constr. Innov.* **2021**, *23*, 265–286. [CrossRef]
24. Cholewa, M.; Minh, L.H.B. PLM Solutions in the Process of Supporting the Implementation and Maintenance of the Circular Economy Concept in Manufacturing Companies. *Sustainability* **2021**, *13*, 10589. [CrossRef]
25. An Advanced Circular and Agile Manufacturing Ecosystem based on Rapid Reconfigurable Manufacturing Process and Individualized Consumer Preferences | KYKLOS 4.0 Project | Fact Sheet | H2020. CORDIS | European Commission. Available online: https://cordis.europa.eu/project/id/872570/it (accessed on 7 August 2023).
26. *ISO 10303-1:2021(En)*; Industrial Automation Systems and Integration—Product Data Representation and Exchange. ISO: Geneva, Switzerland, 2021. Available online: https://www.iso.org/obp/ui/#iso:std:iso:10303:-1:ed-2:v1:en (accessed on 29 August 2023).
27. Farjana, S.H.; Mahmud, M.A.P.; Huda, N. Chapter 1—Introduction to Life Cycle Assessment. In *Life Cycle Assessment for Sustainable Mining*; Farjana, S.H., Mahmud, M.A.P., Huda, N., Eds.; Elsevier: Amsterdam, The Netherlands, 2021; pp. 1–13. [CrossRef]
28. Di Maria, A.; Eyckmans, J.; Van Acker, K. 26—Use of LCA and LCC to Help Decision-Making between Downcycling versus Recycling of Construction and Demolition Waste. In *Advances in Construction and Demolition Waste Recycling*; Pacheco-Torgal, F., Ding, Y., Colangelo, F., Tuladhar, R., Koutamanis, A., Eds.; Woodhead Publishing Series in Civil and Structural Engineering; Woodhead Publishing: Sawston, UK, 2020; pp. 537–558. [CrossRef]
29. Francart, N.; Malmqvist, T. Investigation of Maintenance and Replacement of Materials in Building LCA. *IOP Conf. Ser. Earth Environ. Sci.* **2020**, *588*, 032027. [CrossRef]
30. Meex, E.; Hollberg, A.; Knapen, E.; Hildebrand, L.; Verbeeck, G. Requirements for Applying LCA-Based Environmental Impact Assessment Tools in the Early Stages of Building Design. *Build. Environ.* **2018**, *133*, 228–236. [CrossRef]
31. Sommerville, J.; Craig, N.; Bowden, S. The Standardisation of Construction Snagging. *Struct. Surv.* **2004**, *22*, 251–258. [CrossRef]
32. López-Guerrero, R.E.; Vera, S.; Carpio, M. A Quantitative and Qualitative Evaluation of the Sustainability of Industrialised Building Systems: A Bibliographic Review and Analysis of Case Studies. *Renew. Sustain. Energy Rev.* **2022**, *157*, 112034. [CrossRef]
33. Kragh, M.K.; Jakica, N. 21—Circular Economy in Facades. In *Rethinking Building Skins*; Gasparri, E., Brambilla, A., Lobaccaro, G., Goia, F., Andaloro, A., Sangiorgio, A., Eds.; Woodhead Publishing Series in Civil and Structural Engineering; Woodhead Publishing: Sawston, UK, 2022; pp. 519–539. [CrossRef]
34. Loizou, L.; Barati, K.; Shen, X.; Li, B. Quantifying Advantages of Modular Construction: Waste Generation. *Buildings* **2021**, *11*, 622. [CrossRef]
35. Torres, J.; Garay-Martinez, R.; Oregi, X.; Torrens-Galdiz, J.I.; Uriarte-Arrien, A.; Pracucci, A.; Casadei, O.; Magnani, S.; Arroyo, N.; Cea, A.M. Plug and Play Modular Façade Construction System for Renovation for Residential Buildings. *Buildings* **2021**, *11*, 419. [CrossRef]
36. Vavallo, M.; Arnesano, M.; Revel, G.M.; Mediavilla, A.; Sistiaga, A.F.; Pracucci, A.; Magnani, S.; Casadei, O. Accelerating Energy Renovation Solution for Zero Energy Buildings and Neighbourhoods—The Experience of the RenoZEB Project. *Proceedings* **2019**, *20*, 1. [CrossRef]
37. Jordan, M.; McAlpine Gears Up as British Land's Broadgate Scheme Given Green Light. Building. 2019. Available online: https://www.building.co.uk/news/5097621.article (accessed on 8 August 2023).
38. Polymorphic Information Content (PIC). *Encyclopedia of Genetics, Genomics, Proteomics and Informatics*; Rédei, G.P., Ed.; Springer: Dordrecht, The Netherlands, 2008; p. 1531. [CrossRef]
39. Cambier, C.; Galle, W.; De Temmerman, N. Research and Development Directions for Design Support Tools for Circular Building. *Buildings* **2020**, *10*, 142. [CrossRef]
40. Heinrich, M.; Lang, W. *Materials Passports—Best Practice. Innovative Solutions for a Transition to a Circular Economy in the Built Environment*; Technische Universitat Munchen, BAMB: Munich, Germany, 2019.
41. Mulhall, D.; Ayed, A.-C.; Schroeder, J.; Hansen, K.; Wautelet, T. The Product Circularity Data Sheet—A Standardized Digital Fingerprint for Circular Economy Data about Products. *Energies* **2022**, *15*, 3397. [CrossRef]
42. Rovelli, D.; Brondi, C.; Andreotti, M.; Abbate, E.; Zanforlin, M.; Ballarino, A. A Modular Tool to Support Data Management for LCA in Industry: Methodology, Application and Potentialities. *Sustainability* **2022**, *14*, 3746. [CrossRef]

43. Kedir, F.; Bucher, D.F.; Hall, D.M. *A Proposed Material Passport Ontology to Enable Circularity for Industrialized Construction*; Computing in Construction; ETH: Zürich, Switzerland, 2021; Volume 2, pp. 91–98. [CrossRef]
44. Fouche, D. What You Need to Know about Building Lifecycle Data (A Brief Primer). 2021. Available online: https://www.fohlio.com/blog/need-know-building-lifecycle-data-brief-primer (accessed on 21 August 2023).
45. Alexander, I.; Beus-Dukic, L. Scenarios. In *Discovering Requirements: How to Specify Products and Services*; Wiley: Hoboken, NJ, USA, 2009.
46. Alexander, I.F.; Maiden, N. Scenarios in System Development. In *Scenarios, Stories, Use Cases: Through the Systems Development Life-Cycle*; Wiley: Hoboken, NJ, USA, 2004.
47. Wang, X.; Dunston, P.S. Design, Strategies, and Issues towards an Augmented Reality-Based Construction Training Platform. *J. Inf. Technol. Constr. ITcon* **2007**, *12*, 363–380.
48. Kivrak, S.; Arslan, G. Using Augmented Reality to Facilitate Construction Site Activities. In *Advances in Informatics and Computing in Civil and Construction Engineering*; Mutis, I., Hartmann, T., Eds.; Springer International Publishing: Cham, Switzerland, 2019; pp. 215–221. [CrossRef]
49. Hall, G.E.; Hord, S.M. *Implementing Change: Patterns, Principles, and Potholes*, 2nd ed.; Allyn and Bacon: Boston, MA, USA, 2006.
50. Otaduy, I.; Diaz, O. User Acceptance Testing for Agile-Developed Web-Based Applications: Empowering Customers through Wikis and Mind Maps. *J. Syst. Softw.* **2017**, *133*, 212–229. [CrossRef]
51. Roach, A.T.; Kratochwill, T.R.; Frank, J.L. School-Based Consultants as Change Facilitators: Adaptation of the Concerns-Based Adoption Model (CBAM) to Support the Implementation of Research-Based Practices. *J. Educ. Psychol. Consult.* **2009**, *19*, 300. [CrossRef]
52. Secinaro, S.; Dal Mas, F.; Brescia, V.; Calandra, D. Blockchain in the Accounting, Auditing and Accountability Fields: A Bibliometric and Coding Analysis. *Account. Audit. Account. J.* **2021**, *35*, 168–203. [CrossRef]
53. Silva, F.B.; Reis, D.C.; Mack-Vergara, Y.L.; Pessoto, L.; Feng, H.; Pacca, S.A.; Lasvaux, S.; Habert, G.; John, V.M. Primary Data Priorities for the Life Cycle Inventory of Construction Products: Focus on Foreground Processes. *Int. J. Life Cycle Assess.* **2020**, *25*, 980–997. [CrossRef]
54. WBCSD: World Business Council for Sustainable Development. The EU Digital Product Passport Shapes the Future of Value Chains: What It Is and How to Prepare Now. 2023. Available online: https://www.wbcsd.org/uxdqq (accessed on 8 August 2023).
55. European Commission: Executive Agency for Small and Medium-Sized Enterprises; Volt, J.; Toth, Z.; Glicker, J.; De Groote, M.; Borragán, G.; De Regel, S.; Dourlens-Quaranta, S.; Carbonari, G. *Definition of the Digital Building Logbook: Report 1 of the Study on the Development of a European Union Framework for Buildings' Digital Logbook*; Publications Office of the European Union: Luxembourg, 2020.
56. Jusselme, T.; Rey, E.; Andersen, M. An Integrative Approach for Embodied Energy: Towards an LCA-Based Data-Driven Design Method. *Renew. Sustain. Energy Rev.* **2018**, *88*, 123–132. [CrossRef]

buildings

MDPI

Article

# Exploring the Mechanisms Influencing Users' Willingness to Pay for Green Real Estate Projects in Asia Based on Technology Acceptance Modeling Theory

**Siqin Wang [1,2], Sen Chen [3] and Ken Nah [1,2,*]**

[1] International Design School for Advanced Studies, Hongik University, Seoul 04068, Republic of Korea; wangsiqin@g.hongik.ac.kr
[2] International Design Trend Center, Hongik University, Seoul 04068, Republic of Korea
[3] Graduate Department, Xi'an Physical Education University, Xi'an 710068, China; rockychensen@126.com
* Correspondence: knah@hongik.ac.kr; Tel.: +82-010-6522-7858

**Abstract:** In contrast to the extensive research on the impact of green buildings on character, property values, energy efficiency, and environmental quality, less emphasis has been placed on understanding the factors driving users' selection of green real estate projects, an essential element for the future business sustainability of such buildings. This study investigates the connection between various user motivations, such as perceived usefulness, perceived usability, a sense of responsibility, awareness of consequences, and perceived behavioral control, and their impact on Asian users' willingness to pay for green real estate projects. These users include those who have already invested in and are considering long-term investment in such projects. An empirical methodology, integrating these motivations with a technology acceptance model, is proposed to assess their willingness to pay either through renting or buying. The study applies structural equation modeling to analyze sample data against a hypothetical model, thereby testing the hypotheses. It concludes that, in influencing green real estate projects, the following apply: (1) perceived usefulness and perceived behavioral control significantly shape Asian users' attitudes towards financial commitment; (2) perceived usability directly influences their perceived usefulness; (3) awareness of consequences acts as a mediating factor in the relationship between perceived usability and behavioral control among these users. Moreover, the study offers strategic design recommendations for future green real estate projects, informed by the diverse motivational factors of Asian users, to guide their purchasing and selection decisions.

**Keywords:** green building; green real estate; user research; technology acceptance model; structural equation model; business sustainability

## 1. Introduction

Green building, a global architectural paradigm, emphasizes environmental stewardship and resource efficiency throughout its life cycle and is gaining traction across continents due to its universal appeal and adaptability. In addition to ensuring healthy, comfortable, and energy-efficient spaces, this paradigm integrates green health concepts with sustainable development principles, resonating with international initiatives like the United Nations' Sustainable Development Goals. In its execution, green building employs a holistic and scientific approach, amalgamating an array of advanced technologies, including, but not limited to, greening strategies, natural ventilation, daylighting, energy-efficient building envelopes, solar energy exploitation, eco-friendly materials, and intelligent system management. This convergence aims to foster a symbiotic relationship between technology and the environment, as well as between human culture and architectural practice. Emphasizing a lifecycle perspective, this methodology seeks to minimize the consumption of natural resources and adverse environmental impacts during the design, construction, operational, and decommissioning phases. Concurrently, it aims to amplify the wellbeing and comfort of occupants, thereby contributing to the societal goal of sustainable development. Contrasted with conventional real estate practices, which predominantly prioritize cost efficiency and functionality, often at the expense of environmental considerations and future sustainability, green real estate emerges as a paradigm of enduring commitment to ecological stewardship and resource prudence. This sustainable approach transcends the mere adoption of energy-efficient and eco-friendly architectural designs, material procurement, and construction methodologies. It extends to a comprehensive emphasis on the sustainability of the entire building lifecycle, encompassing judicious water management, stringent air quality control, meticulous waste minimization, and the integration of environmentally benign operational practices. Green real estate, therefore, represents a holistic endeavor to harmonize human habitats with the natural environment, reflecting a profound shift towards responsible and sustainable living spaces [1]. In the context of a burgeoning global green economy, there is an escalating cognizance regarding the influence of buildings on residential, commercial, and natural ecosystems. This awareness is catalyzing a paradigm shift towards sustainable methodologies within the construction sector on a worldwide scale [2]. Globally, nations like the United States and Germany are at the forefront of adopting eco-friendly building technologies, establishing a precedent that is increasingly gaining traction on a global scale. This shift, known as the "going green" movement, is prominently featured in the global green building (GB) initiative. The GB movement advocates for an array of strategies aimed at enhancing built environments and reducing ecological footprints [3]. A notable example is the concept of net-zero energy buildings (ZEB), which is gaining attention for its potential to address critical issues like energy consumption and environmental degradation [4]. In essence, GB involves adopting healthier and more resource-efficient methodologies in construction, renovation, operation, maintenance, and demolition processes [5]. The term "green building" can denote either the physical structure (noun) or the eco-friendly building approach (gerund). Particularly, sustainable low-energy office buildings are designed to exploit their structural and physical attributes, aiming to foster superior living and working conditions with minimal reliance on primary energy sources [6].

The burgeoning interest in green building (GB) within Asia in recent years has significantly propelled the region's construction industry towards sustainable practices, contributing to the global proliferation of green building rating systems. Nations such as China, Singapore, Thailand, Japan, and South Korea have been at the forefront, vigorously advancing national research standards in this field. Their research efforts predominantly revolve around the development and delivery of GB programs, the intricacies of GB certification, energy efficiency measures, and the implementation of cutting-edge technologies [7]. Most renowned green building rating systems evaluate projects against a comprehensive set of eight categories: project management, site optimization, energy conservation, water efficiency, material selection, management of hazardous substances and emissions, storage, and the quality of the indoor environment [8]. Most renowned green building rating systems, such as the Leadership in Energy and Environmental Design (LEED) in the United States and the Building Research Establishment Environmental Assessment Method (BREEAM) in the United Kingdom, evaluate projects against a comprehensive set of categories [9]. These international standards have influenced Asian rating systems and are instrumental in setting a global benchmark for sustainable construction. Building on this momentum, these systems not only set national benchmarks but also award certifications to buildings that meet these standards. Prominent examples include China's Green Building Label (GBL), the Building Environmental Assessment Method (BEAM) in Hong Kong, Japan's Comprehensive Assessment System for Built Environment Efficiency (CASBEE), and the Building Construction Authority Green Mark Scheme in Singapore [10].

Furthermore, the evolution of green real estate projects in Asia showcases a progression towards more mature constructions, particularly those anchored in green building principles. A case in point is the SkyTerrace @ Dawson in Singapore. Launched in 2008 by the Housing and Development Board (HDB) and executed by SCDA, this initiative marked a significant step in advancing public housing design. It incorporates key green building elements such as green roofs, rainwater harvesting, and reuse methods [11]. However, the focus of research has predominantly been on the architectural journey from design to completion rather than on understanding the willingness of users to pay for green real estate designs and the factors that influence this decision-making process.

The subtopic under "GB project delivery and development" is considered a need for researchers to focus on. Green communities aim to conserve natural resources, but it is still being determined whether residents have the motivation to maintain sustainable development independently [12]. Prominent concerns within extant research investigations involve the comparative analysis of green buildings and conventional buildings concerning several critical aspects, including energy efficiency, water efficiency, indoor environmental quality, thermal comfort, health, and productivity [13]. However, there is a notable gap in research on consumer preferences for green buildings, despite numerous studies on their impact on property values, energy efficiency, and environmental quality. This gap is essential to address for advancing sustainability in the built environment. Green buildings are often assessed for their environmental impact during the design, post-construction, and occupancy phases, but these assessments typically lack empirical evidence about the occupants' perceptions and experiences [14].

The present study addresses an under-explored area of GB: the factors influencing users' choices in green real estate projects. While there has been a significant amount of research on various aspects of green building, such as its impact on architectural character, property value, energy efficiency, and environmental quality, this study focuses on the critically important yet relatively neglected topic of user motivation and willingness to pay for such projects. This study integrates several established concepts and theories in technology acceptance, motivation, and decision-making into a comprehensive framework. By incorporating the elements of perceived usefulness, perceived usability, sense of responsibility, consequence awareness, and perceived behavioral control, the study provides a valuable tool for understanding and predicting Asian user attitudes and behaviors towards green real estate projects. The study provides a solid theoretical foundation for understanding and predicting the attitudes and behaviors of Asian users of green real estate projects.

Therefore, this study aims to evaluate the relationship between user motivations—including perceived usefulness, perceived usability, sense of responsibility, awareness of consequences, and perceived behavioral control—and their influence on Asian users' willingness to invest in green real estate design. Acknowledging that investment in green real estate encompasses both renting and buying, it is not feasible to generalize about "consumers". The primary focus is on individuals in Asia who have invested or are considering investing in green real estate projects for long-term use. An empirical methodology, integrating the technology acceptance model (TAM) with structural equation modeling (SEM), is proposed to test the hypotheses. The study is divided into three sections: firstly, we construct a willingness-to-pay model for green buildings based on the TPB and develop a questionnaire from the model's assumptions. Secondly, we fit the sample data and hypothesized model using AMOS 24.0 software, thereby establishing SEM. Lastly, we discuss the impact of various user motivations.

## 2. Model Construction and Underlying Assumptions

### 2.1. Factors Influencing Users' Willingness to Pay

Critical success factors for achieving green buildings and green real estate can be classified into three categories: technological, managerial, and behavioral. Behavioral and cultural factors are also crucial to the development of green buildings [15,16]. Research endeavors pertaining to the performance of green buildings can likewise gain valuable insights through the adoption of a user-centric approach, which places emphasis on the assessment of user requirements and satisfaction levels [17]. User payment intention is an integral part of the promotion and use of green real estate projects. If consumers do not accept green real estate ideas, designs, and materials, achieving the desired "green effect" is problematic.

Over the recent decades, scholars both nationally and internationally have utilized classical theories to explore factors influencing willingness to accept green buildings. A strong correlation exists between pro-environmental attitudes and the notable "forgiveness factor" often associated with green buildings. A study focusing on green housing in Hong Kong identified five key factors impacting consumer decisions in the purchase of green buildings [18]. These include the pricing of green buildings, environmental awareness, the use of green building materials and interiors, personal green consumption habits, and income levels.

Furthermore, Berardi [19] posited that green buildings serve to address a range of societal and economic imperatives. These include educational goals, inclusivity, community cohesion, affordability, economic viability, local economic impact, indoor environmental quality, cultural appreciation, and inspirational aspects. According to Tan and Goh [20], the primary determinants shaping consumers' inclination to acquire a green residential property encompass their attitudes regarding green homes, perceived moral responsibilities, environmental considerations, perceived value, self-identity perceptions, and assessments of financial risk. Tan [21] found that households are motivated by four economic incentives: their beliefs about green homes, a healthy and sustainable environment, energy efficiency, and livability. Closer to the study in this paper, Liu et al. concluded that the key to promoting green homes lies in the attitudes, subjective knowledge, and social trust of the residents.

### 2.2. Theory of Planned Behavior

Understanding the attitudes of potential customers towards green properties is essential, as it directly influences their intention to purchase. The theory of planned behavior (TPB) suggests that an individual's behavioral intention is the most immediate determinant of their action [22]. Within this framework, attitude is identified as a key driver of behavioral intention [23]. As noted by Mohd Suki [24], the attitude towards green products significantly affects the intention to purchase them. This finding is echoed in Costa et al.'s study [25], which found a positive correlation between attitudes towards green products and the willingness to pay for them. Consequently, residents' attitudes towards green-labeled real estate are likely to positively influence their behavioral intention to use these buildings to a certain extent.

Conner and Armitage have provided both empirical and theoretical evidence for six variables within the theory of planned behavior (TPB): belief salience measures, past behaviors/habits, perceived behavioral control (PBC) and self-efficacy, moral norms, self-identity, and affective beliefs. Ajzen [26], drawing on expectation theory, developed the TPB to elucidate the general process of individual behavioral decision-making. Attitude is regarded as an individual's favorable or unfavorable assessment of a specific behavior. The adoption of green real estate is influenced by users' behavioral intentions, which are determined by three factors: attitudes towards green practices, subjective norms, and PBC. However, findings in green consumption research indicate that, due to the positive externalities of such consumption, users' attitudes and subjective norms do not significantly impact their intention to consume green real estate. This study aims to develop an extended model based on the technology acceptance model, offering a more comprehensive explanation of users' willingness to invest in green real estate projects in Asia.

### 2.3. Technology Acceptance Model

The technology acceptance model (TAM) [27,28] is a well-established framework for elucidating user acceptance of information and computer technology, and is extensively applied in interpreting the acceptance of various technological innovations. The adoption of the theory of planned behavior (TPB) in this context is motivated by its dual importance: it is a widely used psychological theory for predicting individual behavioral intentions, and it is frequently cited in green consumerism research. The TPB is based on the causal sequence of belief–attitude–intention–behavior [29].

The TAM's validity and its effective explanation of end-user system usage are well established [30]. According to the TAM, people's beliefs about a technology's usefulness and usability are key predictors of their attitude toward it [31]. This attitude then determines their intention to use the technology, which in turn influences actual usage. The model suggests that the adoption of new technology is driven by the behavioral intention to use it. This intention is shaped by a desire to use the technology, which is influenced by both its perceived usefulness and usability. External factors may also affect these perceptions. Often, the intention to use a technology or service is used as a stand-in for actual usage, especially when direct measurement of usage is challenging [32].

In the model, "perceived usefulness" is defined as the degree to which a user believes that a technology will provide them with economic benefits. Conversely, "perceived usability" pertains to the ease with which an individual can use a specific technology. In this study, "perceived usefulness" relates to a user's view of the effectiveness or utility of a green real estate project compared to conventional real estate. "Perceived usability", on the other hand, refers to the direct benefits a user perceives from owning and using a green real estate property, such as environmentally friendly features. Therefore, this study proposes substituting the "behavioral attitudes" component in the TPB model with "perceived usefulness" and "perceived usability" to assess users' willingness to invest in green real estate projects in Asia. This modification is illustrated in Figure 1.

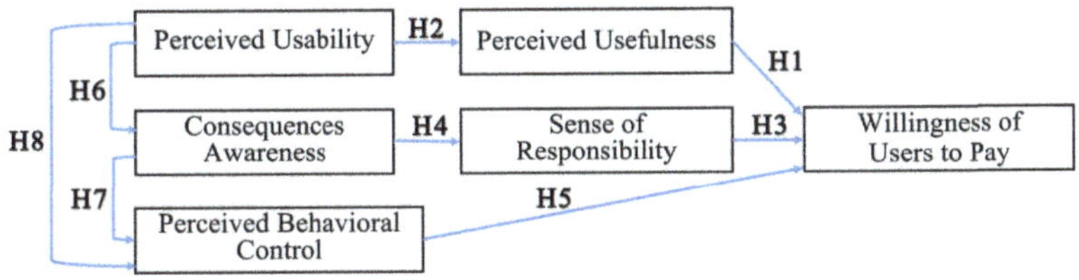


**Figure 1.** Theoretical model of the influence mechanism of users' willingness to invest in green real estate projects.

## *2.4. Hypothesis of the Study*

This study makes the following assumptions about the relationship between the influencing factors of the comprehensive measurement model of users' willingness to pay for green real estate projects in Asia (H1–H8 in Figure 1 represent the hypothesized paths between the latent variables, respectively).

Perceived usefulness in this study is linked to the user experience of green real estate projects. Perceived value (PV) is critical in shaping purchase intentions [33], and in the realm of green purchasing, Chen and Chang [34] discovered that green PV positively influences green purchase intentions. The technology acceptance model (TAM) posits that perceived usefulness and perceived usability collectively form people's attitudes towards technology usage. These attitudes, along with perceived usefulness, affect behavioral intentions, which in turn shape actual behavior. Perceived usefulness is also seen as beneficial at an economic or social level, positively impacting individuals' willingness to act [35]. Existing research has found that perceived usefulness, enjoyment, and price are the three central values influencing the relationship with use, which is reflected explicitly in the GB domain [36]. In comparison to conventional buildings, green buildings' thermal comfort levels have been extensively researched, with a clear link established between comfort and satisfaction. For a green building to exceed conventional buildings in comfort and satisfaction, it must significantly enhance indoor environmental quality [37]. Users' beliefs about the health benefits, quality of life improvements, cost reductions, and energy savings of green real estate are connected to their perceptions of the project's usefulness. As green real estate projects are emerging technologies, their adoption depends on their perceived benefits in production and living environments. Therefore, this study proposes Hypothesis 1.

**Hypothesis 1 (H1).** *Asian users' perceived usefulness of green real estate projects positively affects Asian users' willingness to pay for green real estate projects.*

Perceived usability relates to the internal characteristics of a green real estate project. The International Organization for Standardization's ISO 9241-11 guideline [38] defines "Usability" as "the extent to which a given user can use the product to achieve specific effectiveness, efficiency, and satisfaction goals in a given usage environment". The theory of planned behavior (TPB) operates on the foundational premise that behavioral intention serves as the primary predictor of actual behavior. Behavioral intentions, characterized by the willingness to accept specific actions, are in turn influenced by a composite interplay of individuals' attitudes and perceived usefulness [39]. In a broader context, usability is intrinsically linked to user engagement [40]. For instance, Kibert [41] has advocated for the incorporation of certain standard design elements in green properties, such as natural ventilation, the use of low-toxicity décor and furnishings, enhanced natural daylighting for superior lighting quality, operable windows, fans to facilitate personal control over environmental conditions, and mitigation of exposure to outdoor sounds. These design features not only augment the overall user experience but also contribute positively to user willingness and the time spent using the property, consequently enhancing its usability. The extent to which users take notice of and engage with these features within green real estate projects is a matter that pertains to their perceived usability of such projects. Therefore, this study proposes Hypothesis 2.

**Hypothesis 2 (H2).** *Asian users' perceived usability of green real estate projects positively affects the perceived usefulness of green real estate projects.*

A sense of responsibility among users is linked to their environmental perception and awareness. Social identity theory [42] lays the foundation for understanding the positive relationship between the degree of environmental concern in individuals and their willingness to spend more on green initiatives. Perceived morals (PMs), also known as personal norms or ethical codes, refer to an individual's judgment about the moral rightness or wrongness of performing a certain behavior [43]. This means that individuals may feel a moral duty to act in specific ways when faced with ethical decisions. Growing environmental awareness has led to increased interest in green consumption, significantly enhancing the market value of green products. Liu et al. [44] have underscored the importance of including perceived moral obligations in the TPB framework, particularly regarding sustainable building practices. Chen and Tung [45] argued that perceived moral obligation positively affects consumers' tendencies to choose eco-friendly buildings. Similarly, Barbarossa and Pelsmacker [46] demonstrated the positive impact of moral obligation on consumers' willingness to buy eco-friendly products. Furthermore, according to the normative activation theory (NAT) [47], individuals are more likely to opt for environmentally responsible products if they feel a sense of duty in their decision-making. The stronger the sense of environmental responsibility in individuals, the more likely they are to invest in eco-friendly real estate, even if it means incurring additional costs. Hence, an increased sense of responsibility is likely to positively influence an individual's willingness to financially support green real estate projects.

**Hypothesis 3 (H3).** *Asian users' awareness of an environmental sense of responsibility positively influences Asian users' willingness to pay for green real estate projects.*

Consequence awareness involves an individual's understanding of how their actions impact both their personal and external environments. This awareness is crucial in evaluating the significance and value of investing in green real estate projects. The construction phase in the real estate sector is known for its high resource consumption [48]. Additionally, concepts of benevolence and social altruism link consequence awareness beliefs to concerns about environmental responsibility [49], establishing a connection between consequence awareness and environmental responsibility. Awareness of the potential impacts of one's actions on others moderates the relationship between ethical considerations and behavior. The normative activation theory (NAT) posits that consequence awareness significantly influences the attribution of responsibility [50,51]. Without an understanding of the consequences of their actions, individuals may not feel responsible for those outcomes or consider the efficacy of their actions. Therefore, the more Asian users are aware of the negative impacts of environmental deficiencies, the more likely they are to develop a personal sense of responsibility towards the environment, influencing their attitudes and behaviors towards green real estate.

**Hypothesis 4 (H4).** *Asian users' consequence awareness of green real estate projects positively affects their sense of responsibility in green real estate projects.*

Perceived behavioral control (PBC) relates to how external factors, including environmental features and information interference, affect green real estate projects. It provides a plausible explanation for the complexity of behavioral prediction [52]. Price tolerance for PBC is a good reference point. Smith and Paladino [53] identify project cost as a crucial factor in payment behavior, noting that price and accessibility are major barriers. Green projects typically have higher initial costs than standard buildings, with a study showing that "green label" condos command a 4.7% price premium [54]. Thus, it is essential to analyze the users' perception of control over the green project, taking into account the diminishing influence of users on external factors such as price, location, and amenities, which also relate to the developer's pricing strategy. Additionally, societal and familial information influences PBC. The perceived control of a behavior directly impacts behavioral intention [55], highlighting how perceived constraints shape willingness to invest in green real estate, particularly among Asian users.

**Hypothesis 5 (H5).** *Asian users' perceived behavioral control of green real estate projects positively affects Asian users' willingness to pay for green real estate projects.*

Furthermore, the correlation between perceived usability, consequence awareness, and PBC is also worth exploring. The perceived usefulness of green real estate projects appeals to people's perceived consequence awareness of the environment and themselves, which means that the more people recognize the usefulness of green real estate projects, the stronger the awareness of the consequences. At the same time, the more efficient and effective green real estate projects are in terms of environmental protection, the more consumers will recognize and avoid the negative consequences of ordinary real estate projects. Therefore, a positive correlation exists between perceived usefulness and consequence awareness, and Hypothesis 6 is proposed.

**Hypothesis 6 (H6).** *Asian users' perceived usability of green real estate projects positively affects consequence awareness of green real estate projects.*

The normative activation theory (NAT) links consequence awareness to the recognition of negative effects on oneself or the external environment. A user's heightened awareness of consequences increases their likelihood to invest in green real estate projects and strengthens their perceived behavioral control (PBC) over these projects. Concurrently, the stronger a user's perceived behavioral control is in using and purchasing green real estate, the greater their desire for others to adhere to environmental protection rules, fostering a better ecological environment. Consequently, Hypothesis 7 is proposed.

**Hypothesis 7 (H7).** *Asian users' consequence awareness of green real estate projects positively affects their perceived behavioral control of green real estate projects.*

Users also consider the health benefits of long-term residence in green real estate compared to standard real estate. Based on the research showing that green buildings have led to improvements in users' health and productivity levels [56,57], the personal benefits of holding a green real estate project for an extended period deserve to be taken into account. Popescu et al. [58] additionally posit that energy retrofit measures in green real estate yield advantages that extend beyond enhanced energy efficiency. These measures can also augment the property's intrinsic value, thereby facilitating a reduction in the investment payback period associated with energy efficiency enhancements. Therefore, perceived usability should have a role in PBC, but it needs to be directly realized. The analysis of mediating effects is imperative for assessing both the presence of a variable's mediating role and the extent to which it fulfills this role. Baron and Kenny's [59] causal step approach has been used to test for mediating effects. The independent variable (perceived usability) influences the dependent variable (PBC) through the mediator variable (perceived awareness of consequences), which represents a non-zero mediation effect, and therefore Hypothesis 8 is proposed.

**Hypothesis 8 (H8).** *The perceived usability of green real estate projects by Asian users influences their perceived behavioral control through consequence awareness to create a mediating effect.*

## 3. Methods

### 3.1. Questionnaire Design

This study used a quantitative pairwise research methodology to develop a questionnaire for measuring the willingness to pay of Asian users for green real estate design by reviewing the experience in related literature. The questionnaire consists of two parts. Part 1 investigates the demographic information of the respondents. Part 2 collects data on respondents' willingness to pay, perceived usefulness, perceived usability, sense of responsibility, consequence awareness, and perceived behavioral control. Respondents were asked to rate these items. Each item was rated using a five-point Likert scale (1: strongly disagree, 2: disagree, 3: neither agree nor disagree, 4: agree, and 5: strongly agree).

Specifically, the questionnaire questions were adapted from previous literature, for example, the research questions on perceived usefulness were set regarding the User Experience Usability Metrics (UMUX) [60] and the UMUX-LITE based on factor and item analysis [61]; perceived usefulness research questions were set concerning the computer system usability questionnaire (CSUQ) [62] and the system usability scale (SUS) [63]. The sources of the scales are presented in Table 1. The questionnaire was considered by five experts with more than five years of experience in the real estate development and market management industry. The specific questionnaire can be found in Appendix A.

**Table 1.** Definition of variables and sources of reference scales.

| Research Variable | Operability Definition | Reference Scale |
|---|---|---|
| Willingness of Users to Pay | Asian users willing to rent or buy green real estate projects. | Kang et al. [64] Ajzen and Driver [65] |
| Perceived Usefulness | Asian users perceive green real estate programs as useful based on user experience. | Lewis et al. [61] Finstad [60] |
| Perceived Usability | Asian users perceive the availability of green real estate projects based on their internal characteristics. | Lewis [62] Brooke [63] |
| Sense of Responsibility | Asian users are environmentally responsible and considerate of the green credibility of their real estate brands. | Taufique et al [66] Kaiser et al. [67] |
| Consequences Awareness | Asian users perceive the significance and value of behavior in the future from a personal and social perspective. | Hansla et al. [49] Schwartz [51] |
| Perceived Behavioral Control | The extent to which Asian users are confronted with the consideration of external factors and the degree to which they are disturbed by external information about green real estate. | Terry and O'Leary [68] Ajzen and Madden [55] |

### 3.2. Research Methodology

Since the 1980s, structural equation modeling (SEM) has been used in a wide range of research areas and is effective in psychological and behavioral research [69]. SEM serves as a pivotal analytical technique for investigating the connections between latent and observable variables through the examination of collected data. It stands as a vital tool for delineating the relationships among pivotal factors. Moreover, SEM finds extensive utility in the realm of multivariate data analysis, addressing the limitations inherent in conventional statistical methods. Consequently, this study employed SEM as the analytical approach to assess H1–H7.

This study utilized SPSS 28.0 and AMOS 24.0 software to conduct structural equation modeling (SEM) analysis to test the hypotheses within the theoretical framework. SEM, a widely accepted multivariate analysis technique in the fields of psychology and behavioral research, was chosen for its appropriateness, as endorsed by Hoyle [70]. SEM offers distinct advantages for this research: firstly, it accommodates both latent and observed variables, aligning with the variables under investigation; secondly, it integrates factor and path analyses, facilitating a comprehensive examination of the proposed integrated framework. Furthermore, SEM serves a dual purpose, functioning as a tool for assessing measurement scale reliability and validity in its measurement modeling component and as a means for hypothesis testing in its structural modeling component.

Meanwhile, mediation modeling is an essential research approach for exploring the impact of mediating variables on the relationship between independent and dependent variables [71]. This methodology is grounded in the foundational premise that the independent variable exerts its influence on the dependent variable through an intermediary variable. The evaluation of the mediation model commonly involves the utilization of the bootstrap sampling method, which is a non-parametric statistical technique utilized for estimating the distribution, standard error, and confidence interval of a statistical measure [72]. Accordingly, this study will employ the mediation model to investigate the hypothesis denoted as H8.

### 3.3. Data Collection

The questionnaire was distributed through two channels. The first was an online survey using a "snowball" sampling method, whereby a link to the questionnaire was sent to users aged 18 and over through WeChat, Line, Kakao, and other Asian online platforms, which basically guaranteed that the respondents would be from Asian countries. At first, respondents were randomly selected based on social media. The respondents subsequently disseminated the questionnaire to additional individuals who met the broader criteria of the sample, including their acquaintances and family members. This method ensures the sample's characteristics, but the recovered questionnaires' utilization rate is low.

The second survey method was to distribute the questionnaire through various real estate forums and online through social media platforms and agencies in different Asian countries. The questionnaire survey method, while affording a reasonable degree of control over the sample, is characterized by its time-intensive nature and demands for additional effort. Consequently, this study adopted a hybrid approach, combining both survey methods.

In the questionnaire, all participants were explicitly apprised of the survey's objectives and the policy equating lease and purchase rights, which served as a proxy for user intention to pay. Subsequently, respondents proceeded to complete the questionnaire after providing their informed consent to participate in the research.

The survey lasted three months, from July 2023 to October 2023, and covered people of different genders, ages, and educational backgrounds, most of whom were potential users. They all came from China, Korea, Singapore, Thailand, Japan, and other Asian countries. In this study, 500 self-administered questionnaires distributed online resulted in 385 completed questionnaires (77% response rate). In the data screening phase, the standard deviation for the scores across 33 items was computed, and questionnaires displaying a standard deviation of 0 or containing incomplete data were omitted from the dataset. Subsequently, 356 questionnaires remained eligible for inclusion in the final data analysis. Kline [73] suggested that the sample size must be at least ten times each item's size. With 33 items in this study, the sample size needed to be at least 330, thus meeting the requirement.

### 3.4. *Measurement Model*

#### 3.4.1. Reliability Analysis

Reliability analysis pertains to the extent of consistency observed in various test measurements. In academic research, Cronbach's alpha is the predominant method employed to assess internal consistency. The $\alpha$ coefficient's range spans from 0 to 1, with higher values indicating stronger internal consistency within the scale [74]. If the $\alpha$ coefficient is more significant than 0.7, then the analysis is highly reliable [75]. In this study, SPSS 28.0 software was used for the reliability test. The Cronbach's alpha coefficients for the variables ranged from 0.917 to 0.93, as shown in Table 2. The overall Cronbach's alpha coefficient of the questionnaire was 0.920, which met the overall and local requirements, indicating that the questionnaire had satisfactory reliability.

**Table 2.** Reliability analysis.

| Variable | | Cronbach's Alpha | N of Items |
|---|---|---|---|
| Perceived Usefulness | PUF | 0.917 | 5 |
| Perceived Usability | PUA | 0.915 | 5 |
| Sense of Responsibility | SR | 0.919 | 5 |
| Consequence awareness | CA | 0.907 | 5 |
| Perceived Behavioral Control | PBC | 0.935 | 5 |
| Willingness of Users to Pay | WUP | 0.929 | 5 |

#### 3.4.2. Validation Factor Analysis

The goodness-of-fit test is a statistical procedure employed to assess whether the sample data align with the expected normal distribution [76]. It is also a statistical method of making inferences about observations to determine whether the sample group is truly representative of the whole population [77]. Therefore, they determine the relationship between actual and predicted values in the model. From Table 3, it can be seen that CMIN/DF is 1.478, which is less than the criterion of below 3. Meanwhile, TLI and CFI meet the criterion of 0.9 or more, and RMSEA is 0.037, which is less than 0.08. Each fitting indicator meets the general research criteria, so this model has a good fit.

**Table 3.** Goodness-of-fit test results.

| CMIN | DF | CMIN/DF | TLI | CFI | RMSEA |
|---|---|---|---|---|---|
| 709.344 | 480 | 1.478 | 0.972 | 0.975 | 0.037 |

A survey questionnaire is designed to facilitate measurements of adequate validity, enabling the derivation of meaningful conclusions. Enhanced validity levels correspond to heightened survey authenticity, as highlighted by Taherdoost [78]. Validity assessment primarily encompasses content and structural validity, with the latter encompassing convergent and discriminant validity. To enhance the questionnaire's validity assessment, this study employs AMOS 24.0 to perform a confirmatory factor analysis rooted in exploratory factor analysis. This analysis evaluates the scale's validity through measures such as composite reliability (CR) and average variance extracted (AVE), as proposed by Bagozz [79]. As can be seen from Table 4, the standardized factor loadings of each measure were more significant than 0.6 or more. All CR scores were greater than the threshold value of 0.70 [80], and AVE was more significant than 0.5, indicating good convergent validity.

**Table 4.** Convergent validity test results.

| | | | Estimate | CR | AVE |
|---|---|---|---|---|---|
| PUF1 | ← | PUF | 0.912 | 0.919 | 0.696 |
| PUF2 | ← | PUF | 0.806 | | |
| PUF3 | ← | PUF | 0.776 | | |
| PUF4 | ← | PUF | 0.839 | | |
| PUF5 | ← | PUF | 0.832 | | |
| PUA1 | ← | PUA | 0.750 | 0.917 | 0.689 |
| PUA2 | ← | PUA | 0.842 | | |
| PUA3 | ← | PUA | 0.778 | | |
| PUA4 | ← | PUA | 0.877 | | |
| PUA5 | ← | PUA | 0.893 | | |
| SR1 | ← | SR | 0.763 | 0.921 | 0.701 |
| SR2 | ← | SR | 0.758 | | |
| SR3 | ← | SR | 0.885 | | |
| SR4 | ← | SR | 0.859 | | |
| SR5 | ← | SR | 0.908 | | |
| CA1 | ← | CA | 0.858 | 0.909 | 0.667 |
| CA2 | ← | CA | 0.769 | | |
| CA3 | ← | CA | 0.832 | | |
| CA4 | ← | CA | 0.781 | | |
| CA5 | ← | CA | 0.839 | | |
| PBC1 | ← | PBC | 0.861 | 0.941 | 0.761 |
| PBC2 | ← | PBC | 0.830 | | |
| PBC3 | ← | PBC | 0.887 | | |
| PBC4 | ← | PBC | 0.927 | | |
| PBC5 | ← | PBC | 0.853 | | |
| WUP1 | ← | WUP | 0.848 | 0.930 | 0.627 |
| WUP2 | ← | WUP | 0.823 | | |
| WUP3 | ← | WUP | 0.689 | | |
| WUP4 | ← | WUP | 0.748 | | |
| WUP5 | ← | WUP | 0.893 | | |
| WUP6 | ← | WUP | 0.810 | | |
| WUP7 | ← | WUP | 0.712 | | |
| WUP8 | ← | WUP | 0.790 | | |

### 3.4.3. Correlation Analysis and Differential Validity

In the preceding section, we established the structure of dimensions and their associated items via validity and reliability analyses. The dimension scores were computed as the average of the scores from the questions within each respective dimension, followed by the execution of correlation analysis. Correlation analysis primarily serves to examine the relationships between variables. The correlation coefficient's range falls between $-1$ and 1, where a larger absolute value signifies a more pronounced association between the variables.

In assessing discriminant validity, the comparison involved examining whether the square root of the AVE for a specific construct exceeded the correlations between that construct and others. When the square root of the AVE of a construct surpassed the non-diagonal elements within the corresponding rows and columns of the correlation matrix, it indicated that the indicator was more closely associated with that particular construct than with others. As demonstrated in Table 5, the diagonal entries are represented in bold and signify the square root of the AVE. Given that all diagonal entries are greater than the non-diagonal entries, it suggests that discriminant validity is adequately established for all constructs.

**Table 5.** Correlation analysis and differential validity test results.

| | Correlations | | | | | |
|---|---|---|---|---|---|---|
| | **PUF** | **PUA** | **SR** | **CA** | **PBC** | **WUP** |
| PUF | 0.834 | | | | | |
| PUA | 0.597 ** | 0.830 | | | | |
| SR | 0.211 ** | 0.205 ** | 00.837 | | | |
| CA | 0.401 ** | 0.420 ** | 0.043 | 0.817 | | |
| PBC | 0.494 ** | 0.401 ** | 0.288 ** | 0.464 ** | 0.872 | |
| WUP | 0.367 ** | 0.405 ** | 0.145 ** | 0.387 ** | 0.334 ** | 0.792 |

** Correlation is significant at the 0.01 level (2-tailed).

The above questionnaire's reliability and validity analysis showed that the questionnaire's reliability, validity, and correlation are good enough for structural equation modeling.

## 4. Results

### 4.1. Descriptive Statistical Analysis

This study's intended participants were prospective buyers of environmentally friendly residential properties in the Asian region who were aged 18 or older. To ensure the accuracy of the questionnaire and to address any potential issues related to wording and measurement, a pre-test was administered to a sample of 30 respondents before the main data collection phase. It is important to note that the results from the pre-test were not included in the final sample size used for analysis.

The gender distribution of participants was 150 (42.1%) males and 206 (57.9%) females. The user group was concentrated in the 30–39 (34%) and 40–49 (27%) age ranges. Regarding property ownership, 75.0% of respondents currently own property, while 25.0% do not. The percentage of respondents with experience purchasing green real estate is 36.8%. Out of the total number of respondent users, eight have not used property due to living in self-built houses in the countryside for a long time or not generating commercial activities. The number of users with experience using a green property was a pleasant surprise at 69.9%. For more information, see Table 6's respondents' profiles.

**Table 6.** Basic data statistics.

| Basic Data Statistics | | Frequency | Percentage |
|---|---|---|---|
| Gender | Male | 150 | 42.1% |
| | Female | 206 | 57.9% |
| Age | 18–29 | 50 | 14.0% |
| | 30–39 | 121 | 34.0% |
| | 40–49 | 96 | 27.0% |
| | 50–59 | 57 | 16.0% |
| | 60 years and over | 32 | 9.0% |
| Educational Background | High school diploma | 50 | 14.0% |
| | Bachelor degree | 136 | 38.2% |
| | Master's degree | 92 | 25.8% |
| | Doctor's degree | 64 | 18.0% |
| | Other | 14 | 3.9% |
| Green Real Estate Homeownership | Have not purchased any real estate | 89 | 25.0% |
| | Purchased only regular real estate | 136 | 38.2% |
| | Purchased only green real estate | 28 | 7.9% |
| | Purchased both regular and green real estate | 103 | 28.9% |
| Use of Green Real Estate | Have not used any real estate | 8 | 2.2% |
| | Used only ordinary real estate | 99 | 27.8% |
| | Used only green real estate | 61 | 17.1% |
| | Used both regular and green real estate | 188 | 52.8% |

### 4.2. Structural Equation Modeling (SEM)

SEM is a statistical approach employed to examine the interrelationships among variables by utilizing their covariance matrices. It utilizes a posteriori reasoning to formulate a model that resembles prior research findings. Once the model is established, its viability is assessed through an evaluation of its overall fit and the significance of individual paths within the model. Subsequently, the influence of independent variables on the dependent variable is systematically determined.

Before the empirical test, the structural equation model was constructed, and then AMOS 24.0 was used to perform the calculations using the maximum likelihood method to obtain Figure 2 and Table 7. The goodness of fit was tested according to the criteria of the previous studies [81]. The goodness-of-fit index quantifies the level of congruence between the postulated theoretical model and the empirical data. A higher goodness-of-fit score signifies a greater alignment between the theoretical model and the observed data, as articulated by Hatcher and O'Rourke in 2013 [82]. From Table 7, it can be seen that the CMIN/DF is 1.718, which is less than the standard of less than 3; the TLI and CFI are both above the standard of 0.9; and the RMSEA is 0.045, which is less than the standard of 0.08. The model fitting indexes satisfy the independent level of recommended fitting and the combination rule, so they have favorable fitting ability and excellent fitness.

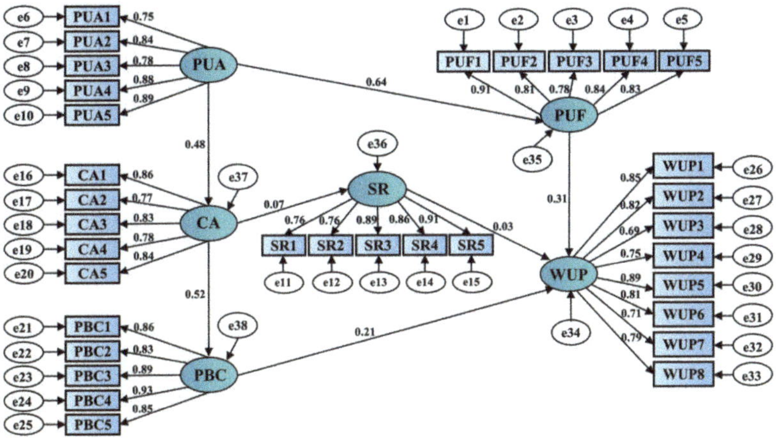


**Figure 2.** Structural equation modeling.

**Table 7.** Measurement model fit indices.

| CMIN | DF | CMIN/DF | TLI | CFI | RMSEA |
|---|---|---|---|---|---|
| 838.616 | 488 | 1.718 | 0.958 | 0.961 | 0.045 |

### 4.3. Model Evaluation

Path coefficients represent standardized iterations of linear regression weights and are employed in SEM to explore potential causal connections among statistical variables [83]. Standardized path coefficients are computed by multiplying the cyclic coefficients with the standard deviations of the corresponding predictor variables, enabling comparisons to assess the relative impact of these variables within the regression model. Alongside standardized path coefficients, studies typically report the $p$-value for the path coefficient test to ascertain its statistical significance. From Table 8, it can be obtained that, in H1–H7, CA is not significant for SR (CR = 1.203, $p = 0.229 > 0.05$) and SR is not significant for WUP (CR = 0.61, $p = 0.542 > 0.05$). As a result, hypotheses H3 and H4 were not valid.

Comparatively, the other paths have $p < 0.05$, and the standardized path coefficients are positive. Thus, H1, H2, H5, H6, and H7 all have significant positive effects [84], and the hypothesis is valid.

**Table 8.** Path coefficients.

| | | | STDEstimate | Estimate | S.E. | C.R. | $p$ |
|---|---|---|---|---|---|---|---|
| CA | ← | PUA | 0.482 | 0.575 | 0.069 | 8.297 | *** |
| PUF | ← | PUA | 0.643 | 1.045 | 0.092 | 11.377 | *** |
| SR | ← | CA | 0.069 | 0.075 | 0.062 | 1.203 | 0.229 |
| PBC | ← | CA | 0.523 | 0.663 | 0.069 | 9.575 | *** |
| WUP | ← | PUF | 0.314 | 0.231 | 0.04 | 5.735 | *** |
| WUP | ← | SR | 0.032 | 0.029 | 0.048 | 0.61 | 0.542 |
| WUP | ← | PBC | 0.208 | 0.164 | 0.042 | 3.872 | *** |

*** $p < 0.001$.

### 4.4. Mediation Effect Test

This study systematically examined the mediating effect of perceived usability on perceived behavioral control by Asian users of green real estate projects using the bootstrap sampling method (Table 9). After 5000 iterations of sampling using the bootstrap sampling method, whether the 95% confidence interval of bias-corrected contained 0 was observed to determine whether the indirect effect was significant. Based on the results of the analysis, it can be seen that the value of the indirect effect of PUA on PBC through CA is 0.252, and the 95% confidence interval does not contain 0, indicating that CA as a mediating variable has a mediating effect in the relationship of PUA on PBC.

The obtained $p$-value of 0.000 falls below the conventional significance threshold of 0.05 for mediation effects, underscoring the statistical significance of the mediating effect, denoted as (PUA→CA→PBC) within this study. This underscores the central role played by the mediating variable (CA) in elucidating the association between the independent variable (PUA) and the dependent variable (PBC). Therefore, H8 is valid.

**Table 9.** Mediation effect test results.

| | STD.Estimate | Lower | Upper | $p$ |
|---|---|---|---|---|
| PUA→ CA→PBC | 0.252 | 0.180 | 0.327 | 0.000 |

### 4.5. Hypothesis Verification

This study aims to establish a theoretical framework of the factors influencing users' willingness to pay for green real estate projects and to draw conclusions by identifying the key factors influencing users' willingness to pay for green real estate through structural equation modeling to provide reference design suggestions for future green real estate project development and design. In the validation analysis of the structural equation model and the experimental results, it was found that the relationship of the theoretical model based on the mechanism of influencing the users' willingness to pay for green real estate projects supported six out of eight hypotheses and did not have a significant positive effect on two hypotheses. The results are shown in Table 10.

This allows for the verification of H1, suggesting that Asian users' perceived usefulness of green real estate projects positively affects Asian users' willingness to pay for green real estate projects. H2 can also be validated to show that Asian users' perceived usability of green real estate projects positively affects the perceived usefulness of green real estate projects. H5 is also valid, which implies that Asian users' perceived behavioral control of green real estate projects positively affects Asian users' willingness to pay for green real estate projects. H6 can be validated to show that Asian users' perceived usability of green real estate projects positively affects consequence awareness of green real estate projects.

H7 is also valid, which means that Asian users' consequence awareness of green real estate projects positively affects perceived behavioral control of green real estate projects. H8 establishes that Asian users' perceived usability of green real estate projects has a mediating effect on perceived behavioral control, and consequence awareness serves as a mediating variable.

**Table 10.** Verification results of hypotheses.

| Hypothesis | Content | Result |
| --- | --- | --- |
| Hypothesis 1 (H1) | Asian users' perceived usefulness of green real estate projects positively affects Asian users' willingness to pay for green real estate projects. | Valid |
| Hypothesis 2 (H2) | Asian users' perceived usability of green real estate projects positively affects the perceived usefulness of green real estate projects. | Valid |
| Hypothesis 3 (H3) | Asian users' awareness of an environmental sense of responsibility positively influences Asian users' willingness to pay for green real estate projects. | Invalid |
| Hypothesis 4 (H4) | Asian users' consequence awareness of green real estate projects positively affects their sense of responsibility in green real estate projects. | Invalid |
| Hypothesis 5 (H5) | Asian users' perceived behavioral control of green real estate projects positively affects Asian users' willingness to pay for green real estate projects. | Valid |
| Hypothesis 6 (H6) | Asian users' perceived usability of green real estate projects positively affects consequence awareness of green real estate projects. | Valid |
| Hypothesis 7 (H7) | Asian users' consequence awareness of green real estate projects positively affects their perceived behavioral control of green real estate projects. | Valid |
| Hypothesis 8 (H8) | The perceived usability of green real estate projects by Asian users influences their perceived behavioral control through consequence awareness to create a mediating effect. | Valid |

In contrast, H3 does not hold, indicating that awareness of environmental responsibility among Asian users does not have a positive influence on the willingness of Asian users to pay for green real estate projects. H4 is invalid, indicating that Asian users' awareness of the consequences of green real estate projects does not positively influence the accountability of green real estate projects.

Figure 2 shows that, among the key influencing factors, perceived usefulness is weighted more heavily, suggesting that Asian users value the long-term benefits to physical and mental health and experience when paying for green real estate projects. Real estate developers should thoroughly consider this factor. In addition, perceived behavioral control is another factor valued by Asian users, indicating that users are less likely to be influenced by external factors when paying for green real estate. Property developers should also take full advantage of this factor to realize the commercial sustainability value of green real estate projects.

## 5. Discussion

A growing body of research focuses on users' and potential users' perceptions, attitudes, and acceptance of green real estate and buildings. Nevertheless, few empirical studies have linked them to user motivation. To address this issue, this study focuses on the relationship between users' motivational factors (perceived usefulness, perceived usability, sense of responsibility, consequence awareness, and perceived behavioral control) and the quantitative relationship between users' intention to pay for green real estate design in Asia, with two main results. One result is that including perceived usefulness, perceived usability, and perceived behavioral control in the original TAM is meaningful in explaining users' attitudes toward green real estate and willingness to pay for green real estate. Another result is that, although consequence awareness, mediated by a sense of responsibility, does not correlate with users' willingness to pay for green real estate, consequence awareness can be the basis for a correlation between perceived usability and perceived behavioral control. They can provide factor-specific theoretical and empirical support for the future promotion of green real estate projects in Asia.

Perceived usability and perceived usefulness are mutually reinforcing processes and critically impact Asian users' willingness to pay for green real estate. Specifically, after perceiving the usability features of a green real estate project, Asian users deepen their understanding of the features through experiential behaviors such as living, working, and visiting, resulting in the perception of usefulness. Similar to the findings of Alshurafat et al. [85], the results of this study imply that perceived usability and perceived usefulness are significant predictors of attitudes toward use and willingness to pay. As a result, Asian users are willing to pay for green real estate based on perceived usefulness, and this further generates the likelihood of action, which is consistent with previous research [86].

The role of perceived behavioral control on users' willingness to pay for green real estate in Asia was also emphasized. Based on the benefits and expectations of homeownership and rental use associated with a green real estate project, users will somewhat discount external factors, the most important of which is price. This is consistent with the results of the current study that users can accept the premium price of green real estate, usually in the range of 7–10% [87]. The arguments of this study are also consistent with other studies on the role of TPB in the willingness to invest in and use green projects [88]. In addition, based on perceived behavioral control, Asian users' willingness to pay for green real estate will be less affected by policies and recommendations. Portnov et al. even suggested that policy incentives for the paying population may produce emotional and opposing responses.

Comparatively, it is essential to note that the relationship between the sense of responsibility and the willingness of Asian users to pay for green real estate is not significant. Users' positive perceptions and attitudes towards a sense of responsibility do not necessarily lead to their willingness to pay for green initiatives in real estate projects. It has long been established that attitudes are weakly related to behavior when broad measures of attitudes are used to predict specific behaviors [89]. Indeed, the relationship between users' attitudes toward social responsibility and paying behavior is complex and depends on product quality, company motivations, and user characteristics [90,91]. Although Asian users may be willing to pay increased costs for green real estate, other factors come into play when paying for it, rather than being influenced by a sense of responsibility, which is in line with Li et al.'s [92] study.

Overall, the theory and its implications, as evidenced by the data results of this study, are threefold. Perceived usefulness and perceived behavioral control are the primary drivers of Asian users' attitudes toward paying behaviors in green real estate projects. This finding emphasizes the importance of these factors in shaping user preferences and choices. Perceived usability plays a crucial role, as it influences perceived usefulness. This highlights the significance of ensuring that green real estate projects are user-friendly and accessible to potential Asian users. Consequence awareness mediates between perceived usability and perceived behavioral control among Asian users. This nuanced finding provides a deeper understanding of how these factors interact and impact user decisions.

Furthermore, it needs to be explored that consequence awareness does not work for the sense of responsibility but does work for perceived behavioral control. It may be because consequence awareness is categorized into perceived consequences for the external environment and perceived consequences for oneself. However, perceived consequences for the external environment had less effect on the sense of responsibility. This corresponds with Howell's [93] suggestion that environmental values influence green consumption. In contrast, the perceived consequences of self have a strong stimulus on the personal will of Asian users, which can make Asian users reduce the possibility of being interfered with by external factors, thus promoting the formation of perceived behavioral control. Meanwhile, the source of perceived consequences is the perceived usability of Asian users based on their knowledge of the characteristics of green real estate projects.

## 6. Conclusions

Grounded in the theory of planned behavior (TPB), this investigation endeavors to elucidate the factors propelling users' willingness to remunerate for green real estate ventures in Asia. Employing a structured questionnaire administered to 356 individuals and subsequent analysis through structural equation modeling, this study discerns pivotal determinants affecting users' financial commitment, notably perceived usefulness, usability, behavioral control, and awareness of consequences. The findings illuminate the quintessential influence of perceived usefulness and behavioral control in molding payment-related attitudes, the foundational role of perceived usability in augmenting perceived usefulness, and the intermediary effects of perceived usability and awareness of consequences on behavioral control.

Moreover, the study furnishes targeted design and marketing stratagems for developers to captivate Asian clientele towards green real estate investments. Amidst a milieu of sustainable development and corporate metamorphoses within the real estate sector, the research underscores the necessity to transcend conventional energy conservation and eco-friendly practices. Instead, it advocates for integrating user-centric green innovations such as intelligent home systems and ecologically mindful interface designs. Furthermore, the strategic infusion of culturally resonant design elements, like Feng Shui or Vastu Shastra, is recognized as instrumental in augmenting project allure through a synthesis of effective communication and cultural reverence.

This study contributes to the field by presenting an empirical methodology to investigate users' motivation and willingness to pay for green real estate projects. The method combines established models, such as the TAM, with factors particularly relevant to green real estate. This approach ensures the study's theoretical framework is well-suited to the research questions. In addition to the academic contribution, the study provides practical implications by suggesting a reference design for future developers of green real estate projects targeting Asian users. This provides actionable insights from the theoretical framework and bridges the gap between academic research and practical applications in green real estate development.

Notwithstanding its insightful contributions to understanding the fiscal propensities towards green real estate in Asia, the study acknowledges limitations in its scope: primarily, its focus on a subset of users either with or contemplating green investments does not encapsulate the entire spectrum of potential users. Future inquiries should endeavor to widen the ambit of investigation, delving into a more heterogeneous array of demographics and locales to enrich the comprehension of user willingness to pay. This expansion will enhance the research's relevance and transferability. Additionally, the reliance on cross-sectional data prompts a call for future studies to harness financial disclosures of publicly traded entities, stock data, and national panel statistics to perpetuate and deepen the investigative narrative. Through such an enriched lens, a more nuanced appreciation of the global tendencies and preferences in green real estate can be achieved, furthering the quest for sustainable development.

**Author Contributions:** Conceptualization, S.W. and S.C.; data curation, S.W.; formal analysis, S.W.; funding acquisition, S.W.; investigation, S.W.; methodology, S.W.; project administration, S.W. and K.N.; software, S.W.; validation, S.W.; visualization, S.W. and S.C.; writing—original draft, S.W.; writing—review & editing, S.W. and S.C.; resources, K.N.; supervision, K.N. All authors have read and agreed to the published version of the manuscript.

**Funding:** This research received no external funding.

**Data Availability Statement:** Data are contained within the article.

**Conflicts of Interest:** The authors declare no conflict of interest.

## Appendix A. Questionnaire

Questionnaire on Mechanisms Influencing User Willingness to Pay for Green Real Estate Design in Asia.

**The Purpose of This Research**

The purpose of this study is to assess the impact of user motivation and intention to pay for green real estate design in Asia and to examine whether purchase intention affects their willingness to purchase these real estate projects. The target respondents of this questionnaire are mainly those who have paid or are considering paying for green real estate projects in Asia to use them in the long term. This study aims to sample 450 people of different backgrounds. It is up to you to decide whether to participate or not. You can withdraw your participation at any time without giving any reason or facing any negative consequences.

**Confidentiality**

All the information that I collect about you during the course of the research will be kept strictly confidential and will only be accessible to my supervisor and assessors of the dissertation. Once you have given consent, any personal data to represent your credibility as a source will be anonymized. Personal data will be held securely for the duration of this research project and destroyed once the dissertation has been assessed.

1. Have you read the consent form and are willing to continue to participate in the questionnaire?

   Yes or no.

   Before starting the questionnaire, please answer the following guiding questions:

2. Have you paid for or are you considering paying for a green real estate project and using it for the long term in Asia?

- Yes, please continue to answer the following questionnaire.
- No, you have finished the questionnaire.

**Questionnaire on Mechanisms Influencing User Willingness to Pay for Green Real Estate in Asia**

| | Items | Options | | |
|---|---|---|---|---|
| **Personal Basic Information** | Gender | 1 | Male | ○ |
| | | 2 | Female | ○ |
| | Age | 1 | 18-29 | ○ |
| | | 2 | 30-39 | ○ |
| | | 3 | 40-49 | ○ |
| | | 4 | 50-59 | ○ |
| | | 5 | 60 years and over | ○ |
| | Educational Background | 1 | High School Diploma | ○ |
| | | 2 | Bachelor degree | ○ |
| | | 3 | Master's degree | ○ |
| | | 4 | Doctor's degree | ○ |
| | | 5 | Other | ○ |
| | Green Real Estate Homeownership | 1 | Have not purchased any real estate | ○ |
| | | 2 | Purchased only regular real estate | ○ |
| | | 3 | Purchased only green real estate | ○ |
| | | 4 | Purchased both regular and green real estate | ○ |
| | Use of Green Real Estate | 1 | Have not used any real estate | ○ |
| | | 2 | Used only ordinary real estate | ○ |
| | | 3 | Used only green real estate | ○ |
| | | 4 | Used both regular and green real estate | ○ |

| **Category** | **Measurement items** | | **1** | **2** | **3** | **4** | **5** |
|---|---|---|---|---|---|---|---|
| Willingness of Users to Pay | WUP1 | I would like to buy or rent a green real estate. | ○ | ○ | ○ | ○ | ○ |
| | WUP2 | I would definitely buy or rent a green real estate. | ○ | ○ | ○ | ○ | ○ |
| | WUP3 | I would make an effort to buy or rent green real estate. | ○ | ○ | ○ | ○ | ○ |
| | WUP4 | I want to work or live in a green real estate. | ○ | ○ | ○ | ○ | ○ |
| | WUP5 | I think it is the right choice to buy or rent a green real estate instead of a regular one. | ○ | ○ | ○ | ○ | ○ |
| | WUP6 | I'm willing to raise my budget for green real estate. | ○ | ○ | ○ | ○ | ○ |
| | WUP7 | I no longer consider paying for regular real estate as opposed to green real estate. | ○ | ○ | ○ | ○ | ○ |
| | WUP8 | I hope to convince family and friends to buy or rent green real estate as well. | ○ | ○ | ○ | ○ | ○ |
| Perceived Usefulness | PUF1 | I think buying or renting green real estate is good for my health. | ○ | ○ | ○ | ○ | ○ |
| | PUF2 | I think buying or renting green real estate improves my quality of life. | ○ | ○ | ○ | ○ | ○ |
| | PUF3 | I think green real estate is more comfortable to experience than regular real estate. | ○ | ○ | ○ | ○ | ○ |
| | PUF4 | I think buying or renting green real estate can help me save money on energy consumption. | ○ | ○ | ○ | ○ | ○ |
| | PUF5 | I think the user experience of green real estate was an important factor in my choosing it. | ○ | ○ | ○ | ○ | ○ |
| Perceived Usability | PUA1 | I think buying or renting a green real estate allows access to better environmental amenities. | ○ | ○ | ○ | ○ | ○ |
| | PUA2 | I think buying or renting a green real estate allows me to enjoy better natural light and ventilation. | ○ | ○ | ○ | ○ | ○ |
| | PUA3 | I think buying or renting green real estate allows for | ○ | ○ | ○ | ○ | ○ |

| | | | | | | | |
|---|---|---|---|---|---|---|---|
| | | better control of environmental conditions and exposure to outdoor sounds. | | | | | |
| | PUA4 | I think buying or renting green real estate reduces exposure to harmful building materials. | O | O | O | O | O |
| | PUA5 | I think the internal characteristics of green real estate were an important factor in my choosing it. | O | O | O | O | O |
| Sense of Responsibility | SR1 | I think buying or renting a green real estate allows you to be responsible for your surroundings. | O | O | O | O | O |
| | SR2 | I believe that buying or renting green real estate can contribute to the sustainability of society | O | O | O | O | O |
| | SR3 | I think buying or renting a green real estate allows me to be a neighbor to people who are equally responsible. | O | O | O | O | O |
| | SR4 | I would like to be involved in any project on environmental and social responsibility. | O | O | O | O | O |
| | SR5 | My knowledge and awareness of the need to protect the environment has led me to choose to buy or rent a green real estate. | O | O | O | O | O |
| Consequences Awareness | CA1 | I think green real estate has greater appreciation than regular real estate. | O | O | O | O | O |
| | CA2 | I think green real estate has more long-term residential value than regular real estate. | O | O | O | O | O |
| | CA3 | I believe that health and productivity levels are enhanced by living and working in a green real estate. | O | O | O | O | O |
| | CA4 | I think green real estate offers a higher return on investment than regular real estate. | O | O | O | O | O |
| | CA5 | I think buying or renting green real estate can bring long-term meaning and value to my life. | O | O | O | O | O |
| Perceived Behavioral Control | PBC1 | I can accept a premium over regular real estate for the same level of green real estate. | O | O | O | O | O |
| | PBC2 | Accept that lowering the requirements for amenities and lots surrounding a green real estate project is acceptable. | O | O | O | O | O |
| | PBC3 | It is acceptable to reduce the requirements for internal factors such as house types for green real estate projects. | O | O | O | O | O |
| | PBC4 | I don't think buying or renting green real estate will be influenced by what others recommend or discourage. | O | O | O | O | O |
| | PBC5 | I don't think buying or renting a green real estate property will be influenced by social views and policies. | O | O | O | O | O |
| (1: strongly disagree, 2: disagree, 3: neither agree nor disagree, 4: agree and 5: strongly agree) | | | | | | | |

## References


1. Hou, Y.; Chen, S.; Yao, Z.; Huang, Q.; Shen, X.; Cao, L.; Cheng, J.; Gui, F.; Zhang, Y.; Wang, X. Green Building Consumption Perception and Its Impact on Fitness Service Purchasing Intentions: An Extended Institutional Analysis and Development Decision-Making Model Analysis. *Buildings* **2023**, *13*, 2536. [CrossRef]
2. Weniger, A.; Del Rosario, P.; Backes, J.G.; Traverso, M. Consumer Behavior and Sustainability in the Construction Industry—Relevance of Sustainability-Related Criteria in Purchasing Decision. *Buildings* **2023**, *13*, 638. [CrossRef]
3. Lu, W.; Chen, X.; Peng, Y.; Liu, X. The effects of green building on construction waste minimization: Triangulating 'big data'with 'thick data'. *Waste Manag.* **2018**, *79*, 142–152. [CrossRef] [PubMed]
4. Li, D.H.; Yang, L.; Lam, J.C. Zero energy buildings and sustainable development implications—A review. *Energy* **2013**, *54*, 1–10. [CrossRef]
5. Li, Y.; Li, M.; Sang, P.; Chen, P.H.; Li, C. Stakeholder studies of green buildings: A literature review. *J. Build. Eng.* **2022**, *54*, 104667. [CrossRef]
6. Kozusznik, M.W.; Maricutoiu, L.P.; Peiró, J.M.; Vîrgă, D.M.; Soriano, A.; Mateo-Cecilia, C. Decoupling office energy efficiency from employees' well-being and performance: A systematic review. *Front. Psychol.* **2019**, *10*, 293. [CrossRef]

7. Darko, A.; Chan, A.P. Critical analysis of green building research trend in construction journals. *Habitat Int.* **2016**, *57*, 53–63. [CrossRef]
8. Gou, Z.; Lau, S.S.Y. Contextualizing green building rating systems: Case study of Hong Kong. *Habitat Int.* **2014**, *44*, 282–289. [CrossRef]
9. Song, Y.; Lau, S.K.; Lau, S.S.Y.; Song, D. A comparative study on architectural design-related requirements of green building rating systems for new buildings. *Buildings* **2023**, *13*, 124. [CrossRef]
10. Lu, W.; Chi, B.; Bao, Z.; Zetkulic, A. Evaluating the effects of green building on construction waste management: A comparative study of three green building rating systems. *Build. Environ.* **2019**, *155*, 247–256. [CrossRef]
11. SCDA. SkyTerrace @ Dawson. 2008. Available online: https://www.scdaarchitects.com/architecture/skyterracedawson (accessed on 1 September 2023).
12. Hostetler, M.; Noiseux, K. Are green residential developments attracting environmentally savvy homeowners? *Landsc. Urban Plan.* **2010**, *94*, 234–243. [CrossRef]
13. Zuo, J.; Zhao, Z.Y. Green building research—Current status and future agenda: A review. *Renew. Sustain. Energy Rev.* **2014**, *30*, 271–281. [CrossRef]
14. Zhang, Y.; Altan, H. A comparison of the occupant comfort in a conventional high-rise office block and a contemporary environmentally-concerned building. *Build. Environ.* **2011**, *46*, 535–545. [CrossRef]
15. Cole, R.J.; Brown, Z.; McKay, S. Building human agency: A timely manifesto. *Build. Res. Inf.* **2010**, *38*, 339–350. [CrossRef]
16. Deuble, M.P.; de Dear, R.J. Green occupants for green buildings: The missing link? *Build. Environ.* **2012**, *56*, 21–27. [CrossRef]
17. Kim, M.J.; Oh, M.W.; Kim, J.T. A method for evaluating the performance of green buildings with a focus on user experience. *Energy Build.* **2013**, *66*, 203–210. [CrossRef]
18. Lan, S.H.; Sheng, T.C. The Study on Key Factors of Influencing Consumers' Purchase of Green Buildings: Application of Two-stage Fuzzy Analytic Hierarchy Process. *Int. Bus. Res.* **2014**, *7*, 49. [CrossRef]
19. Berardi, U. Clarifying the new interpretations of the concept of sustainable building. *Sustain. Cities Soc.* **2013**, *8*, 72–78. [CrossRef]
20. Tan, W.L.; Goh, Y.N. The role of psychological factors in influencing consumer purchase intention towards green residential building. *Int. J. Hous. Mark. Anal.* **2018**, *11*, 788–807. [CrossRef]
21. Tan, T.H. Satisfaction and motivation of homeowners towards green homes. *Soc. Indic. Res.* **2014**, *116*, 869–885. [CrossRef]
22. Ajzen, I. The theory of planned behavior. *Organ. Behav. Hum. Decis. Process.* **1991**, *50*, 179–211. [CrossRef]
23. Conner, M.; Armitage, C.J. Extending the theory of planned behavior: A review and avenues for further research. *J. Appl. Soc. Psychol.* **1998**, *28*, 1429–1464. [CrossRef]
24. Mohd Suki, N. Green product purchase intention: impact of green brands, attitude, and knowledge. *Br. Food J.* **2016**, *118*, 2893–2910. [CrossRef]
25. Costa, C.S.R.; da Costa, M.F.; Maciel, R.G.; Aguiar, E.C.; Wanderley, L.O. Consumer antecedents towards green product purchase intentions. *J. Clean. Prod.* **2021**, *313*, 127964. [CrossRef]
26. Ajzen, I. The theory of planned behaviour: Reactions and reflections. *Psychol. Health* **2011**, *26*, 1113–1127. [CrossRef]
27. Davis, F.D. Perceived usefulness, perceived ease of use, and user acceptance of information technology. *MIS Q.* **1989**, *13*, 319–340. [CrossRef]
28. Davis, F.D.; Bagozzi, R.P.; Warshaw, P.R. User acceptance of computer technology: A comparison of two theoretical models. *Manag. Sci.* **1989**, *35*, 982–1003. [CrossRef]
29. Bagozzi, R.P. Attitudes, intentions, and behavior: A test of some key hypotheses. *J. Personal. Soc. Psychol.* **1981**, *41*, 607. [CrossRef]
30. Wu, J.H.; Chen, Y.C.; Lin, L.M. Empirical evaluation of the revised end user computing acceptance model. *Comput. Hum. Behav.* **2007**, *23*, 162–174. [CrossRef]
31. Lah, U.; Lewis, J.R.; Šumak, B. Perceived usability and the modified technology acceptance model. *Int. J. Hum. Comput. Interact.* **2020**, *36*, 1216–1230. [CrossRef]
32. Bhattacherjee, A.; Sanford, C. The intention—Behaviour gap in technology usage: The moderating role of attitude strength. *Behav. Inf. Technol.* **2009**, *28*, 389–401. [CrossRef]
33. Wu, H.C. An empirical study of the effects of service quality, perceived value, corporate image, and customer satisfaction on behavioral intentions in the Taiwan quick service restaurant industry. *J. Qual. Assur. Hosp. Tour.* **2013**, *14*, 364–390. [CrossRef]
34. Chen, Y.S.; Chang, C.H. Enhance green purchase intentions: The roles of green perceived value, green perceived risk, and green trust. *Manag. Decis.* **2012**, *50*, 502–520. [CrossRef]
35. Karahanna, E.; Straub, D.W. The psychological origins of perceived usefulness and ease-of-use. *Inf. Manag.* **1999**, *35*, 237–250. [CrossRef]
36. Yen, Y.S. Channel integration affects usage intention in food delivery platform services: The mediating effect of perceived value. *Asia Pac. J. Mark. Logist.* **2023**, *35*, 54–73. [CrossRef]
37. Paul, W.L.; Taylor, P.A. A comparison of occupant comfort and satisfaction between a green building and a conventional building. *Build. Environ.* **2008**, *43*, 1858–1870. [CrossRef]
38. *ISO 9241-210:2019* ; Ergonomic Requirements for Office Work with Visual Display Terminals (VDTs): Guidance on Usability. International Organization for Standardization: Geneva, Switzerland, 1998.
39. Montano, D.E.; Kasprzyk, D. Theory of reasoned action, theory of planned behavior, and the integrated behavioral model. *Health Behav. Theory Res. Pract.* **2015**, *70*, 231.

40. Han, S.; Teng, A.K.; Lin, S.Y.; Demiris, G.; Zaslavsky, O.; Chen, A.T. Examining Engagement and Usability in an Online Discussion Platform for Older Adults: Findings From Pilot Studies. *CIN Comput. Inform. Nurs.* **2023**, *41*, 665–672. [CrossRef] [PubMed]
41. Kibert, C.J. *Sustainable Construction: Green Building Design and Delivery*; John Wiley & Sons: Hoboken, NJ, USA, 2016.
42. Tajfel, H.; Turner, J.C. The social identity theory of intergroup behavior. In *Political Psychology*; Psychology Press: London, UK, 2004; pp. 276–293.
43. Chen, M.F. The impacts of perceived moral obligation and sustainability self-identity on sustainability development: A theory of planned behavior purchase intention model of sustainability-labeled coffee and the moderating effect of climate change skepticism. *Bus. Strategy Environ.* **2020**, *29*, 2404–2417. [CrossRef]
44. Liu, M.T.; Liu, Y.; Mo, Z. Moral norm is the key: An extension of the theory of planned behaviour (TPB) on Chinese consumers' green purchase intention. *Asia Pac. J. Mark. Logist.* **2020**, *32*, 1823–1841. [CrossRef]
45. Chen, M.F.; Tung, P.J. Developing an extended theory of planned behavior model to predict consumers' intention to visit green hotels. *Int. J. Hosp. Manag.* **2014**, *36*, 221–230. [CrossRef]
46. Barbarossa, C.; De Pelsmacker, P. Positive and negative antecedents of purchasing eco-friendly products: A comparison between green and non-green consumers. *J. Bus. Ethics* **2016**, *134*, 229–247. [CrossRef]
47. Harland, P.; Staats, H.; Wilke, H.A. Situational and personality factors as direct or personal norm mediated predictors of pro-environmental behavior: Questions derived from norm-activation theory. *Basic Appl. Soc. Psychol.* **2007**, *29*, 323–334. [CrossRef]
48. Patel, C.; Chugan, P.K. Measuring awareness and preferences of real estate developers for green buildings over conventional buildings. In *Consumer Behaviour and Emerging Practices in Marketing*; Aagja, J., Awasthi, A.K., Jain, S., Eds.; Institute of Management, Nirma University, Himalaya Publishing House: Mumbai, India, 2013; pp. 332–341.
49. Hansla, A.; Gamble, A.; Juliusson, A.; Gärling, T. The relationships between awareness of consequences, environmental concern, and value orientations. *J. Environ. Psychol.* **2008**, *28*, 1–9. [CrossRef]
50. De Groot, J.I.; Steg, L. Morality and prosocial behavior: The role of awareness, responsibility, and norms in the norm activation model. *J. Soc. Psychol.* **2009**, *149*, 425–449. [CrossRef] [PubMed]
51. Schwartz, S.H. Awareness of consequences and the influence of moral norms on interpersonal behavior. *Sociometry* **1968**, *31*, 55–369. [CrossRef]
52. Lim, W.M.; Weissmann, M.A. Toward a theory of behavioral control. *J. Strateg. Mark.* **2023**, *31*, 185–211. [CrossRef]
53. Smith, S.; Paladino, A. Eating clean and green? Investigating consumer motivations towards the purchase of organic food. *Australas. Mark. J.* **2010**, *18*, 93–104. [CrossRef]
54. Shimizu, C. Sustainable measures and economic value in green housing. *Open House Int.* **2013**, *38*, 57–63. [CrossRef]
55. Ajzen, I.; Madden, T.J. Prediction of goal-directed behavior: Attitudes, intentions, and perceived behavioral control. *J. Exp. Soc. Psychol.* **1986**, *22*, 453–474. [CrossRef]
56. Pan, Y.; Yin, R.; Huang, Z. Energy modeling of two office buildings with data center for green building design. *Energy Build.* **2008**, *40*, 1145–1152. [CrossRef]
57. Gou, Z.; Lau, S.S.Y.; Chen, F. Subjective and objective evaluation of the thermal environment in a three-star green office building in China. *Indoor Built Environ.* **2012**, *21*, 412–422. [CrossRef]
58. Popescu, D.; Bienert, S.; Schützenhofer, C.; Boazu, R. Impact of energy efficiency measures on the economic value of buildings. *Appl. Energy* **2012**, *89*, 454–463. [CrossRef]
59. Baron, R.M.; Kenny, D.A. The moderator—Mediator variable distinction in social psychological research: Conceptual, strategic, and statistical considerations. *J. Personal. Soc. Psychol.* **1986**, *51*, 1173. [CrossRef] [PubMed]
60. Finstad, K. The usability metric for user experience. *Interact. Comput.* **2010**, *22*, 323–327. [CrossRef]
61. Lewis, J.R.; Utesch, B.S.; Maher, D.E. UMUX-LITE: When there's no time for the SUS. In Proceedings of the SIGCHI Conference on Human Factors in Computing Systems, Paris, France, 27 April–2 May 2013; pp. 2099–2102.
62. Lewis, J.R. Measuring perceived usability: The CSUQ, SUS, and UMUX. *Int. J. Hum. Comput. Interact.* **2018**, *34*, 1148–1156. [CrossRef]
63. Brooke, J. Sus: A 'quick and dirty' usability. *Usability Eval. Ind.* **1996**, *189*, 189–194.
64. Kang, K.H.; Stein, L.; Heo, C.Y.; Lee, S. Consumers' willingness to pay for green initiatives of the hotel industry. *Int. J. Hosp. Manag.* **2012**, *31*, 564–572. [CrossRef]
65. Ajzen, I.; Driver, B. Contingent value measurement: On the nature and meaning of willingness to pay. *J. Consum. Psychol.* **1992**, *1*, 297–316. [CrossRef]
66. Taufique, K.; Siwar, C.; Talib, B.; Chamhuri, N. Measuring consumers' environmental responsibility: A synthesis of constructs and measurement scale items. *Curr. World Environ.* **2014**, *9*, 27. [CrossRef]
67. Kaiser, F.G.; Ranney, M.; Hartig, T.; Bowler, P.A. Ecological behavior, environmental attitude, and feelings of responsibility for the environment. *Eur. Psychol.* **1999**, *4*, 59. [CrossRef]
68. Terry, D.J.; O'Leary, J.E. The theory of planned behaviour: The effects of perceived behavioural control and self-efficacy. *Br. J. Soc. Psychol.* **1995**, *34*, 199–220. [CrossRef] [PubMed]
69. Shah, R.; Goldstein, S.M. Use of structural equation modeling in operations management research: Looking back and forward. *J. Oper. Manag.* **2006**, *24*, 148–169. [CrossRef]

70. Hoyle, R.H. The structural equation modeling approach: Basic concepts and fundamental issues. In *Structural Equation Modeling: Concepts, Issues, and Applications*; Sage Publications, Inc.: Thousand Oaks, CA, USA, 1995.
71. Preacher, K.J.; Hayes, A.F. Asymptotic and resampling strategies for assessing and comparing indirect effects in multiple mediator models. *Behav. Res. Methods* **2008**, *40*, 879–891. [CrossRef] [PubMed]
72. Johnson, R.W. An introduction to the bootstrap. *Teach. Stat.* **2001**, *23*, 49–54. [CrossRef]
73. Kline, R.B.; Santor, D.A. Principles & practice of structural equation modelling. *Can. Psychol.* **1999**, *40*, 381.
74. Hair, J.F.; Ringle, C.M.; Sarstedt, M. PLS-SEM: Indeed a silver bullet. *J. Mark. Theory Pract.* **2011**, *19*, 139–152. [CrossRef]
75. Taber, K.S. The use of Cronbach's alpha when developing and reporting research instruments in science education. *Res. Sci. Educ.* **2018**, *48*, 1273–1296. [CrossRef]
76. Anderson, T.W.; Darling, D.A. A test of goodness of fit. *J. Am. Stat. Assoc.* **1954**, *49*, 765–769. [CrossRef]
77. D'Agostino, R. *Goodness-of-Fit-Techniques*; Routledge: Oxford, UK, 2017.
78. Taherdoost, H. Validity and reliability of the research instrument; how to test the validation of a questionnaire/survey in a research. How to test the validation of a questionnaire/survey in a research. *SSRN Electron. J.* **2016**, *5*, 28–36.
79. Bagozzi, R.P. Evaluating structural equation models with unobservable variables and measurement error: A comment. *J. Mark. Res.* **1981**, *18*, 375–381. [CrossRef]
80. Fornell, C.; Larcker, D.F. Evaluating structural equation models with unobservable variables and measurement error. *J. Mark. Res.* **1981**, *18*, 39–50. [CrossRef]
81. Bandalos, D.L. The effects of item parceling on goodness-of-fit and parameter estimate bias in structural equation modeling. *Struct. Equ. Model.* **2002**, *9*, 78–102. [CrossRef]
82. Hatcher, L.; O'Rourke, N. *A Step-by-Step Approach to Using SAS for Factor Analysis and Structural Equation Modeling*; Sas Institute: Cary, NC, USA, 2013.
83. Wright, S. Path coefficients and path regressions: Alternative or complementary concepts? *Biometrics* **1960**, *16*, 189–202. [CrossRef]
84. Loehlin, J.C. *Latent Variable Models: An Introduction to Factor, Path, and Structural Equation Analysis*; Psychology Press: London, UK, 2004.
85. Alshurafat, H.; Al-Mawali, H.; Al Shbail, M.O. The influence of technostress on the intention to use blockchain technology: The perspectives of Jordanian auditors. *Dev. Learn. Organ. Int. J.* **2023**, *37*, 24–27. [CrossRef]
86. Zheng, S.; Cheng, Y.; Ju, Y. Understanding the intention and behavior of renting houses among the young generation: Evidence from Jinan, China. *Sustainability* **2019**, *11*, 1507. [CrossRef]
87. Portnov, B.A.; Trop, T.; Svechkina, A.; Ofek, S.; Akron, S.; Ghermandi, A. Factors affecting homebuyers' willingness to pay green building price premium: Evidence from a nationwide survey in Israel. *Build. Environ.* **2018**, *137*, 280–291. [CrossRef]
88. Lee, S.S.; Kim, Y.; Roh, T. Pro-environmental behavior on electric vehicle use intention: Integrating value-belief-norm theory and theory of planned behavior. *J. Clean. Prod.* **2023**, *418*, 138211. [CrossRef]
89. Weigel, R.H.; Newman, L.S. Increasing attitude-behavior correspondence by broadening the scope of the behavioral measure. *J. Personal. Soc. Psychol.* **1976**, *33*, 793. [CrossRef]
90. Nadeem, W.; Alimamy, S.; Ashraf, A.R. Navigating through difficult times with ethical marketing: Assessing consumers' willingness-to-pay in the sharing economy. *J. Retail. Consum. Serv.* **2023**, *70*, 103150. [CrossRef]
91. Malik, S.Y.; Hayat Mughal, Y.; Azam, T.; Cao, Y.; Wan, Z.; Zhu, H.; Thurasamy, R. Corporate social responsibility, green human resources management, and sustainable performance: Is organizational citizenship behavior towards environment the missing link? *Sustainability* **2021**, *13*, 1044. [CrossRef]
92. Li, Q.; Long, R.; Chen, H. Differences and influencing factors for Chinese urban resident willingness to pay for green housings: Evidence from five first-tier cities in China. *Appl. Energy* **2018**, *229*, 299–313. [CrossRef]
93. Howell, R.A. It's not (just) "the environment, stupid!" Values, motivations, and routes to engagement of people adopting lower-carbon lifestyles. *Glob. Environ. Chang.* **2013**, *23*, 281–290. [CrossRef]

buildings

MDPI

Article

# Coevolution Mechanism of Remanufacturer–Construction Enterprise–Public in Construction and Demolition Waste Resource Utilization Projects under Green Value Co-Creation

**Beiyu Yi, Yuxi Zou, Weihong Chen, Zihan Li, Yuman He, Zican Jin, Hao Zhang and Xingwei Li ***

College of Architecture and Urban-Rural Planning, Sichuan Agricultural University, Chengdu 611830, China; 202108254@stu.sicau.edu.cn (B.Y.); 2023325024@stu.sicau.edu.cn (Y.Z.); 202008146@stu.sicau.edu.cn (W.C.); 202208085@stu.sicau.edu.cn (Z.L.); 202207928@stu.sicau.edu.cn (Y.H.); 202208206@stu.sicau.edu.cn (Z.J.); zhanghao2@stu.sicau.edu.cn (H.Z.)

* Correspondence: xwl@sicau.edu.cn

**Abstract:** The utilization of resources plays a crucial role in mitigating the environmental pollution issue that improper disposal of construction and demolition waste (CDW) causes. However, the slow growth of the recycled building materials market limits the development of CDW resource utilization. Green value co-creation among remanufacturers, construction enterprises, and the public in CDW resource utilization projects is an effective way to address the issue. This study, based on the theory of value co-creation, uses the evolutionary game method to construct an evolutionary game model for CDW resource utilization projects. The main conclusions are as follows: (1) When the degree of green value co-creation is 0.1 or 0.5, the remanufacturer, the construction enterprise, and the public cannot maintain a state of green value co-creation; when the degree of green value co-creation is 0.9, the remanufacturer, the construction enterprise, and the public in the CDW resource utilization project finally reach a stable state of green value co-creation. (2) When the degree of green value co-creation is 0.5, enhancing the green value co-creation willingness of the remanufacturer or the public can lead other CDW resource utilization project stakeholders to participate in green value co-creation. This study contributes to the promotion of stakeholder cooperation in CDW resource utilization projects, thus providing implications for the promotion of CDW resources.

**Keywords:** green value co-creation; construction and demolition waste resource utilization; coevolution; remanufacturer; construction enterprise; public

## 1. Introduction

The United Nations reports that global waste generation exceeds 10 billion tons annually [1]. As a major source of solid waste [2], improper disposal of construction and demolition waste (CDW) not only leads to increased costs, decreased profits, and overconsumption of raw materials for construction companies, but it also forces the public to face health issues and safety risks [3]. Therefore, how to efficiently handle the large amount of CDW generated has become an urgent problem.

Resource utilization is an effective way of dealing with the CDW [4]. Studies have confirmed the positive economic, social, and environmental effects of CDW resource utilization compared to landfills [5]. However, construction enterprises are reluctant to pay for reproductions due to issues such as low social acceptance and a lack of assurance of product quality [6]. This weakens the enthusiasm of remanufacturers and hinders the development of the CDW recycling market [2]. Therefore, it is necessary to take effective measures to improve the willingness of construction enterprises to pay for recycled building materials.

Green value co-creation improves product quality and customer acceptance through interactive feedback among CDW resource utilization project stakeholders [7], thus promoting the sustainable development of the CDW recycling market. Based on the participation of stakeholders in CDW resource utilization projects, green value co-creation incorporates environmental concepts and encompasses green co-production and green use value [8]. On the one hand, green innovation is the key to green co-production and affects product quality and the willingness of beneficiaries to engage in green value co-creation [9]. However, due to the complexity and uncertainty of innovation activities [10], remanufacturers, as the main body of innovation in CDW recycling, may not choose independent innovation. On the other hand, use is the premise of realizing green use value. However, construction enterprises, as the main users of recycled building materials, are reluctant to use recycled materials without significant advantages in quality and price [2]. Construction enterprises will only consider using recycled building materials for certain purposes, such as creating a green image for the enterprise or obtaining government incentives [2]. In addition, the public is an important driving force for enterprises to consider environmental concepts. Their environmental governance behavior can not only coordinate with the government's environmental governance but also guide companies to regulate their behavior [11]. However, practical factors such as the difficulty of obtaining enterprise information and the high costs of participation continue to hinder public participation in environmental governance [12]. Therefore, further research is necessary to understand how the behaviors of remanufacturers, construction enterprises, and the public in CDW resource utilization projects affect green value co-creation.

Once remanufacturers, construction enterprises, and the public meet the basic requirements for participation, they can jointly create green value through interactive feedback at the stages of production and consumption [13]. The degree of those interactive feedback behaviors among remanufacturers, construction enterprises, and the public in CDW resource utilization projects is expressed as the degree of green value co-creation. The core of assessing the degree of green value co-creation is to evaluate the degree of active participation of project stakeholders in the creation process in terms of both green co-production and green use value. Re et al. assessed the degree of value co-creation using a questionnaire with several indicators, providing a methodology for quantitatively assessing the degree of green value co-creation [7]. The higher the degree of green value co-creation, the more outputs it generates [7]. Specifically, the richer, deeper, and more frequent the interactive feedback behaviors among remanufacturers, construction enterprises, and the public, the more benefits green value co-creation generates. However, the enhancement and deepening of interactive feedback behaviors mean an increase in associated costs. This could potentially impede or even undermine the green value co-creation behaviors of remanufacturers, construction enterprises, and the public in CDW resource utilization projects [14]. Research on green value co-creation primarily concentrates on its positive impact on green innovation [15], enterprise performance, and other aspects [16]. However, there is a dearth of research that delves into the impact of varying levels of green value co-creation on the behaviors of co-creation subjects. To encourage stable green value co-creation among remanufacturers, construction enterprises, and the public in CDW resource utilization projects, it is important to understand how different levels of green value co-creation change the ways that the three groups act.

To obtain green value co-creation for CDW resource utilization project stakeholders, the innovative choices of remanufacturers directly affect the quality and price of recycled building materials, thus influencing the choice of materials by construction enterprises. Construction enterprises' use of materials reflects the market demand for recycled building materials, which in turn influences remanufacturers' innovative choices. In addition, corporate behavior impacts the environment. The state of the environment influences public environmental governance behavior, which in turn shapes corporate behavior [17]. The decision making processes of remanufacturers, construction enterprises, and the public are interdependent and intertwined. Thus, exploring the evolution of the behaviors of remanufacturers, construction enterprises, and the public within the same framework is analyzing their coevolution, or dynamic interaction [18]. Clarifying the coevolution mechanism among remanufacturers, construction enterprises, and the public in CDW resource utilization projects is essential for facilitating CDW resource utilization [19]. However, the coevolution of remanufacturers, construction enterprises, and the public in CDW resource utilization projects under the green value co-creation perspective has not received enough attention. The evolutionary game approach is suitable for analyzing competition, learning, and adaptation among limited rational subjects and is widely used to study the evolution of stakeholder behavior in CDW resource utilization projects [20]. Therefore, the evolutionary game method can be used to explore the behavioral evolution among the three subjects.

Based on this, this paper aims to provide targeted suggestions for promoting CDW resource utilization by deepening the understanding of the coevolution among remanufacturers, construction enterprises, and the public in CDW resource utilization projects under green value co-creation. Specifically, this paper focuses on answering the following two scientific questions: First, what is the impact of the degree of green value co-creation among remanufacturers, construction enterprises, and the public in CDW resource utilization projects on realizing the green value co-creation steady state? Second, how does the willingness of remanufacturers, construction enterprises, and the public to participate in CDW resource utilization projects affect green value co-creation?

In addition, the contributions of this paper are as follows: (1) Innovatively consider the green value co-creation behavior of remanufacturers, construction enterprises, and the public in the adoption of CDW recycled building materials. This study not only enriches the research on green value co-creation but also establishes a theoretical foundation for encouraging green value co-creation among remanufacturers, construction enterprises, and the public. (2) This study introduces the theory of value co-creation into CDW resource utilization projects for the first time. It shows the mechanism of coevolution among remanufacturers, construction enterprises, and the public in CDW resource utilization projects under green value co-creation and provides new evidence to support the growth of CDW resource utilization.

The rest of the paper is structured as follows: Section 2 provides relevant literature reviews. Section 3 constructs a tripartite evolutionary game model of green value co-creation among a remanufacturer, a construction enterprise, and the public. Section 4 analyzes the asymptotic stability of the remanufacturer, the construction enterprise, and the public strategies, and it identifies the stability conditions for the different evolutionary stable strategies (ESSs). Section 5 carries out numerical simulations and discusses the effect of the relevant parameters on the evolutionary paths. Section 6 summarizes the conclusions and limitations of this paper.

## 2. Literature Review

This section reviews the literature from three aspects: CDW resource utilization projects, green value co-creation, and coevolution.

### 2.1. CDW Resource Utilization Projects

CDW resource utilization projects refer to using reasonable technology or management tools to recover useful substances in CDW [21]. Recycling CDW can reduce raw material consumption and conserve energy [22]. It can also effectively alleviate the problems of CDW occupying land resources and polluting the environment [23]. Therefore, many CDW resource utilization projects have emerged and received extensive attention from scholars.

Among them, the operation mode of CDW resource utilization projects and the coordination and cooperation of stakeholders have become research hotspots. On the one hand, the adoption of the public–private partnership (PPP) model, based on the characteristics of CDW resource utilization projects, can reduce the government's financial pressure and, at the same time, enhance the investment willingness of enterprises [24]. Although CDW resource utilization PPP projects have the advantages mentioned above, the key to their success lies in reasonable risk and benefit allocation [25]. On the other hand, stakeholders' coevolution will affect the smooth implementation of CDW resource utilization projects. The public, construction enterprises, and remanufacturers are the core stakeholders of the CDW resource utilization projects [26]. Previous research has used system dynamics models [27] and evolutionary game models [21] to examine the relationships among remanufacturers, construction companies, and the public in great detail. These analyses reveal that the interactions among the three, as well as their respective costs and benefits, all affect the CDW resource utilization effect.

Above all, an effective management solution is a way to promote CDW resource utilization. To foster coevolution among stakeholders in CDW resource utilization projects, they should pay particular attention to the distribution of costs and benefits. If the construction enterprises are reluctant to use the remanufactured products, then the remanufacturers' resource behavior will lose motivation. Therefore, it is necessary to take valid measures to improve the willingness of construction enterprises to pay for recycled building materials. Unfortunately, the existing research mainly focuses on the generation and recycling stages of CDW. How to improve the willingness of construction enterprises to pay for recycled building materials has not received enough attention.

### 2.2. Green Value Co-Creation

Value co-creation is consumers' active participation in cooperation and creating value with enterprises at one or more stages of production and consumption [28]. In recent years, research on value co-creation has gradually expanded from the dual value co-creation of enterprises and consumers to the multi-value co-creation among stakeholders [29]. In addition, green value co-creation has been proposed in the context of higher demands for green sustainability, that is, considering green concepts based on value co-creation [30].

There are two components of green value co-creation: green co-production and green value in use [30]. On the one hand, green co-production is a collaborative process among enterprises in green innovation, including three elements: knowledge sharing, equity, and interaction [30]. Voluntary interactive feedback among enterprises and stakeholders on demand and green awareness can help enterprises solve dynamic problems, clarify customer needs, and attract more customers [31]. On the other hand, green value in use refers to the customer's experience in using, maintaining, and repairing green innovative products, including three elements: experience, personalization, and interpersonal relationships [32]. Feedback from customers provides enterprises with ideas for improving their products and opportunities to increase customer satisfaction. Based on the close relationship between green value co-creation and green innovation, green value co-creation is considered the decisive factor in green innovation [8]. Through quantitative analysis, the positive relationship between green value co-creation and green innovation has been confirmed by research [15]. Additionally, one study confirmed that green value co-creation positively affects enterprise performance, which further suggests that green value co-creation plays an important role in enterprise development [16].

Above all, in the process of green value co-creation, enterprises can not only promote enterprise innovation but also improve enterprise performance through interactive feedback with stakeholders. Unfortunately, research on green value co-creation is inadequate [33]. Green value co-creation, in particular, not only helps to improve product quality, but it can also promote consumer acceptance and adoption of products. While it effectively propels the growth of the market for recycled building materials, it remains unclear how to encourage stakeholders in the CDW resource utilization project to co-create green value.

### 2.3. Coevolution

Coevolution is a pattern of interactions between populations characterized by multilevelness, multidirectional causality, nonlinearity, feedback, and path dependence [34]. First, multilevelness shows that coevolution may occur between enterprises [35], between enterprises and consumers [36], and between enterprises and the social environment [37]. The analysis of coevolution should be integrated at the micro, meso, and macro levels, and one level should not be analyzed separately while ignoring the others [38]. Secondly, multidirectional causality refers to the change in organizations in the environment due to mutual influence. The change in one organization is the result of interactions among other organizations and environmental changes [39]. Thirdly, nonlinearity arises because the organization and the environment influence each other, and the environment is changeable. Therefore, the process of coevolution is complex and changeable. Fourthly, feedback is reflected in the interaction between the organization and the environment. Evolution is a result of feedback [40]. Finally, feedback determines the characteristics of coevolution path dependence. Path dependence reflects the system's evolutionary tendencies under the feedback mechanism [40].

The concept of coevolution has been applied to the fields of ecological economy and organizational management. In the field of organizational management, coevolution provides a way to explore how organizations interact with each other, adapt to, and influence the environment [41]. To clarify how organizations adapt to and influence the environment, scholars have carried out a series of studies. On the one hand, the study of coevolution between enterprises confirms that the micro-behavior of enterprises will cause macro-changes in the relationship between enterprises [42]. On the other hand, the interaction between enterprise and consumer demonstrates that customer-based innovation can increase enterprises' profits while facilitating market dominance development [36]. In addition, the coevolution of enterprises and the political environment shows not only that coevolution occurs in the subjects' learning process but that power and influence also lead to coevolution [43].

Above all, a wealth of research has been conducted on the coevolution between the organization and the environment. It reveals the bidirectional relationship between the organization and the environment, which is significant for promoting organizational development. However, the majority of the existing studies concentrate on the coevolution of enterprise to enterprise, enterprise to consumer, or enterprise to environment, leaving a dearth of studies that simultaneously consider the coevolution of enterprise, consumer, and environment. Furthermore, the mutual influence of remanufacturers, construction enterprises, and the public under green value co-creation reflects the coevolution among enterprises, consumers, and the environment. No research has revealed the coevolution mechanism among the three under green value co-creation.

## 3. Methodology

### 3.1. Problem Description

To promote the development of CDW resource utilization, remanufacturers, construction enterprises, and other stakeholders in CDW resource utilization projects have started to participate in green value co-creation. For instance, a remanufacturer collaborated with a construction enterprise to carry out a CDW resource utilization project in Hangzhou, which further enhanced the market acceptance and recognition of recycled building materials by strengthening ties with the public [44]. To encourage green value co-creation among the stakeholders in CDW resource utilization projects, the government has implemented various incentives and constraints for enterprises and the public. These measures have motivated enterprises to participate in CDW resource utilization to a certain extent. However, economic incentives and constraints alone may not be sufficient to participate with a high degree of autonomy [21]. As the promoter of corporate green behavior, the public's environmental governance behavior can motivate enterprises to implement green actions from a reputational perspective. Enterprises' green actions can bring positive environmental benefits to the public. However, the cost of supervision and the difficulty of information acquisition hinder the public's enthusiasm for environmental governance. As the main body of green use, the use behavior of construction enterprises reflects the demand for recycled building materials, which in turn influences the willingness of remanufacturers to innovate. Remanufacturers, the primary source of green innovation, can adopt different innovation strategies. Under different strategies, remanufacturers encounter different risks and costs, leading to variations in the quality and prices of recycled building materials, thereby influencing the willingness of construction enterprises to utilize these materials. Figure 1 illustrates the game relationship among the three players.

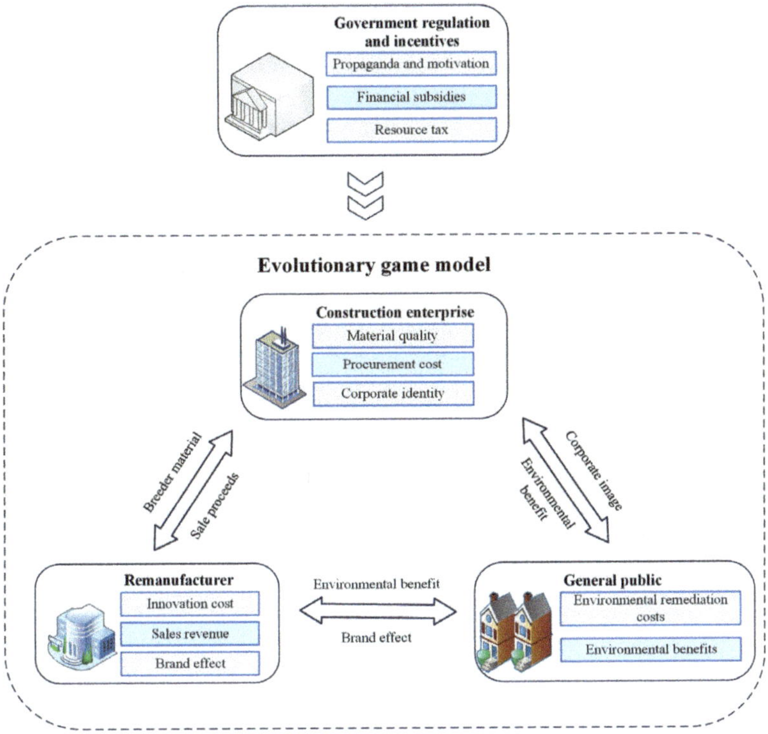


**Figure 1.** The game relationship among the remanufacturer, the construction enterprise, and the public in the CDW resource utilization project.

### 3.2. Model Assumption

Based on the analysis, this paper makes the following assumptions. Table 1 describes the corresponding parameter settings and instructions.

**Table 1.** Parameter settings and descriptions.

| Unit | Parameter | Parameter Description |
|---|---|---|
| Government | $S_1$ | Subsidies for the construction enterprise to positively use recycled building materials [45] |
| | $S_2$ | Subsidies for the independent innovation of the remanufacturer [46] |
| | $Q$ | Incentives for public participation in environmental governance [47] |
| | $T$ | Resource taxes levied on the use of raw materials by the construction enterprise [2] |
| Remanufacturer | $C_1$ | Independent innovation costs [48] |
| | $C_2$ | Imitation innovation costs [49] |
| | $\Delta R$ | The incremental benefits of independent innovation [50] |
| | $b_i$ | The utilization rate of recycled building materials produced by independent innovation or imitation innovation, where $i = 1$ denotes the utilization rate of recycled building materials produced by independent innovation and $i = 2$ denotes the utilization rate of recycled building materials produced by imitation innovation [2] $(0 < b_2 < b_1 < 1)$ |
| | $E_2$ | The environmental benefits of independent innovation [51] |
| | $R_2$ | The brand effect of independent innovation [52] |
| Construction enterprise | $P_1$ | The price of purchasing raw materials [50] |
| | $P_2$ | The price of purchasing recycled building materials [53] |
| | $\Delta C$ | When the three parties join the green value co-creation, the communication costs between the construction enterprise and the remanufacturer [54] |
| | $R_1$ | Image benefits for construction enterprises when positively using recycled building materials [55] |
| | $E_1$ | The environmental benefits of positive use of recycled building materials [56] |
| Public | $C_p$ | Total costs of participation in environmental governance [57] |
| | $\theta$ | The degree of green value co-creation [7] $(0 < \theta < 1)$ |

**Assumption 1.** *To promote CDW resource utilization, the government will guide enterprises and the public to positively participate in CDW resource utilization. At the enterprise level, on the one hand, the government encourages enterprises to make efforts to promote CDW resource utilization by providing them with subsidies. On the other hand, the government promotes the rational use of resources and encourages the use of recycled building materials by imposing taxes on resource products [2]. At the public level, the government points out that organizations and individuals have the right and duty to protect the environment. The government will encourage the public to participate in environmental governance through public opinion propaganda and other means [58].*

**Assumption 2.** *The remanufacturer has two innovation strategies: independent innovation and imitative innovation. Independent innovation is an autonomous enterprise activity for technological breakthroughs and new product development; imitation innovation generally manifests in the purchase of patents [49]. Independent and imitation innovation investments are $C_1$ and $C_2$, respectively. Compared with imitation innovation, although independent innovation is costly and risky [59], targeted innovation can improve product quality, reduce losses during the use of materials, and increase the utilization rate of recycled building materials $b_i$ [2]. The utilization of recycled building materials $b_i$ is relative to the utilization of raw materials. In addition, the enterprise that chooses independent innovation can convert its innovation activities into brand effects $R_2$ [60]. Moreover, the independent innovation of the enterprise can improve the waste recycling rate and save resources, thus bringing incremental benefits $\Delta R$ to the enterprise [50]. Due to the uncertainty and difficulty in quantifying the innovation risk of both independent and imitation innovation, this study does not consider the relevant risk parameters to simplify the model.*

**Assumption 3.** *After meeting the minimum use ratio, the construction enterprise can independently choose whether to continue to use recycled building materials, that is to say, it is up to the construction enterprise to decide whether to positively use recycled building materials. The construction enterprise's attitude towards the use of materials directly determines its material procurement cost, where the raw materials' procurement cost is $P_1$ and the recycled building materials' procurement cost is $P_2$. To simplify the model, this study does not consider other green building techniques that the construction enterprise would adopt and only discusses the construction enterprise's material use choices for sand, gravel, and clay. If the construction enterprise positively uses recycled building materials, it will not only receive government subsidies $S_1$ but also enhance its corporate image $R_1$ [61]. However, it is worth noting that the construction enterprise that positively uses recycled building materials needs to cope with the risks associated with material quality compared to negative use. While the construction enterprise that utilizes recycled building materials negatively does not face this risk, it must comply with government regulations by paying the resource tax T.*

**Assumption 4.** *The public has two strategies, including participating and not participating in environmental governance. The public with increasing concern about environmental issues may be willing to pay $C_p$ to support construction and demolition waste separation and recycling programs, environmental education, and the promotion of recycled product consumption [62]. The public's environmental governance behaviors have the potential to encourage enterprises to adopt green behaviors [63], thereby bringing environmental benefits $E_1$ and $E_2$ to the public. However, the public may still not choose the participation strategy due to constraints such as the difficulty of obtaining information about the enterprises and the cost of monitoring and reporting [12].*

**Assumption 5.** *When the remanufacturer innovates independently, the construction enterprise positively uses recycled building materials, and the public participates in environmental governance, the three parties reach a green value co-creation leading to additional benefits and costs. Both green value co-creation outcomes and costs are positively related to the degree of green value co-creation $\theta$ [7]. The degree of green value co-creation is a measure of the scope of green value co-creation based on the mechanism and form of green value co-creation that indicates the efforts made by the three parties for green value co-creation. The outputs of green value co-creation include improvement in product quality [64], attraction of new customers [65], and enhancement in brand loyalty [66]. The additional costs of green value co-creation include those related to incentives and feedback: specifically, the additional innovation costs $\theta C_1$ paid by the remanufacturer for targeted innovation* *by paying more attention to feedback during the innovation process; the costs $\theta\Delta C$ of positive feedback and communication between the construction enterprise and the remanufacturer; and the additional costs $\theta C_p$ paid by the public for paying attention and feedback.*

**Assumption 6.** *It is supposed that the probability of the remanufacturer choosing independent innovation is x and the probability of imitating innovation is $1 - x(0 \leq x \leq 1)$; the probability of the construction enterprise choosing to use the recycled building materials positively is y and the probability of using the recycled building materials negatively is $1 - y(0 \leq y \leq 1)$; the probability of the public choosing to participate in the environmental governance is z and the probability of not participating in the environmental governance is $1 - z(0 \leq z \leq 1)$. Here, x, y, and z are functions of time.*

Based on the research assumptions and model parameters, the benefits to the remanufacturer, the construction enterprise, and the public are calculated under different combinations of strategies. For example, when the remanufacturer innovates independently, the construction enterprise positively uses recycled building materials, and the public participates in environmental governance, $-(1+\theta)C_p + (1+\theta)E_1 + (1+\theta)E_2 + Q$ is the benefit equation for the public. That is, when three parties are involved in green value co-creation, the benefits to the public consist of the costs of public participation, the environmental benefits of corporate behavior, and government incentives. Table 2 describes the benefits for each stakeholder under the eight strategies.

**Table 2.** The payoff matrix of the three-party game.

| Public | Construction Enterprise | Remanufacturer | |
|---|---|---|---|
| | | **Independent Innovation ($x$)** | **Imitation Innovation ($1-x$)** |
| Participation ($z$) | Positive use ($y$) | $-(1+\theta)C_1 + P_2/(b_1 + b_1\theta) + (1+\theta)R_2 + S_2$ $+(1+\theta)\Delta R$<br>$-P_2/(b_1 + b_1\theta) - \theta\Delta C + (1+\theta)R_1 + S_1$<br>$-(1+\theta)C_p + (1+\theta)E_1 + (1+\theta)E_2 + Q$ | $-C_2 + P_2/b_2$<br>$-P_2/b_2 + R_1 + S_1$<br>$-C_p + E_1 + Q$ |
| | Negative use ($1-y$) | $-C_1 + S_2 + R_2 + \Delta R$<br>$-P_1 - T$<br>$-C_p + E_2 + Q$ | $-C_2$<br>$-P_1 - T$<br>$-C_p + Q$ |
| Non-participation ($1-z$) | Positive use ($y$) | $-C_1 + P_2/b_1 + S_2 + \Delta R$<br>$-P_2/b_1 + S_1$<br>$E_1 + E_2$ | $-C_2 + P_2/b_2$<br>$-P_2/b_2 + S_1$<br>$E_1$ |
| | Negative use ($1-y$) | $-C_1 + S_2 + \Delta R$<br>$-P_1 - T$<br>$E_2$ | $-C_2$<br>$-P_1 - T$<br>0 |

## 4. Model Analysis

### *4.1. Replicated Dynamic Equations*

#### 4.1.1. The Replicated Dynamic Equation for the Remanufacturer, the Construction Enterprise, and the Public

According to the payoff matrix, the expected payoffs of the remanufacturer's independent and imitative innovation strategies are shown in Equations (1) and (2), respectively, and the average expected payoff is shown in Equation (3):

$$U_{R0} = \Delta R + yz\Delta R\theta - (1+yz\theta)C_1 + \frac{y(1+\theta-z\theta)P_2}{(1+\theta)b_1} + z(1+y\theta)R_2 + S_2 \quad (1)$$

$$U_{R1} = \frac{yP_2}{b_2} - C_2 \quad (2)$$

$$\overline{U_R} = x\left[\Delta R + yz\Delta R\theta - (1+yz\theta)C_1 + \frac{y(1+\theta-z\theta)P_2}{(1+\theta)b_1} + z(1+y\theta)R_2 + S_2\right] + (1-x)\left(\frac{yP_2}{b_2} - C_2\right) \quad (3)$$

Based on Equations (1)–(3), the replicated dynamic equation of the remanufacturer is obtained as shown in Equation (4):

$$F(x) = \frac{dx}{dt} = x(U_{R0} - \overline{U_R}) \\ = (1-x)x\left[\Delta R + yz\Delta R\theta - (1+yz\theta)C_1 + C_2 + \frac{y(1+\theta-z\theta)P_2}{(1+\theta)b_1} - \frac{yP_2}{b_2} + z(1+y\theta)R_2 + S_2\right] \quad (4)$$

Based on the payoff matrix, the expected payoffs for the construction enterprise's positive and negative use of recycled building material strategies are shown in Equations (5) and (6), respectively, and the average expected payoff is shown in Equation (7):

$$U_{E0} = -xz\theta\Delta C + \frac{x[-1+(-1+z)\theta]P_2}{(1+\theta)b_1} + \frac{(-1+x)P_2}{b_2} + z(1+x\theta)R_1 + S_1 \quad (5)$$

$$U_{E1} = -T - P_1 \quad (6)$$

$$\overline{U_E} = y\left[-xz\theta\Delta C + \frac{x(-1-\theta+z\theta)P_2}{(1+\theta)b_1} + \frac{(-1+x)P_2}{b_2} + z(1+x\theta)R_1 + S_1\right] + (1-y)(-T-P_1) \quad (7)$$

According to Equations (5)–(7), the replicated dynamic equation of the construction enterprise is obtained as shown in Equation (8):

$$F(y) = \frac{dy}{dt} = y(U_{E0} - \overline{U_E}) \\ = (1-y)y\left[T - xz\theta\Delta C + P_1 + \frac{x(-1-\theta+z\theta)P_2}{(1+\theta)b_1} + \frac{(-1+x)P_2}{b_2} + z(1+x\theta)R_1 + S_1\right] \quad (8)$$

According to the payoff matrix, the expected payoffs for public participation in environmental governance and non-participation in environmental governance strategies are shown in Equations (9) and (10), respectively, and the average expected payoff is shown in Equation (11):

$$U_{P0} = Q - (1+xy\theta)C_P + y(1+x\theta)E_1 + x(1+y\theta)E_2 \quad (9)$$

$$U_{P1} = yE_1 + xE_2 \quad (10)$$

$$\overline{U_P} = Qz - (1+xy\theta)zC_P + y(1+zx\theta)E_1 + x(1+yz\theta)E_2 \quad (11)$$

According to Equations (9)–(11), the replicated dynamic equation of the public is obtained as shown in Equation (12):

$$F(z) = \frac{dz}{dt} = z(U_{P0} - \overline{U_P}) \\ = (1-z)z[Q - (1+xy\theta)C_P + xy\theta(E_1+E_2)] \quad (12)$$

#### 4.1.2. Analysis of the Replicated Dynamic Equation of the Remanufacturer

The derivation of the replicated dynamic equation of the remanufacturer, namely, Equation (4), yields:

$$F'(x) = (1-2x)\left[\Delta R + yz\Delta R\theta - (1+yz\theta)C_1 + C_2 + \frac{y(1+\theta-z\theta)P_2}{(1+\theta)b_1} - \frac{yP_2}{b_2} + z(1+y\theta)R_2 + S_2\right]$$

To facilitate the analysis of the stability of the remanufacturer, let $J(z) = \Delta R + yz\Delta R\theta - (1+yz\theta)C_1 + C_2 + \frac{y(1+\theta-z\theta)P_2}{(1+\theta)b_1} - \frac{yP_2}{b_2} + z(1+y\theta)R_2 + S_2$. Let $J(z) = 0$. Then, $z^* = \frac{-\Delta R + C_1 - C_2 - \frac{yP_2}{(1+\theta)b_1} - \frac{y\theta P_2}{(1+\theta)b_1} + \frac{yP_2}{b_2} - S_2}{y\Delta R\theta - y\theta C_1 - \frac{y\theta P_2}{(1+\theta)b_1} + (1+y\theta)R_2}$. Solve for the first-order partial derivative of $J(z)$ with respect to z, $J'(z) = y\Delta R\theta - y\theta C_1 - \frac{y\theta P_2}{(1+\theta)b_1} + (1+y\theta)R_2$.

When $y\Delta R\theta - y\theta C_1 - \frac{y\theta P_2}{(1+\theta)b_1} + (1+y\theta)R_2 > 0$, $J(z)$ is an increasing function; otherwise, it is a decreasing function.

(1) When $J(z)$ is an increasing function:

If $z > z^*$, then $J(z) > 0$, $F(1) = 0$, $F'(1) < 0$, and the remanufacturer opts for the independent innovation strategy. If $z = z^*$, then $J(z) = 0$, $F(x) \equiv 0$, $F'(x) \equiv 0$, and any strategy of the remanufacturer is stable. If $z < z^*$, then $J(z) < 0$, $F(0) = 0$, $F'(0) < 0$, and the remanufacturer opts for the imitate innovation strategy.

(2) When $J(z)$ is a decreasing function:

If $z > z^*$, then $J(z) < 0$, $F(0) = 0$, $F'(0) < 0$, and the remanufacturer opts for the imitate innovation strategy. If $z = z^*$, then $J(z) = 0$, $F(x) \equiv 0$, $F'(x) \equiv 0$, and any strategy of the remanufacturer is stable. If $z < z^*$, then $J(z) > 0$, $F(1) = 0$, $F'(1) < 0$, and the remanufacturer opts for the independent innovation strategy.

Above all, the incremental revenue of independent innovation, the cost of independent innovation, the price of recycled building materials, the utilization rate of recycled building materials, the brand benefit of independent innovation, the degree of green value co-creation, and the probability of strategic choice of the public and construction enterprise also affect the innovation choice of the remanufacturer. Concretely speaking, when $y\Delta R\theta - y\theta C_1 - \frac{y\theta P_2}{(1+\theta)b_1} + (1+y\theta)R_2 > 0$ and the public's willingness to participate in environmental governance is high, the remanufacturer tends to independent innovation. When the public's willingness to participate in environmental governance is low, the remanufacturer opts for imitation innovation. When the public's willingness to participate in environmental governance is of critical value, the remanufacturer's innovation choice is uncertain. When $y\Delta R\theta - y\theta C_1 - \frac{y\theta P_2}{(1+\theta)b_1} + (1+y\theta)R_2 < 0$, the situation is the opposite.

#### 4.1.3. Analysis of the Replicated Dynamic Equation of the Construction Enterprise

The derivation of the replicated dynamic equation of the construction enterprise, namely, Equation (8), yields:

$$F'(y) = (1-2y)\left[T - xz\theta\Delta C + P_1 + \frac{x(-1-\theta+z\theta)P_2}{(1+\theta)b_1} + \frac{(-1+x)P_2}{b_2} + z(1+x\theta)R_1 + S_1\right]$$

For the purpose of analyzing the stability of the construction enterprise, let $L(x) = T - xz\theta\Delta C + P_1 + \frac{x(-1-\theta+z\theta)P_2}{(1+\theta)b_1} + \frac{(-1+x)P_2}{b_2} + z(1+x\theta)R_1 + S_1$. Let $L(x) = 0$ to obtain $x^* = \frac{-T - P_1 + \frac{P_2}{b_2} - zR_1 - S_1}{-z\theta\Delta C - \frac{P_2}{b_1} + \frac{z\theta P_2}{(1+\theta)b_1} + \frac{P_2}{b_2} + z\theta R_1}$.

The derivation of $L(x)$ obtains $L'(x) = -z\theta\Delta C - \frac{P_2}{b_1} + \frac{z\theta P_2}{(1+\theta)b_1} + \frac{P_2}{b_2} + z\theta R_1$. Thus, $L(x)$ is an increasing function when $-z\theta\Delta C - \frac{P_2}{b_1} + \frac{z\theta P_2}{(1+\theta)b_1} + \frac{P_2}{b_2} + z\theta R_1 > 0$ and a decreasing function when the opposite is true.

(1) When $L(x)$ is an increasing function:

When $x > x^*$, we obtain $L(x) > 0$, $F(1) = 0$, $F'(1) < 0$, and at this time the stable strategy of the construction enterprise is to positively use recycled building materials. When $x = x^*$, we obtain $L(x) = 0$, $F(y) \equiv 0$, $F'(y) \equiv 0$, and then any strategy of the construction enterprise is stable. When $x < x^*$, we obtain $L(x) < 0$, $F(0) = 0$, $F'(0) < 0$, and then the stable strategy of the construction enterprise is to use recycled materials negatively.

(2) When $L(x)$ is a decreasing function:

When $x > x^*$, we obtain $L(x) < 0$, $F(0) = 0$, $F'(0) < 0$, and at this point, the construction enterprise's stable strategy is to use recycled building materials negatively. When $x = x^*$, we obtain $L(x) = 0$, $F(y) \equiv 0$, $F'(y) \equiv 0$, and at this time any strategy of the construction enterprise is stable. When $x < x^*$, we obtain $L(x) > 0$, $F(1) = 0$, $F'(1) < 0$, and the stable strategy of the construction enterprise at this time is to use recycled building materials positively.

In summary, the decision making of the construction enterprise is influenced by the price of recycled building materials, the utilization rate of recycled building materials, the communication costs, the degree of green value co-creation, and the probability of strategic choices between the remanufacturer and the public. Specifically, when $-z\theta\Delta C - \frac{P_2}{b_1} + \frac{z\theta P_2}{(1+\theta)b_1} + \frac{P_2}{b_2} + z\theta R_1 > 0$ is satisfied and the remanufacturer's willingness to innovate independently is high, the construction enterprise will choose to use the recycled building materials positively; on the contrary, when the remanufacturer's willingness to innovate independently is low, the construction enterprise will choose to use the recycled building materials negatively; in particular, when the probability of the remanufacturer's innovation independently is at a certain value, the construction enterprise's choice is uncertain. The opposite is true when $-z\theta\Delta C - \frac{P_2}{b_1} + \frac{z\theta P_2}{(1+\theta)b_1} + \frac{P_2}{b_2} + z\theta R_1 < 0$.

#### 4.1.4. Analysis of the Replicated Dynamic Equation of the Public

The derivation of the replicated dynamic equation of the public, namely, Equation (12), yields $F'(z) = (1-2z)[Q - (1+xy\theta)C_P + xy\theta(E_1+E_2)]$. To facilitate the analysis of the stability of the public, let $Q(y) = Q - (1+xy\theta)C_P + xy\theta(E_1+E_2)$. Let $Q(y) = 0$ to obtain $y^* = \frac{Q - C_P}{x\theta(C_P - E_1 - E_2)}$. The derivation of $Q(y)$ yields $Q'(y) = x\theta(E_1 + E_2 - C_P)$. It follows that $Q(y)$ is an increasing function when $E_1 + E_2 - C_P > 0$ and a decreasing function when the opposite is true.

(1) When $Q(y)$ is an increasing function:

When $y > y^*$, we obtain $Q(y) > 0$, $F(1) = 0$, $F'(1) < 0$, and the stable strategy of the public is to participate in environmental governance. When $y = y^*$, we obtain $Q(y) = 0$, $F(z) \equiv 0$, $F'(z) \equiv 0$, and at this point any strategy of the public is stable. When $y < y^*$, we obtain $Q(y) < 0$, $F(0) = 0$, $F'(0) < 0$, and at this time the public's stable strategy is not to participate in environmental governance.

(2) When $Q(y)$ is a decreasing function:

When $y > y^*$, we obtain $Q(y) < 0$, $F(0) = 0$, $F'(0) < 0$, and at this time, the stable strategy of the public is not to participate in environmental governance. When $y = y^*$, we obtain $Q(y) = 0$, $F(z) \equiv 0$, $F'(z) \equiv 0$, and any strategy of the public is stable. When $y < y^*$, we obtain $Q(y) > 0$, $F(1) = 0$, $F'(1) < 0$, and at this point, the public's stable strategy is to participate in environmental governance.

The analysis shows that the public's participation behaviors are influenced by their participation costs and the behavior of the enterprises. When $E_1 + E_2 - C_P > 0$ is satisfied and the willingness of the construction enterprise to positively use recycled building materials is not high enough, the public will choose not to participate regardless of the strategy chosen by the remanufacturer; on the contrary, if the willingness of the construction enterprise to positively use recycled building materials is high enough, the public will choose the participation strategy; when the probability of the construction enterprise positively using recycled building materials is at a certain value, the public's decision making is uncertain. However, when $E_1 + E_2 - C_P < 0$, the opposite is true.

The replicated dynamic system obtained from the analysis is shown in Equation (13).

$$\begin{cases} F(x) = (1-x)x\left[\Delta R + yz\Delta R\theta - (1+yz\theta)C_1 + C_2 + \dfrac{y(1+\theta-z\theta)P_2}{(1+\theta)b_1} - \dfrac{yP_2}{b_2} + z(1+y\theta)R_2 + S_2\right] \\ F(y) = (1-y)y\left[T - xz\theta\Delta C + P_1 + \dfrac{x(-1-\theta+z\theta)P_2}{(1+\theta)b_1} + \dfrac{(-1+x)P_2}{b_2} + z(1+x\theta)R_1 + S_1\right] \\ F(z) = (1-z)z[Q - (1+xy\theta)C_P + xy\theta(E_1+E_2)] \end{cases} \quad (13)$$

## 4.2. Evolutionary Stable Strategy

The replicated dynamic equation reflects the changes in the direction and speed of the subjects' behavioral choices. The equilibrium point is obtained when the replicated dynamic equations are equal to 0. It can be calculated according to Equation (14):

$$\begin{cases} F(x) = (1-x)x\left[\Delta R + yz\Delta R\theta - (1+yz\theta)C_1 + C_2 + \dfrac{y(1+\theta-z\theta)P_2}{(1+\theta)b_1} - \dfrac{yP_2}{b_2} + z(1+y\theta)R_2 + S_2\right] = 0 \\ F(y) = (1-y)y\left[T - xz\theta\Delta C + P_1 + \dfrac{x(-1-\theta+z\theta)P_2}{(1+\theta)b_1} + \dfrac{(-1+x)P_2}{b_2} + z(1+x\theta)R_1 + S_1\right] = 0 \\ F(z) = (1-z)z[Q - (1+xy\theta)C_P + xy\theta(E_1+E_2)] = 0 \end{cases} \quad (14)$$

By solving Equation (13), eight pure strategic equilibrium points, $E_1$ (0,0,0), $E_2$ (1,0,0), $E_3$ (0,1,0), $E_4$ (0,0,1), $E_5$ (0,1,1), $E_6$ (1,1,0), $E_7$ (1,0,1), and $E_8$ (1,1,1), are obtained.

Ritzberger and Weibull showed that it is the strict Nash equilibrium and the pure strategy Nash equilibrium that become asymptotically stable solutions to the evolving game system and therefore this paper did not consider mixed strategy solutions [67]. Thus, this section only analyzes eight pure strategy equilibrium points. According to the Friedman method, the stability of the equilibrium point of the evolutionary game can be judged by the local stability analysis method of the Jacobi matrix [68]. When all eigenvalues are negative, the point is ESS.

The Jacobi matrix of the tripartite evolutionary game system is shown in Equation (15):

$$J = \begin{bmatrix} \frac{\partial F(x)}{\partial x} & \frac{\partial F(x)}{\partial y} & \frac{\partial F(x)}{\partial z} \\ \frac{\partial F(y)}{\partial x} & \frac{\partial F(y)}{\partial y} & \frac{\partial F(y)}{\partial z} \\ \frac{\partial F(z)}{\partial x} & \frac{\partial F(z)}{\partial y} & \frac{\partial F(z)}{\partial z} \end{bmatrix} = \begin{bmatrix} (1-2x)\left[\frac{(1+\theta-\theta z)yP_2}{(1+\theta)b_1} - \frac{yP_2}{b_2} + \Delta R + yz\Delta R\theta - (1+yz\theta)C_1 + C_2 + z(1+y\theta)R_2 + S_2\right] & (1-x)x\left[-z\theta C_1 + z\theta(\Delta R + R_2) - \frac{P_2}{b_2} + \frac{(1+\theta-\theta z)P_2}{(1+\theta)b_1}\right] & (1-x)x\left[y\theta\Delta R + R_2 + y\theta\left(-C_1 - \frac{P_2}{b_1+\theta b_1} + R_2\right)\right] \\ (1-y)y\left[-z\theta\Delta C + \frac{(-1-\theta+z\theta)P_2}{(1+\theta)b_1} + \frac{P_2}{b_2} + z\theta R_1\right] & (1-2y)\left[T - xz\theta\Delta C + P_1 + \frac{x(-1-\theta+z\theta)P_2}{(1+\theta)b_1} + \frac{(-1+x)P_2}{b_2} + z(1 + x\theta)R_1 + S_1\right] & (1-y)y\left[-x\theta\Delta C + R_1 + x\theta\left(\frac{P_2}{b_1+\theta b_1} + R_1\right)\right] \\ (1-z)zy\theta(-C_P + E_1 + E_2) & (1-z)zx\theta(-C_P + E_1 + E_2) & (1-2z)\left[Q - (1 + xy\theta)C_P + xy\theta(E_1 + E_2)\right] \end{bmatrix} \quad (15)$$

By putting the pure strategy equilibrium points into the Jacobi matrix, the eigenvalues can be obtained, as shown in Table 3.

This paper examines the equilibrium points and Jacobi matrix eigenvalues of the system in Table 3. Finding the positive and negative eigenvalues of each equilibrium point cannot be determined. It shows that the game's strategy is constantly changing due to a variety of factors.

$E_8$ (1,1,1) is the most ideal result. In this situation, the remanufacturer opts for independent innovation, the construction enterprise opts for positively using recycled building materials, and the public opts for participating in environmental governance. This state can create more value for all parties while promoting the popularization and application of recycled building materials, thus promoting the development of CDW resource utilization.

**Table 3.** Equilibrium points and Jacobi matrix eigenvalues.

| Equilibrium Point | Eigenvalue | | |
|---|---|---|---|
| | $\lambda_1$ | $\lambda_2$ | $\lambda_3$ |
| $E_1(0,0,0)$ | $-C_1+C_2+S_2+\Delta R$ | $\frac{-P_2+b_2(T+P_1+S_1)}{b_2}$ | $Q-C_p$ |
| $E_2(1,0,0)$ | $C_1-C_2-S_2-\Delta R$ | $\frac{-P_2}{b_1}+T+P_1+S_1$ | $Q-C_p$ |
| $E_3(0,1,0)$ | $\frac{P_2}{b_1}-\frac{P_2}{b_2}-C_1+C_2+S_2+\Delta R$ | $-\frac{-P_2+b_2(T+P_1+S_1)}{b_2}$ | $Q-C_p$ |
| $E_4(0,0,1)$ | $-C_1+C_2+S_2+R_2+\Delta R$ | $\frac{-P_2+b_2(T+P_1+S_1+R_1)}{b_2}$ | $-Q+C_p$ |
| $E_5(0,1,1)$ | $\frac{P_2}{(1+\theta)b_1}-\frac{P_2}{b_2}-(1+\theta)C_1$ $+C_2+S_2+(1+\theta)R_2+(1+\theta)\Delta R$ | $\frac{P_2}{b_2}-T-P_1-S_1-R_1$ | $-Q+C_p$ |
| $E_6(1,1,0)$ | $-\frac{P_2}{b_1}+\frac{P_2}{b_2}+C_1-C_2-S_2-\Delta R$ | $\frac{P_2}{b_1}-T-P_1-S_1$ | $Q-(1+\theta)C_p+$ $\theta(E_1+E_2)$ |
| $E_7(1,0,1)$ | $C_1-C_2-S_2-R_2-\Delta R$ | $\frac{-P_2}{(1+\theta)b_1}+T+P_1+S_1$ $-\theta\Delta C+(1+\theta)R_1$ | $-Q+C_p$ |
| $E_8(1,1,1)$ | $-\frac{P_2}{(1+\theta)b_1}+\frac{P_2}{b_2}+(1+\theta)C_1$ $-C_2-S_2-(1+\theta)R_2-(1+\theta)\Delta R$ | $\frac{P_2}{(1+\theta)b_1}-T-P_1-S_1$ $+\theta\Delta C-(1+\theta)R_1$ | $-Q+(1+\theta)C_p$ $-\theta(E_1+E_2)$ |

To realize the stable state of $E_8$ (1,1,1), all its eigenvalues should be negative:

$$\begin{cases} -\frac{P_2}{(1+\theta)b_1}+\frac{P_2}{b_2}+(1+\theta)C_1-C_2-S_2-(1+\theta)R_2-(1+\theta)\Delta R<0 \\ \frac{P_2}{(1+\theta)b_1}-T-P_1-S_1+\theta\Delta C-(1+\theta)R_1<0 \\ -Q+(1+\theta)C_p-\theta(E_1+E_2)<0 \end{cases}$$

That means when the other two parties are involved, the remanufacturer will choose to innovate independently if the benefits of independent innovation are higher than those of imitation. Similarly, the construction enterprise and the public will only choose the green value co-creation strategy, namely, using recycled building materials positively and taking part in environmental governance, if green value co-creation can bring higher benefits.

## 5. Numerical Simulation and Discussion

This paper focuses on how the three parties can achieve a steady state of green value co-creation through evolutionary game model analysis. The green value co-creation of the remanufacturer, the construction enterprise, and the public in the CDW resource utilization project are affected by the willingness of the parties and the degree of green value co-creation. This study uses MATLAB R2022b to numerically simulate and analyze the evolution of the system, making it easier to understand how the strategies of the remanufacturer, the construction enterprise, and the public worked together in the CDW resource utilization project.

Firstly, in Section 5.1, this paper considers the effects of different degrees of green value co-creation ($\theta$) and initial willingness combinations of the remanufacturer, the construction enterprise, and the public on the steady state of system evolution. This helps to clarify the overall influence of multiple factors on the evolutionary trend of the system and provides a basis for further analysis on how to facilitate the evolution of the system to the desired state (1,1,1).

Secondly, in Section 5.2, this paper further analyzes how the degree of green value co-creation ($\theta$) among the remanufacturer, the construction enterprise, and the public of the CDW resource utilization project affects the steady state of green value co-creation among the three. The optimal degree of green value co-creation to maintain green value co-creation homeostasis among the three is then obtained.

Finally, based on the analyses in Sections 5.1 and 5.2, Section 5.3 aims to explore how the initial willingness of green value co-creation among the remanufacturer, the construction enterprise, and the public affects the state of green value co-creation among the three. Specifically, this paper explore the impact of changing the initial willingness of individual subjects on the system evolution outcome, particularly when the system evolution outcome is influenced by the combination of initial willingness but the degree of green value co-creation has not yet reached an optimal level. This helps to clarify which subject's initial willingness enhancement is better suited to motivating other subjects to participate in green value co-creation. This provides ideas for promoting green value co-creation among remanufacturers, construction companies, and the public.

Based on this, this paper determines the initial values of the parameters through literature surveys and expert interviews, as shown in Table 4.

**Table 4.** Initial parameter settings.

| Parameters | Values | Parameters | Values |
|---|---|---|---|
| $S_1$ | 2 [56] | $C_2$ | 3 [49] |
| $S_2$ | 2 [69] | $\Delta R$ | 6 [48] |
| $Q$ | 2 [47] | $b_1$ | 1.0 [2] |
| $C_P$ | 3 [70] | $E_2$ | 2.88 [71] |
| $P_1$ | 7 [50] | $T$ | 1 |
| $P_2$ | 6 [50] | $E_1$ | 2.88 |
| $\Delta C$ | 5 [54] | $b_2$ | 0.55 |
| $R_1$ | 5 [55] | $R_2$ | 7 |
| $C_1$ | 10 [51] | | |

### *5.1. The Effect of the Degree of Green Value Co-Creation and the Combination of Initial Willingness on the Evolutionary Steady State of the System*

In this section, $\theta$ = 0.1, 0.5, and 0.9 are used to denote the low, medium, and high degrees of green value co-creation, respectively [72]. And different initial willingness combinations are taken to evolve 50 times under low, medium, and high degrees [73], to explore the effect of the degree of green value co-creation $\theta$ and the combination of the initial willingness of the remanufacturer, the construction enterprise, and the public on the system evolution. Figure 2 describes how the degree of green value co-creation and the initial willingness combination affect the system's evolutionary steady state.

As can be seen from Figure 2a, the system has no stable point, i.e., the system evolution cannot reach a steady state. As can be seen from Figure 2b,c, the system converges to a steady state (1,1,1) under some combinations of initial strategies, and the rest of the combinations have no steady state. This suggests that different combinations of the initial willingness of the remanufacturer, the construction enterprise, and the public lead to different evolutionary outcomes at medium and high levels of green value co-creation. This is consistent with the findings of Dou et al. [74], which together validate the butterfly effect in value co-creation [75]. Furthermore, as the degree of green value co-creation among the remanufacturer, the construction enterprise, and the public increases, the system transitions from a non-steady state to a steady state (1,1,1), and this motivates more initial willingness combinations to evolve towards (1,1,1). This means that the increase in the degree of green value co-creation among the stakeholders in the CDW resource utilization project helps the system reach the ideal state of three-party green value co-creation.

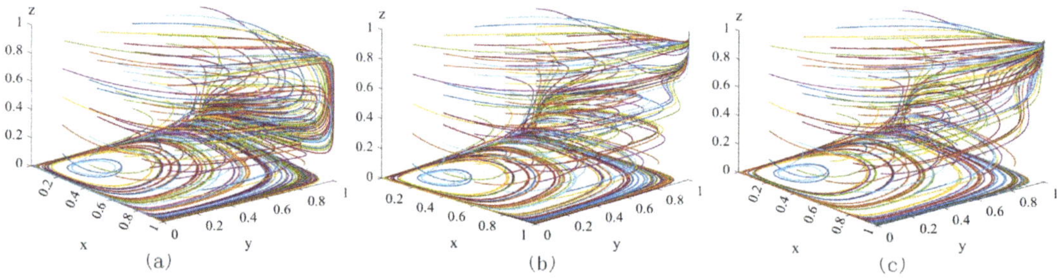


**Figure 2.** The effect of the degree of green value co-creation and initial willingness combination on the stable state of system evolution. In particular, subfigures (**a–c**) represent the results of the system's green value co-creation, respectively.

### *5.2. The Impact of the Degree of Green Value Co-Creation on the Evolutionary Steady State of the System*

This section considers the effects of different degrees of green value co-creation on the evolutionary steady state of the system. The initial willingness of the game subjects is set to 0.5 to eliminate the influence of the subjects' green value co-creation preference. Figure 3 describes the effect of the degree of green value co-creation on the evolution of the behavior of the remanufacturer, the construction enterprise, and the public.

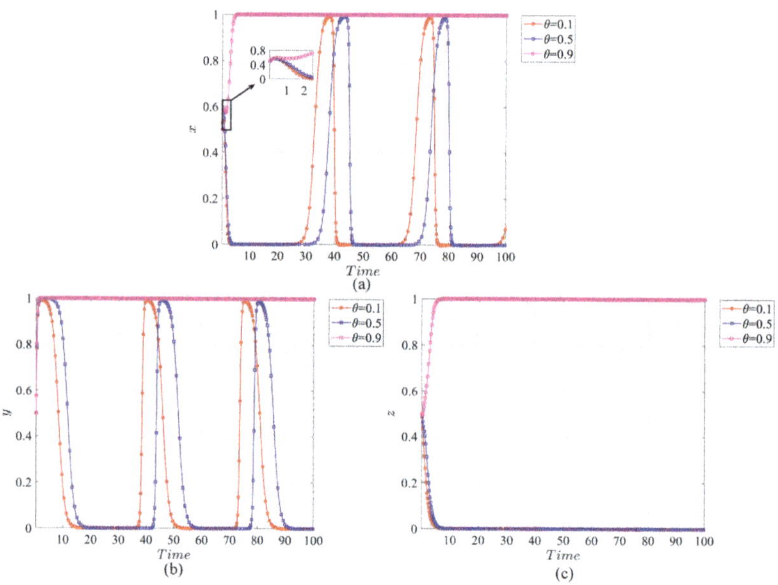


**Figure 3.** The effect of the degree of green value co-creation on the behavioral evolution of the game subjects. In particular, subfigures (**a–c**) represent the evolutionary paths of the remanufacturer, the construction enterprise, and the public when $\theta$ = 0.1, 0.5, and 0.9, respectively.

As can be seen from Figure 3, when $\theta = 0.1$ or $\theta = 0.5$, $x$ and $y$ fail to reach a steady state, while $z$ converges to 0; when $\theta = 0.9$, $x$, $y$, and $z$ all converge to 1. That is, under the medium and low degrees of green value co-creation, the strategic choices of the remanufacturer and the construction enterprise show cyclical oscillations, and the public will ultimately choose not to participate in the environmental governance; under the high degree of green value co-creation, the three main parties all choose green value co-creation behavior. This suggests that a high degree of green value co-creation is more conducive to a stable state of green value co-creation among the three parties, in line with the findings of Li et al. [76]. Li et al. stated that a high degree of value co-creation between manufacturers and retailers maximizes profits. The public's consistent refusal to participate at medium and low levels may be due to their insufficient awareness of the environmental benefits and government incentives to offset their participation costs.

By analyzing Figure 3a,b, it is evident that the strategic choices of the remanufacturer and the construction enterprise are uncertain and show cyclical oscillations. This may be because when the remanufacturer shows a willingness to independently innovate, the construction enterprise will positively use recycled building materials to create a green corporate image. However, innovation is a long-term activity, and there is a lag in the construction enterprise's recognition of the remanufacturer's innovative activities. This largely affects the willingness of the remanufacturer to independently innovate, in turn affecting the willingness of the construction enterprise to use them.

### *5.3. The Impact of Initial Willingness on the Evolutionary Steady State of the System*

In this section, three initial strategy scenarios of the subjects are considered at a medium degree of green value co-creation [77]. The initial state of each strategy is 0.1, 0.5, and 0.9, respectively.

#### 5.3.1. The Impact of Changes in the Initial Willingness of the Remanufacturer on the Evolutionary Steady State of the System

The impact of the initial willingness of the remanufacturer to innovate independently on the strategic choices of the remanufacturer, the construction enterprise, and the public, respectively, is shown in Figure 4.

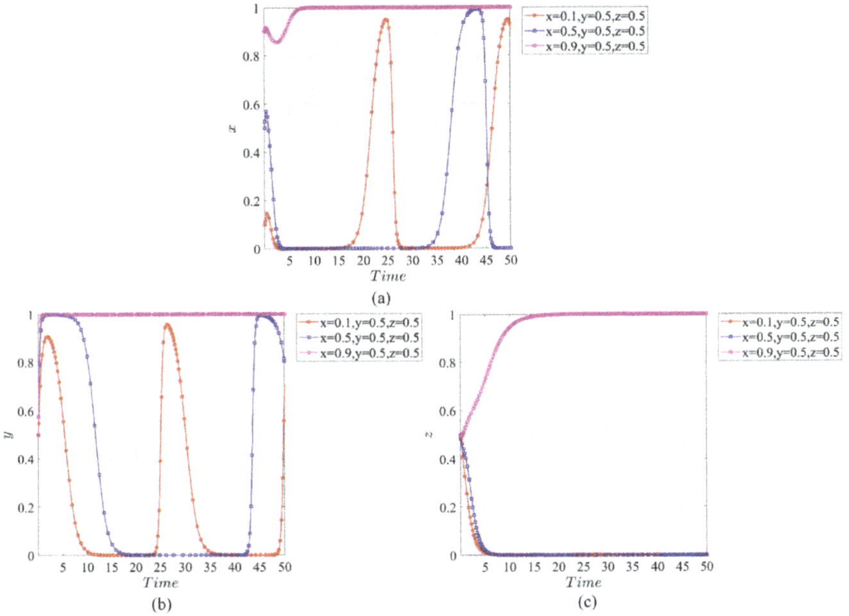


**Figure 4.** The impact of the initial willingness of the remanufacturer on the strategy choices of the game subjects. In particular, subfigures (**a–c**) represent the effects of the initial willingness of the remanufacturer on the evolution of the decision making of the remanufacturer, the construction enterprise, and the public, respectively.

As can be seen from Figure 4, the change in the initial willingness of the remanufacturer has an effect on the evolution of the system. That is, when the remanufacturer's willingness to innovate independently is 0.9, the remanufacturer, the construction enterprise, and the public all converge to the stable state of adopting green value co-creation behavior; when the remanufacturer's initial willingness to innovate independently is 0.1 or 0.5, the decision making of the remanufacturer and the construction enterprise shows cyclical oscillations, and the higher the remanufacturer's initial willingness to innovate independently, the slower the frequency of the oscillations. And eventually, the public converges on the stable state of not participating in environmental governance. This suggests that the increase in green value co-creation willingness of the remanufacturer can drive the construction enterprise and the public to participate in green value co-creation.

#### 5.3.2. The Impact of Changes in the Initial Willingness of the Construction Enterprise on the Evolutionary Steady State of the System

The impact of the initial willingness of the construction enterprise to actively use recycled building materials on the strategic choices of the remanufacturer, the construction enterprise, and the public is shown in Figure 5.

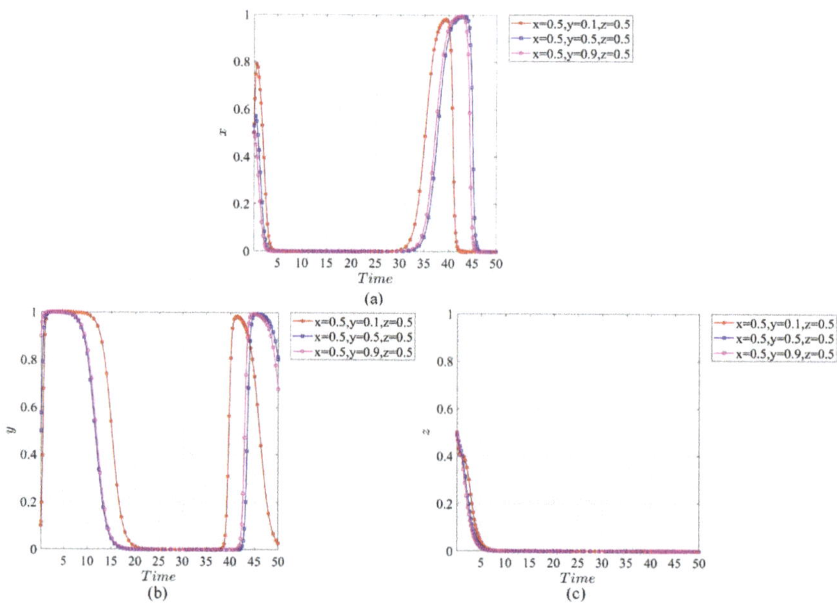


**Figure 5.** The impact of the initial willingness of the construction enterprise on the strategy choices of game subjects. In particular, subfigures (**a–c**) represent the effects of the construction enterprise's initial willingness on the evolution of the decisions of the remanufacturer, the construction enterprise, and the public, respectively.

In Figure 5, we see that the effect of changes in the initial willingness of the construction enterprise on the evolution of the system. Specifically, regardless of the initial willingness of the construction enterprise to use recycled building materials, the decisions of the remanufacturer and the construction enterprise always show periodic oscillations, and the public always converges to a steady state of non-participation in environmental governance. Moreover, the change in the initial willingness of the construction enterprise to positively use reproductions has little effect on the rate of system evolution. When the construction enterprise's initial willingness to positively use recycled building materials is medium or high, the system evolution speed and trend are almost identical. This suggests that changing the green value co-creation willingness of the construction enterprise has little effect on the green value co-creation willingness of the remanufacturer and the public. This may be because the use behavior of the construction enterprise provides a free-rider opportunity for the remanufacturer and the public, and the remanufacturer and the public are more inclined to obtain something for nothing. Therefore, the willingness of the construction enterprise to use the product does not affect the decision making of the remanufacturer or the public.

### 5.3.3. The Impact of Changes in Initial Public Willingness on the Evolutionary Steady State of the System

The impact of the initial willingness of the public to participate in environmental governance on the strategic choices of the remanufacturer, the construction enterprise, and the public is shown in Figure 6.

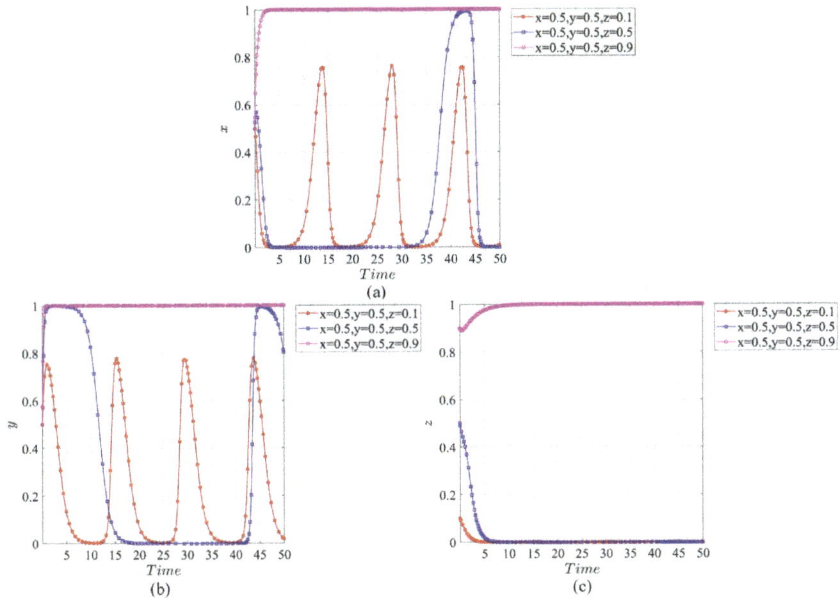


**Figure 6.** The impact of public initial willingness on the strategy choices of game subjects. In particular, subfigures (**a**–**c**) represent the effects of initial public willingness on the evolution of the decision making of the remanufacturer, the construction enterprise, and the public, respectively.

As can be seen from Figure 6a,b, when the initial willingness of public participation is 0.1 or 0.5, the strategy choices of the remanufacturer and the construction enterprise show periodic oscillation, with the frequency of oscillation increasing with a lower initial willingness. When the initial willingness of public participation is 0.9, the strategies of both the remanufacturer and the construction enterprise change to adopt green value co-creation behaviors, i.e., independent innovation and positive use of recycled building materials. As can be seen from Figure 6c, when the initial willingness of public participation is 0.1 and 0.5, the public ultimately chooses the non-participation strategy, and when the initial willingness of public participation is 0.5, the convergence to the non-participation strategy is slower. In contrast, when the initial willingness of the public to participate is 0.9, the public strategy choice shifts to participation. This indicates that the change in public initial willingness can affect the strategy choices of the game subjects. Specifically, the higher the initial willingness of the public to participate, the more favorable it is for the three parties to reach a stable state of green value co-creation. In other words, the increase in the public's willingness to participate in green value co-creation can lead to the participation of the remanufacturer and the construction enterprise in green value co-creation.

The results in Figures 4 and 6 are mostly the same, which suggests that making the public and remanufacturer more willing to participate in green value co-creation has a positive effect on the system's evolution to a steady state where all three parties opt for green value co-creation behaviors. That is, the selection of green value co-creation behaviors by the remanufacturer and the public can effectively promote the participation of other subjects in green value co-creation. This, together with the findings of Shi et al. [78], confirms the positive effect of the green value co-creation behavior of the remanufacturer on system evolution. On this basis, this study also highlights the positive effect of public green value co-creation behavior on system evolution. This may be because increased public green awareness can lead to enterprise transformation and green actions, which contribute to green value co-creation [79].

## 6. Conclusions and Limitations

Based on the theory of value co-creation, this paper constructs an evolutionary game model of the CDW resource utilization project among a remanufacturer, a construction enterprise, and the public using the evolutionary game method. The impacts of the degree of green value co-creation and the initial willingness combination of the remanufacturer, the construction enterprise, and the public on the evolutionary behavior of their collaborative governance of CDW are investigated. The main conclusions are as follows.

(1) When the degree of green value co-creation is 0.1 or 0.5, the remanufacturer, the construction enterprise, and the public cannot maintain a state of green value co-creation. The decision making of the remanufacturer and the construction enterprise shows cyclical oscillations, and the public finally chooses not to participate in environmental governance. When the degree of green value co-creation is 0.9, the remanufacturer, the construction enterprise, and the public of the CDW resource utilization project finally reach a stable state of green value co-creation. Therefore, improving the degree of green value co-creation can help to achieve a stable state of green value co-creation among the remanufacturer, the construction enterprise, and the public.

(2) When the degree of green value co-creation is 0.5, increasing the green value co-creation willingness of the remanufacturer or the public can lead other CDW resource utilization projects' stakeholders to participate in green value co-creation. However, increasing the construction enterprise's green value co-creation willingness has little effect on the system evolution trend and does not change the system evolution result. Therefore, it is a feasible way to encourage other stakeholders to participate in green value co-creation by increasing the green value co-creation willingness of remanufacturers or the public.

This paper, considering green value co-creation among remanufacturers, construction enterprises, and the public in CDW resource utilization projects for the first time, provides new evidence from CDW resource utilization for studies related to green value co-creation. Meanwhile, in the practical sense, the management insights of this paper are as follows:

(1) Enhance the degree of green value co-creation among remanufacturers, construction enterprises, and the public in CDW resource utilization projects. For example, during the production stage, enterprises can strengthen their ties with stakeholders by distributing questionnaires or conducting interviews, thereby involving them in the product development stage and collaborating on the product's design. During the consumption stage, enterprises can target product improvements by offering incentives or discounts to customers who provide feedback. A collaborative process at one or more stages of production or consumption would allow for a stronger green value co-creation relationship between remanufacturers, construction enterprises, and the public.

(2) Enhance the positivity of remanufacturers and the public to co-create green value. At the remanufacturer level, on the one hand, the government should adopt a reasonable incentive model and policy guarantee for remanufacturers' innovation. On the other hand, through a series of effective measures, the government can alleviate remanufacturers' concerns about the failure of innovations related to CDW resource utilization. Specifically, the government should establish or improve the relevant technical standards and operational norms to ensure the innovations are measurable and comparable. This will not only improve the quality of innovative products but also increase market acceptance and reduce the losses suffered by remanufacturers due to the fear that innovation results will not meet the standards. Encourage research organizations to establish a close partnership with remanufacturers so that research results can be quickly transformed into practical applications. At the same time, remanufacturers can also provide feedback on market demand to scientific research institutes to guide the direction of scientific research to be closer to the actual demand. In the way, remanufacturers are encouraged to actively participate in independent innovation. At the public level, firstly, the education of the public should be strengthened to increase their awareness of environmental protection and encourage their participation in the supervision and management of CDW resources. Secondly, strengthen the disclosure of enterprise information, give full play to the role of various network platforms, and reduce the difficulty of public information access. In addition, the government should establish and improve the public supervision and reporting mechanism, increase the incentives for public participation in supervision, and reduce the cost of public participation in environmental governance.

The conclusions of this paper provide suggestions for promoting green value co-creation among the remanufacturer, the construction enterprise, and the public in CDW resource utilization projects, but this paper still suffers from the following limitations. Firstly, innovation risk is an important factor influencing enterprises' innovation choices, but this paper only considers market risk and does not take into account risks such as innovation failure. Secondly, the data used in this study were taken from the relevant literature, and future research could draw from actual cases to obtain more realistic conclusions. Finally, this paper assumes that the influence of the degree of green value co-creation on both the costs and benefits of green value co-creation is linear, but the actual situation is more complicated. Future research can use non-linear regression, neural networks, and other methods to explore the relationship between the degree of green value co-creation and the costs and benefits of green value co-creation to more accurately understand the synergistic evolution mechanism of the stakeholders of CDW resource utilization projects under green value co-creation.

**Author Contributions:** Methodology, validation, formal analysis, investigation, resources, data curation, writing—original draft, writing—review and editing, visualization, B.Y.; conceptualization, methodology, writing—original draft, supervision, project administration, X.L.; writing—review and editing, Y.Z., W.C., Z.L., Y.H., Z.J. and H.Z. All authors have read and agreed to the published version of the manuscript.

**Funding:** This work was supported by the National Natural Science Foundation of China (grant number 72204178), the Sichuan Science and Technology Program, Natural Science Foundation of Sichuan, China (grant number 2023NSFSC1053), and the National College Students Innovation and Entrepreneurship Training Plan (grant number 202410626004).

**Data Availability Statement:** Data are contained within the article.

**Conflicts of Interest:** The authors declare no conflicts of interest.

## References

1. United Nations Home Page. Available online: https://www.un.org/en/actnow/facts-and-figures (accessed on 25 May 2024).
2. Su, Y.; Si, H.; Chen, J.; Wu, G. Promoting the sustainable development of the recycling market of construction and demolition waste: A stakeholder game perspective. *J. Clean. Prod.* **2020**, *277*, 122281. [CrossRef]

3. Tafesse, S.; Girma, Y.E.; Dessalegn, E. Analysis of the socio-economic and environmental impacts of construction waste and management practices. *Heliyon* **2022**, *8*, e09169. [CrossRef] [PubMed]
4. Ma, L.; Zhang, L. Evolutionary game analysis of construction waste recycling management in China. *Resour. Conserv. Recycl.* **2020**, *161*, 104863. [CrossRef]
5. Yuan, F.; Shen, L.; Li, Q. Emergy analysis of the recycling options for construction and demolition waste. *Waste Manag.* **2011**, *31*, 2503–2511. [CrossRef] [PubMed]
6. Zhao, W.; Leeftink, R.B.; Rotter, V.S. Evaluation of the economic feasibility for the recycling of construction and demolition waste in China—The case of Chongqing. *Resour. Conserv. Recycl.* **2010**, *54*, 377–389. [CrossRef]
7. Re, B.; Magnani, G. Value co-creation processes in the context of circular entrepreneurship: A quantitative study on born circular firms. *J. Clean. Prod.* **2023**, *392*, 135883. [CrossRef]
8. Yousaf, Z. Go for green: Green innovation through green dynamic capabilities: Accessing the mediating role of green practices and green value co-creation. *Environ. Sci. Pollut. Res.* **2021**, *28*, 54863–54875. [CrossRef] [PubMed]
9. Yen, C.-H.; Teng, H.-Y.; Tzeng, J.-C. Innovativeness and customer value co-creation behaviors: Mediating role of customer engagement. *Int. J. Hosp. Manag.* **2020**, *88*, 102514. [CrossRef]
10. Zou, H.; Qin, H.; He, D.; Sun, J. Research on an Enterprise Green Innovation Ecosystem From the Vulnerability Perspective: Evolutionary Game and Simulation. *IEEE Access* **2021**, *9*, 140809–140823. [CrossRef]
11. Zhang, Z.; Ling, D.; Tian, W.; Zhou, C.; Song, M.; Fang, S. Public participation and outgoing audit of natural resources: Evidence from tripartite evolutionary game in China. *Environ. Res.* **2023**, *236*, 116734. [CrossRef]
12. Öge, K. Transparent autocracies: The Extractive Industries Transparency Initiative (EITI) and civil society in authoritarian states. *Extr. Ind. Soc.* **2017**, *4*, 816–824. [CrossRef]
13. Re, B.; Magnani, G. Value co-creation in circular entrepreneurship: An exploratory study on born circular SMEs. *J. Bus. Res.* **2022**, *147*, 189–207. [CrossRef]
14. Liu, C.; Guo, H.; Li, Z.; Gao, X.; Li, S. Coevolution of multi-game resolves social dilemma in network population. *Appl. Math. Comput.* **2019**, *341*, 402–407. [CrossRef]
15. Ashraf, S.F.; Li, C.; Wattoo, M.U.; Murad, M.; Mahmood, B. Green horizons: Unleashing green innovation through green business strategies and competencies. *Bus. Strategy Environ.* **2024**, *33*, 4233–4251. [CrossRef]
16. Tian, H.-H.; Huang, S.-Z.; Cheablam, O. How green value co-creation mediates the relationship between institutional pressure and firm performance: A moderated mediation model. *Bus. Strategy Environ.* **2023**, *32*, 3309–3325. [CrossRef]
17. Chen, Y.; Zhang, J.; Tadikamalla, P.R.; Gao, X. The Relationship among Government, Enterprise, and Public in Environmental Governance from the Perspective of Multi-Player Evolutionary Game. *Int. J. Environ. Res. Public Health* **2019**, *16*, 3351. [CrossRef]
18. Li, X.; He, J. Evolutionary mechanism of green product certification behavior in cement enterprises: A perspective of herd behavior. *Environ. Technol. Innov.* **2024**, *33*, 103508. [CrossRef]
19. Liu, J.; Teng, Y. Evolution game analysis on behavioral strategies of multiple stakeholders in construction waste resource industry chain. *Environ. Sci. Pollut. Res.* **2023**, *30*, 19030–19046. [CrossRef]
20. Long, H.; Liu, H.; Li, X.; Chen, L. An Evolutionary Game Theory Study for Construction and Demolition Waste Recycling Considering Green Development Performance under the Chinese Government's Reward–Penalty Mechanism. *Int. J. Environ. Res. Public Health* **2020**, *17*, 6303. [CrossRef]
21. Wang, J.; Song, Y.; Wang, W.; Wang, S.; Guo, F.; Lu, J. Marine Construction Waste Recycling Mechanism Considering Public Participation and Carbon Trading: A Study on Dynamic Modeling and Simulation Based on Sustainability Policy. *Sustainability* **2022**, *14*, 10027. [CrossRef]
22. Tam, V.W.Y. Comparing the implementation of concrete recycling in the Australian and Japanese construction industries. *J. Clean. Prod.* **2009**, *17*, 688–702. [CrossRef]
23. Sun, Y.; Gu, Z. Implementation of Construction Waste Recycling under Construction Sustainability Incentives: A Multi-Agent Stochastic Evolutionary Game Approach. *Sustainability* **2022**, *14*, 3702. [CrossRef]
24. Xu, S.; Yin, W.; Li, X.; Chen, W.; Yi, B.; Wang, Y. Evolutionary mechanism of green technology innovation behavior in the operation period of construction and demolition waste recycling public–private partnership projects. *Manag. Decis. Econ.* **2023**, *44*, 4637–4650. [CrossRef]
25. Wang, Z.; Zhou, Y.; Jin, X.; Zhao, N.; Sun, J. Risk allocation and benefit distribution of PPP projects for construction waste recycling: A case study of China. *Eng. Constr. Archit. Manag.* **2022**, *30*, 3927–3956. [CrossRef]
26. Shao, Z.; Li, M.; Yu, D.; Han, C.; Meng, L. Collaborative Evolution Mechanism and Simulation of Construction Waste Recycling Stakeholders Based on Social Network. *Buildings* **2022**, *12*, 2255. [CrossRef]
27. Wang, Y.; Li, H.; Xue, L.; Gou, W. The Evolution of the Construction Waste Recycling System and the Willingness to Use Recycled Products in China. *Sustainability* **2022**, *14*, 12541. [CrossRef]
28. Ranjan, K.R.; Read, S. Value co-creation: Concept and measurement. *J. Acad. Mark. Sci.* **2016**, *44*, 290–315. [CrossRef]
29. Pera, R.; Occhiocupo, N.; Clarke, J. Motives and resources for value co-creation in a multi-stakeholder ecosystem: A managerial perspective. *J. Bus Res.* **2016**, *69*, 4033–4041. [CrossRef]
30. Chang, C.-H. Do green motives influence green product innovation? The mediating role of green value co-creation. *Corp. Soc. Responsib. Environ. Manag.* **2019**, *26*, 330–340. [CrossRef]

31. Fisher, D.; Smith, S. Cocreation is chaotic: What it means for marketing when no one has control. *Mark. Theory* **2011**, *11*, 325–350. [CrossRef]
32. Yahya, S.; Khan, A.; Farooq, M.; Irfan, M. Integrating green business strategies and green competencies to enhance green innovation: Evidence from manufacturing firms of Pakistan. *Environ. Sci. Pollut. Res.* **2022**, *29*, 39500–39514. [CrossRef] [PubMed]
33. Han, M.; Xu, B. Distance with Customers Effects on Green Product Innovation in SMEs: A Way through Green Value Co-Creation. *Sage Open* **2021**, *11*, 21582440211061539. [CrossRef]
34. Lewin, A.Y.; Volberda, H.W. Prolegomena on Coevolution: A Framework for Research on Strategy and New Organizational Forms. *Organ. Sci.* **1999**, *10*, 519–534. [CrossRef]
35. Madhok, A.; Liu, C. A coevolutionary theory of the multinational firm. *J. Int. Manag.* **2006**, *12*, 1–21. [CrossRef]
36. Harrington, J.E.; Chang, M.-H. Co-evolution of firms and consumers and the implications for market dominance. *J. Econ. Dyn. Control* **2005**, *29*, 245–276. [CrossRef]
37. Dijksterhuis, M.S.; Van den Bosch, F.A.J.; Volberda, H.W. Where Do New Organizational Forms Come from? Management Logics as a Source of Coevolution. *Organ. Sci.* **1999**, *10*, 569–582. [CrossRef]
38. Duarte, R.G.; Rodrigues, S.B. Co-evolution of Industry Strategies and Government Policies: The Case of the Brazilian Automotive Industry. *BAR Braz. Adm. Rev.* **2017**, *14*, e160100. [CrossRef]
39. Janzen, D.H. When is it coevolution? *Evolution* **1980**, *34*, 611–612. [CrossRef] [PubMed]
40. Abatecola, G.; Breslin, D.; Kask, J. Do organizations really co-evolve? Problematizing co-evolutionary change in management and organization studies. *Technol. Forecast. Soc. Change* **2020**, *155*, 119964. [CrossRef]
41. Prim, A.L.; de Freitas, K.A.; Paiva, E.L.; Kumar, M. The development of quality capabilities in Brazilian breweries: A Co-evolutionary approach. *Int. J. Prod. Econ.* **2023**, *256*, 108717. [CrossRef]
42. Sofitra, M.; Takahashi, K.; Morikawa, K. The coevolution of interconnected relationship strategies in supply networks. *Int. J. Prod. Res.* **2015**, *53*, 6919–6936. [CrossRef]
43. Child, J.; Rodrigues, S.B.; Tse, K.K.-T. T The Dynamics of Influence in Corporate Co-Evolution. *J. Manag. Stud.* **2012**, *49*, 1246–1273. [CrossRef]
44. Construction Waste Resource Utilization Case, Waiting for You to Praise! Available online: https://jst.zj.gov.cn/art/2023/12/7/art_1569971_58934520.html (accessed on 29 May 2024).
45. Geng, X.; Lv, L.; Wang, Y.; Sun, R.; Wang, X. Evolutionary Game Research on Green Construction Considering Consumers' Preference under Government Supervision. *Int. J. Environ. Res. Public Health* **2022**, *19*, 16743. [CrossRef] [PubMed]
46. Fan, R.; Wang, Y.; Chen, F.; Du, K.; Wang, Y. How do government policies affect the diffusion of green innovation among peer enterprises?—An evolutionary-game model in complex networks. *J. Clean. Prod.* **2022**, *364*, 132711. [CrossRef]
47. Wang, X.; Lai, C.; Li, H.; Zhang, Z. A tripartite game analysis of public participation in environmental regulation of ionic rare earth mining areas. *Resour. Policy* **2023**, *81*, 103319. [CrossRef]
48. Shi, Y.; Li, Y. An Evolutionary Game Analysis on Green Technological Innovation of New Energy Enterprises under the Heterogeneous Environmental Regulation Perspective. *Sustainability* **2022**, *14*, 6340. [CrossRef]
49. Du, P.; Tian, K.; Wang, Y. Independent Innovation Incentive Mechanism of the National Independent Innovation Demonstration Zone of China Based on Evolutionary Game. *Complexity* **2022**, *2022*, 2894487. [CrossRef]
50. Guo, L.; Zhang, Q.; Wu, J.; Santibanez Gonzalez, E.D.R. An evolutionary game model of manufacturers and consumers' behavior strategies for green technology and government subsidy in supply chain platform. *Comput. Ind. Eng.* **2024**, *189*, 109918. [CrossRef]
51. Zhao, R.; Peng, L.; Zhao, Y.; Feng, Y. Coevolution mechanisms of stakeholder strategies in the green building technologies innovation ecosystem: An evolutionary game theory perspective. *Environ. Impact Assess. Rev.* **2024**, *105*, 107418. [CrossRef]
52. Wang, M.; Li, Y.; Cheng, Z.; Zhong, C.; Ma, W. Evolution and equilibrium of a green technological innovation system: Simulation of a tripartite game model. *J. Clean. Prod.* **2021**, *278*, 123944. [CrossRef]
53. Guo, F.; Wang, J.; Song, Y. How to promote sustainable development of construction and demolition waste recycling systems: Production subsidies or consumption subsidies? *Sustain. Prod. Consum.* **2022**, *32*, 407–423. [CrossRef]
54. Kong, X.; Xu, Q.; Zhu, T. Dynamic Evolution of Knowledge Sharing Behavior among Enterprises in the Cluster Innovation Network Based on Evolutionary Game Theory. *Sustainability* **2020**, *12*, 75. [CrossRef]
55. Zhu, Z.; Wang, X.; Liu, L.; Hua, S. Green sensitivity in supply chain management: An evolutionary game theory approach. *Chaos Solitons Fractals* **2023**, *173*, 113595. [CrossRef]
56. Hu, Q.; Xiong, F.; Shen, G.Q.; Liu, R.; Wu, H.; Xue, J. Promoting green buildings in China's multi-level governance system: A tripartite evolutionary game analysis. *Build. Environ.* **2023**, *242*, 110548. [CrossRef]
57. Sun, M.; Gao, X.; Li, J.; Jing, X. Research on Evolutionary Game of Water Environment Governance Behavior from the Perspective of Public Participation. *Int. J. Environ. Res. Public Health* **2022**, *19*, 14732. [CrossRef] [PubMed]
58. Qiao, X.; Lee, H.; Shen, Q.; Li, Y. Research on the Tripartite Evolutionary Game of Zero-Waste City Construction in China. *Sustainability* **2023**, *15*, 10507. [CrossRef]
59. Zhou, C.; He, J.; Li, Y.; Chen, W.; Zhang, Y.; Zhang, H.; Xu, S.; Li, X. Green Independent Innovation or Green Imitation Innovation? Supply Chain Decision-Making in the Operation Stage of Construction and Demolition Waste Recycling Public-Private Partnership Projects. *Systems* **2023**, *11*, 94. [CrossRef]
60. Zameer, H.; Wang, Y.; Yasmeen, H. Transformation of firm innovation activities into brand effect. *Mark. Intell. Plan.* **2019**, *37*, 226–240. [CrossRef]

61. Jiang, S.; Wei, X.; Jia, J.; Ma, G. Toward sustaining the development of green residential buildings in China: A tripartite evolutionary game analysis. *Build. Environ.* **2022**, *223*, 109466. [CrossRef]
62. Wu, J.; Chang, I.-S.; Yilihamu, Q.; Zhou, Y. Study on the practice of public participation in environmental impact assessment by environmental non-governmental organizations in China. *Renew. Sustain. Energy Rev.* **2017**, *74*, 186–200. [CrossRef]
63. Du, Y.; Li, Z.; Du, J.; Li, N.; Yan, B. Public environmental appeal and innovation of heavy-polluting enterprises. *J. Clean. Prod.* **2019**, *222*, 1009–1022. [CrossRef]
64. Grönroos, C.; Gummerus, J. The service revolution and its marketing implications: Service logic vs service-dominant logic. *Manag. Serv. Qual. Int. J.* **2014**, *24*, 206–229. [CrossRef]
65. Piligrimiene, Z.; Dovaliene, A.; Virvilaite, R. Consumer Engagement in Value Co-Creation: What Kind of Value it creates for Company? *Eng. Econ.* **2015**, *26*, 452–460. [CrossRef]
66. Jaakkola, E.; Alexander, M. The Role of Customer Engagement Behavior in Value Co-Creation: A Service System Perspective. *J. Serv. Res.* **2014**, *17*, 247–261. [CrossRef]
67. Ritzberger, K.; Weibull, J.W. Evolutionary Selection in Normal-Form Games. *Econometrica* **1995**, *63*, 1371–1399. [CrossRef]
68. Friedman, D. Evolutionary Games in Economics. *Econometrica* **1991**, *59*, 637–666. [CrossRef]
69. Li, Y.; Shi, Y. Dynamic Game Analysis of Enterprise Green Technology Innovation Ecosystem under Double Environmental Regulation. *Int. J. Environ. Res. Public Health* **2022**, *19*, 11047. [CrossRef] [PubMed]
70. Chen, W.; Hu, Z.-H. Analysis of Multi-Stakeholders' Behavioral Strategies Considering Public Participation under Carbon Taxes and Subsidies: An Evolutionary Game Approach. *Sustainability* **2020**, *12*, 1023. [CrossRef]
71. Xu, L.; Tang, H.; Liu, L.; Liu, L. An equilibrium analysis on the tripartite evolutionary game of garbage classification recycling. *Front. Environ. Sci.* **2023**, *11*, 1185932. [CrossRef]
72. Zheng, S.; Zhang, Y. Analyzing the evolutionary game of subsidies' strategy in the digitization of marine ranch: A theoretical framework. *Front. Mar. Sci.* **2024**, *11*, 1376256. [CrossRef]
73. Wang, S.; Su, H.; Hou, Q. Evolutionary game study on multi-agent value co-creation of service-oriented digital transformation in the construction industry. *PLoS ONE* **2023**, *18*, e0285697. [CrossRef] [PubMed]
74. Dou, R.; Nan, G.; Wei, Z.; Hsu, C.-Y. Value co-creation in group enterprises: An evolutionary game theory-based analysis. *Int. J. Prod. Res.* **2024**, 1–25, *early access*. [CrossRef]
75. Liu, W.; Long, S.; Wei, S.; Xie, D.; Wang, J.; Liu, X. Smart logistics ecological cooperation with data sharing and platform empowerment: An examination with evolutionary game model. *Int. J. Prod. Res.* **2022**, *12*, 4295–4315. [CrossRef]
76. Li, G.; Shi, X.; Yang, Y.; Lee, P.K.C. Green Co-Creation Strategies among Supply Chain Partners: A Value Co-Creation Perspective. *Sustainability* **2020**, *12*, 4305. [CrossRef]
77. Zhu, C.; Fan, R.; Lin, J.; Chen, R.; Luo, M. How to promote municipal household waste management by waste classification and recycling? A stochastic tripartite evolutionary game analysis. *J. Environ. Manag.* **2023**, *344*, 118503. [CrossRef] [PubMed]
78. Shi, X.; Li, G.; Dong, C.; Yang, Y. Value Co-Creation Behavior in Green Supply Chains: An Empirical Study. *Energies* **2020**, *13*, 3902. [CrossRef]
79. Gao, L.; Yan, A.; Yin, Q. An evolutionary game study of environmental regulation strategies for marine ecological governance in China. *Front. Mar. Sci.* **2022**, *9*, 1048034. [CrossRef]

buildings

MDPI

Article

# Facilitating Green Transition in Small- and Medium-Sized Building Material Enterprises: Collaborative Support via Green Patent Pledge Financing Guarantees

**Zhu Mei [1], Qiaomei Zhou [2,*], Jingjing Zhang [1,*] and Junjie Mao [1]**

[1] School of Management, Jiangsu University, Zhenjiang 212013, China
[2] School of Intellectual Property, Jiangsu University, Zhenjiang 212013, China
* Correspondence: zqm@ujs.edu.cn (Q.Z.); jingjingzhang@ujs.edu.cn (J.Z.)

**Abstract:** Purpose: This study aims to analyze the interactions within the intellectual property pledge financing framework involving SMBMEs, banks, and third-party intermediaries, focusing on factors that promote sustainable cooperation. Methods: The research employs an evolutionary game model to simulate and analyze strategic interactions among the stakeholders, considering bounded rationality and asymmetric information. Results: Key findings include the positive correlation between SMBMEs' reputation and timely repayment, the adverse effects of patent devaluation on cooperation, and the reasonable loan interest rates for facilitating GPPFG cooperation. The study also highlights the need for a transparent information platform and reasonable breach of contract compensation. Conclusions: The paper concludes that an efficient GPPFG mechanism is crucial for SMBMEs' financial support and green transformation. It suggests that third-party intermediaries play a significant role in risk reduction and innovation facilitation. The study acknowledges limitations and calls for future research to explore technological innovations and improve intermediary service systems for SMBMEs' sustainable development.

**Keywords:** green transformation; SMEs in building materials; green patent financing; sustainable development; innovation funding

## 1. Introduction

Amid escalating global climate change and environmental degradation, the consensus on green and sustainable development has increased internationally. As a pivotal segment of the building industry, the green transformation of small- and medium-sized building materials enterprises (SMBMEs) is particularly urgent. This urgency is not only a response to governmental strategies for energy conservation and emissions reduction but also an inevitable choice for the long-term development of enterprises themselves. Firstly, green transformation aids small- and medium-sized building materials enterprises in enhancing resource utilization efficiency and reducing energy consumption and pollution emissions, which are directly related to the enterprises' cost control and environmental responsibilities. Secondly, the introduction of green products and technologies can bolster the market competitiveness of these enterprises, meeting modern consumers' demand for eco-friendly and low-carbon products. Furthermore, with the continuous strengthening of environmental regulations by the government, enterprises can only avoid potential policy risks through green transformation, successfully passing ecological audits, and ensuring continuous operation. Lastly, green transformation also represents a significant manifestation of corporate social responsibility, helping to elevate the corporate image and strengthen the trust of investors and consumers, thereby attracting more social capital and talent.

In the course of transformation, SMBMEs ubiquitously confront the predicament of capital shortage [1], encompassing not only substantial initial investments, the pressure of working capital turnover, but also the enduring financial support required for technological research and development and market promotion, so traditional financing channels alone cannot meet their capital needs [2–4]. In light of this, the exploration and application of green patent pledge financing guarantee (GPPFG) cooperation mechanism emerge as a potentially innovative financing model [5], aimed at addressing the capital bottleneck in the green transformation process. The essence of GPPFG cooperation lies in transforming the potential economic value of green patents (GPs) into direct financing capability, a mechanism that not only mitigates financial institutions' risks but also serves as an effective solution to the financing difficulties for small and medium enterprises [6]. Patent pledge financing refers to a financing method where an enterprise obtains a loan from a bank by legally owning GPs as collateral after evaluation. The GPPFG for SMBMEs implies a mechanism in which SMBMEs utilize their owned green technology patents as collateral to secure financing from financial institutions. This method aims to encourage and support enterprises in the development and application of new green, environmentally friendly technologies and materials. The core of the cooperation in GPPFG lies in transforming the potential economic value of GPs into direct financing capability.

Despite the significant advantages of GPPFG in enhancing corporate financing capabilities and reducing financing costs, current research indicates that the efficiency of this cooperative mechanism is not high [7], with many limiting factors and practical challenges present [8]. The complexity of patent evaluation and the insufficient intellectual property management capabilities of SMBMEs result in higher financing risks. Furthermore, guarantee organizations and financial institutions have low recognition of GPs as collateral, and their concerns about the financial risks associated with GPs make them more cautious in providing guarantees or loans [9], limiting the amount and conditions of loans. How to avoid risks, promote financing guarantee cooperation, and enhance the enthusiasm of guarantee organizations and financial institutions are pressing issues that need to be addressed [10].

In light of the aforementioned context, to give full play to the role of intellectual property rights in value realization and empowerment of green transformation in the field of corporate financing, this paper aims to investigate and construct an efficient GPPFG cooperation mechanism to provide robust financial support for SMBMEs, thereby promoting their green transformation and sustainable development. Through in-depth dissection, model analysis, and strategic proposition, this paper aspires to offer theoretical support and practical guidance for the GPPFG of SMBMEs, to inform related policy-making, and thus contribute to sustainable development.

## 2. Literature Review

### 2.1. Intellectual Property Pledge Financing

Intellectual property is a broad concept that encompasses all property rights created by human intellectual activity, including patents, trademarks, copyrights, etc. Patents, as an important form of intellectual property, can provide enterprises with a means to access funding through their pledge. A review of the literature on intellectual property pledge financing allows for a deeper understanding of the application of intellectual property pledge financing in practice, the challenges faced, and future developmental trends. This provides a theoretical foundation and empirical cases for the study of patent pledge financing, aiding in the identification and optimization of key issues within patent pledge financing. The academic community has conducted extensive research on intellectual property pledge financing, primarily covering the current features, influencing factors, value assessment and risks, model construction, and experience learning of intellectual property pledge financing.

Research has shed light on the current state and characteristics of intellectual property pledge financing. For instance, Nikitenko et al. (2017) explored the modern banking loan model using patent intellectual property as collateral and highlighted the unique aspects of loan collateral registration in some European countries [11]. Qian (2013) reviewed the mechanisms and risk factors of intellectual property supporting finance, noting that the assessment methods still focus on static value [12]. Yuan (2013) and Rinitami Njatrijani (2020) pointed out the shortcomings in the legislation and practice of intellectual property pledge financing in China and Indonesia, respectively [13,14].

Scholars have researched the influencing factors of intellectual property pledge financing from various perspectives. Yang (2018) and Qian (2011) explored the relationship between cooperative behavior of high-tech enterprises and market factors from a game theory perspective [15,16]; Dong (2023), Wang (2009), and Santoso (2021) pointed out that the easy depreciation and poor liquidity of intellectual property are challenging issues restraining business development [17–19]. Additionally, Feng (2020), Wang (2011), Raymond (2009), and Panda (2019) analyzed the mismatch between current legal systems and their practical application from a legal perspective [20–22]. Yan (2021), based on specific social networks, analyzed the factors affecting the trust in intellectual property pledge loans through experimental research [23]; Chen (2018) reviewed the relevant literature, identifying the main challenges faced by small- and medium-sized enterprises in intellectual property financing as talent shortage, poor quality, and obstructed financing channels [24].

From the perspective of value and risk assessment, Su (2023) applied interval theory, the analytic hierarchy process, and the set-valued statistical method to innovate the assessment method for pledged intellectual property, effectively reducing the risks of IPPF [2]. Zhao et al. (2020) were the first to use the analytic hierarchy process and fuzzy comprehensive evaluation method to build a predictive model for the financing amount of patent pledge in pharmaceutical enterprises [25]. Feng (2021) conducted an in-depth study of the pledge value of Yunnan Baiyao's intellectual property by combining three intellectual property value assessment methods [26]. Ge (2021) and Wang (2020) provided new solutions for risk analysis of tech-oriented enterprises through fuzzy analytic hierarchy process [10,27]. Zhang (2013), from the perspective of commercial banks, established a risk evaluation index system for IPPF business for small and medium tech enterprises using the analytic hierarchy process [28]. Furthermore, Zhang (2017), through surveys, analysis, and expert interviews, clarified three key indicators of IPPF risk assessment and their weights using the analytic hierarchy process [29]. These studies not only enrich the theoretical framework of IPPF but also offer important references for practical application.

In the aspect of model construction, Wan (2023) developed a two-tier supply chain model from the perspective of the supply chain, concerning decisions on IPPF under three scenarios: without financing, without government subsidies, and with government subsidies [6]. Su (2013) innovated IPPF loan models under the premise of satisfying risk control and enterprise capital needs using the CreditRisk+ model [30]. Ren (2012) analyzed the difficulties of IPPF in Shaanxi Province and proposed targeted financing guarantee models [31]. Cheng (2013) summarized domestic and foreign IPPF models into four types: single market, government-guided market, government-directed, and government-executed [32]. Additionally, Feng (2012) emphasized strengthening the role of market mechanisms while valuing government guidance to construct an effective IPPF mechanism [33]; Wu (2012) suggested a multi-party cooperation approach to establish IPPF guarantee models from market and government dimensions [34]. Li (2012) discussed operation environments suitable for intellectual property financing in Shanxi [35], while Chen et al. (2016) explored the evolutionary game model between tech-oriented SMEs and banks through simulation studies [36]. However, Chen (2011) noted that certain IPPF models rely too heavily on the government, neglecting the financing capacity of SMEs, resulting in low acceptance [37]. Lastly, research by Zhou (2023) indicated that China's high-tech industry management efficiency needs improvement and requires the introduction of a professional intellectual property platform [38].

Scholars have also paid attention to learning from advanced experiences. Long (2009) and Li (2020) investigated Japan's and the United States' experiences in IPPF, respectively, offering insights for China [39,40]. Marcello (2023) focused on Indonesia's potential to leverage intellectual property assets to increase entrepreneurial capital [41]. Xie (2010) suggested that China could learn from the U.S. experience to improve relevant laws, policies, and assessment mechanisms [42].

### 2.2. SMEs Green Transformation and Patent Pledge Financing

Direct research on the green transformation and patent pledge financing of SMBMEs is rare. However, reviewing the literature on the green transformation and patent pledge financing of SMEs can provide theoretical support and empirical data for this study. Cheng et al. (2024) studied the strategies and tools implemented by SMEs in the Taiwanese metal industry for green operations, pointing out that by adopting green supply chain management and environmental management systems, enterprises can effectively reduce environmental pollution and improve energy efficiency, thus achieving a green transformation [43]. Mazur (2012) found through case studies that small businesses can reduce their environmental impact through the adoption of innovative green technologies and improved product design, and that this transformation also helps enterprises discover new business opportunities and enhance brand image [44]. Baranova (2017) discussed the environmental capabilities of SMEs in the East Midlands region of the UK in the transition to a low-carbon economy. The research shows that although SMEs face many challenges, such as funding constraints and lack of technology, enterprises can effectively achieve green transformation by developing environmental management capabilities and utilizing government support [45].

Early research on patent pledge financing focused on patents as an important intangible asset and their potential value in corporate financing, such as the economic value of patents and their impact on corporate performance [46]. Gradually, the research field expanded to examine how patents, as collateral, could facilitate enterprises to obtain financing, exploring the mechanisms and effects. Mann (2018) examined the role of patent collateral in financing innovation among public firms, questioning whether stronger creditor rights to patents primarily encourage or discourage financing [47]. As the knowledge economy rose and innovative enterprises increased, the patent asset pledge financing method received more attention. Scholars began to delve deeper into the operational modes, risk assessment, legal protection, and market reactions of patent pledge financing. Zhao and Liu (2020) analyzed eight indexes considering both the pledge patent value and the pledger's credit value, proposing a prediction model for patent pledge financing [25]. Yang et al. (2023) explored how the age of firms affects their ability to obtain patent pledge financing, with findings suggesting variance based on the age of startups and their experience [48]. Zhang et al. (2024) explored the impact of multidimensional characteristics of the knowledge base on patent pledge financing within a systematic theoretical framework. In particular, they examined how patent pledge financing affects corporate innovation and how it can be better developed through legal and institutional environments [49]. Palangkaraya (2023) investigated whether patent pledges facilitate innovation, using matched control groups of not-pledged patents to analyze the effects [50]. With the development of big data and artificial intelligence technologies, researchers have begun to use more refined data analysis methods to explore the influencing factors and actual effects of patent pledge financing. For example, by evaluating the quality of patents, the credit record of enterprises, and the overall market environment, they constructed a more precise model for assessing the risks of pledge financing. Jiang et al. (2023) explored predicting financial distress by mining semantic features in patent texts, highlighting a methodology that, while straightforward, misses capturing finer details [51].

Scholars in the academic community have made significant contributions to the advancement and development of businesses through research in intellectual property pledge financing, green transition of small- and medium-sized enterprises (SMEs), and patent pledge financing, exploring the current state, influencing factors, model construction, and lessons learned in the field of intellectual property pledge financing. Faced with the demand for green transformation, some innovative SMEs in the construction materials industry have begun to explore using intellectual property as collateral for financing. By leveraging their patent rights as collateral security, these enterprises can obtain loans from financial institutions, thus introducing a new channel for financing. However, existing research rarely discusses the issues related to GPPFG cooperation for SMEs in the building materials industry, which is not conducive to promoting the green transformation and sustainable development of these enterprises.

Enhancing the enthusiasm of financial institutions for lending and the willingness of guarantee institutions to provide guarantees through the synergistic coupling among participating entities is key to solving the financing difficulties faced by SMBMEs. Against this backdrop, this paper utilizes evolutionary game theory as a foundation, under the premise of considering the characteristics of SMBMEs and the cooperative characteristics of the GPPFG. It seeks to find a stable cooperative state among SMBMEs, commercial banks, and third-party intermediary platforms. This study constructs a GPPFG cooperation model and explores the impact of changes in key factors on the strategic behaviors of the three parties. Finally, in light of the current stage of green transformation development of SMBMEs, it formulates corresponding recommendations for the GPPFG.

In summary, compared to previous scholarly research, this paper primarily differs in the following three aspects: Firstly, it considers the role of third-party intermediary platforms in evaluating, guaranteeing, and defaulting on patent transactions. By linking SMBMEs with commercial banks through the platform, a tripartite evolutionary game model is constructed to analyze the stability of the game participants' strategies and the impact of variable changes. Secondly, it utilizes the first method of Lyapunov to conduct a stability analysis of the pure strategy equilibrium points in the replicator dynamic system and derives their strategy combinations. Lastly, it employs numerical simulation analysis in MATLAB to verify the model's analysis effectiveness, ensuring the research process's scientific rigor. Based on the analysis results, countermeasures and suggestions are proposed to advance the GPPFG.

## 3. Materials and Methods

Evolutionary game theory, which originated from biological evolutionary theory, centers on simulating the evolution of strategies in the process of natural selection. It posits that under conditions of bounded rationality and information, individuals optimize their strategies through imitation and learning. This theory emphasizes the dynamic learning and adaptation of strategies, differing from classical game theory which assumes complete rationality and information and is more aligned with real-world economic conditions. Two key concepts of evolutionary game theory are "Evolutionary stable strategy" and "replicator dynamics". An evolutionary stable strategy describes a state where the dominant strategy of a population is resistant to new strategies, while replicator dynamics simulate the changes in the distribution of population strategies over time. If a strategy yields higher returns than the population average, it will be adopted by more individuals; otherwise, it will be eliminated.

In this study, evolutionary game theory aids in analyzing the strategic choices and interactions among SMBMEs, banks, and third-party intermediary platforms under conditions of bounded rationality and asymmetric information. In the analysis of the game model, we proceeded with simulation analysis. As an essential tool in evolutionary game analysis, simulation can assist researchers in gaining a deeper understanding of the dynamics of the game, predicting potential outcomes, and providing solutions to practical issues. This paper intends to simulate the complex interactions among the three parties on the MATLAB platform and test the model's sensitivity to various parameter changes, thereby identifying the key factors that affect the stability and efficiency of GPPFG cooperation.

In the process of GPPFG cooperation, which centers around GPs as the core, multiple participating entities are involved beyond the financial transactions between SMBMEs and banks. These include guarantee, insurance, evaluation, and operational entities among GP intermediary service organizations. GP intermediary service platforms are conducive to reducing financing guarantee risks, and the effectiveness of risk reduction depends on the cooperative efficiency of all parties involved. Therefore, this paper constructs a theoretical framework and game model for GP pledge financing guarantees that introduces a third-party intermediary service platform, completing the marketization loop of GPPFG.

As shown in Figure 1, the GPPFG model includes entities such as SMBMEs, banks, third-party intermediary platforms, and the government. The third-party intermediary platform is composed of important institutional departments involved in the GPPFG process, forming a systemic intermediary service platform for the GPPFG, which includes guarantee institutions, insurance companies, GP evaluation agencies, and GP operation organizations. Additionally, the government is responsible for maintaining, guiding, and supervising the order of GP operations. From the perspective of the functions of participating entities, guarantee institutions enhance the possibility for SMBMEs to obtain financing from banks, lower the financing threshold for SMBMEs, and actively promote the development of the GPPFG business. Insurance companies also significantly share the risk burden of the GPPFG institutions.

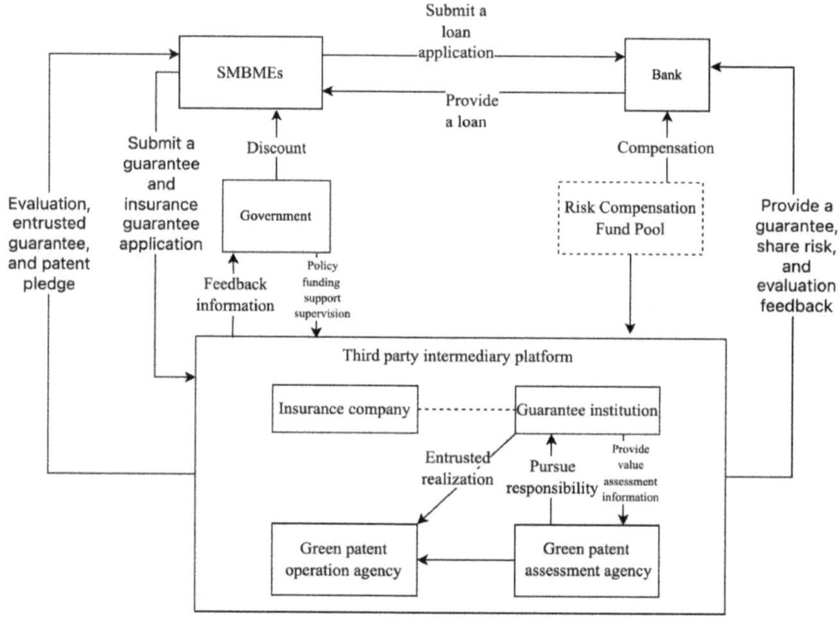


**Figure 1.** Diagram of GPPFG model involving third-party intermediary platforms.

At the same time, guarantee and insurance companies, GP evaluation agencies, and operation organizations together establish a contractual-type intermediary service platform. This enables the relevant entities involved in the GPPFG business to break down information barriers, mitigate risks caused by information asymmetry, and reduce the future cost of business development. Government departments within this pledge financing guarantee model only play a supporting, guidance, and routine supervisory role, and do not directly participate in the specific business of financing guarantees. The aforementioned "SMBME + bank + third-party intermediary platform (guarantee + insurance + evaluation + operation)" GPPFG model includes guarantee companies, insurance companies, GP evaluation firms, and GP operation agencies. This constructs a complete model chain, from financing application to valuation, guarantee, insurance, and eventual realization, ensuring the formation of a good cycle in the GPPFG market.

### 3.1. Model Assumptions

The participation process of third-party intermediary platforms in GPPFG mainly includes three main entities: SMBMEs, banks, and third-party intermediary platforms. SMBMEs with financing needs apply to banks for GP pledge financing loans and, to enhance their credibility, request pre-loan assessments and guarantees as well as insurance from third-party intermediary platforms. In the event of default or bad debts by the enterprise, the involved parties repay the losses proportionally according to the contract. Upon receiving an application, banks entrust GP appraisal agencies to thoroughly assess the value of the intellectual property. The decision to issue a loan is based on the guarantee ability of the guarantee company, the insurance provided by the insurance company, and the patent value assessment provided by the third-party institution. The third-party intermediary platform is responsible for the evaluation, insurance, and disposal of the defaulted collateral, addressing the two key challenges in GP pledge financing guarantee: value assessment and collateral disposal in default situations, ensuring the stable operation of the overall model and risk reduction. The model follows these assumptions:

Assumption 1: In the GPPFG model, the decisions of the game participants are boundedly rational and do not fully comply with the hypothesis of economic man; the SMBME may exhibit opportunistic behavior.

Assumption 2: SMBMEs, banks, and third-party intermediary platforms all make decisions aimed at maximizing their benefits. The guarantee institutions and insurance companies within the third-party intermediary platforms offer guarantee and insurance underwriting services to the SMBMEs, from which they must gain certain profits during the financing process.

Assumption 3: To leverage a larger amount of financing guarantee, SMBMEs use the help of guarantees to initiate bank loans. If an enterprise defaults, the guarantee institutions will use their funds to compensate for the majority of the losses, and the pledged intellectual property of the enterprise will be disposed of by the guarantee institution to repay the debt.

Assumption 4: Government departments merely play a supportive role in this model and do not directly intervene in the operation of GP pledge financing business. The choices of the entities evolve naturally in this environment.

The three participating entities randomly choose their game strategies, resulting in eight possible combinations of strategies. They are as follows: (On-time repayment, Cooperation, Cooperation), (On-time repayment, Cooperation, Non-cooperation), (On-time repayment, Non-cooperation, Cooperation), (On-time repayment, Non-cooperation, Non-cooperation), (Overdue default, Cooperation, Cooperation), (Overdue default, Cooperation, Non-cooperation), (Overdue default, Non-cooperation, Cooperation), (Overdue default, Non-cooperation, Non-cooperation).

### 3.2. Model Parameter Settings

In the process of the GPPFG, the loan amount requested by SMBMEs is denoted as L, and the value assessment of their GP is V. The success rate and return rate of the enterprise projects after obtaining the financing loans are a and b, respectively; the transaction fees paid by the enterprise to the third-party intermediary platform are f, including guarantee fees, insurance fees, and assessment fees. The potential change rate of the value of the GP during the guarantee period is n; if the enterprise defaults and cannot repay the principal and interest, it has to bear the corresponding loss of goodwill denoted as D. The financing interest rate offered by the bank is r, and the cost invested by the bank in the operation of GPPFG business is C1; if the enterprise repays the principal and interest on time, the bank gives a credit reward S to the enterprise. If the investment project fails and the enterprise cannot repay on time, the bank bears a loss proportion denoted as p. For the third-party intermediary platform, the cost invested during the promotion of the business is C2, and the proportion of the cooperation payment paid by the bank when the enterprise investment project is successful is m; when the enterprise fails to repay the financed loan, the third-party intermediary platform spends a cost C3 in trading and transferring the pledged default GP. After reaching a cooperation agreement between the commercial bank and the third-party intermediary platform, if one party fails to fulfill its obligations, it must provide compensation for breach of contract denoted as g to the other party. The specific settings of the parameters of the game model are shown in Table 1.

**Table 1.** Settings of model parameters.

| Parameter | Definitions | Parameter | Definitions |
|---|---|---|---|
| L | The financing amount applied for by SMBMEs | C3 | The cost spent by the third-party intermediary platform in the process of GP trading and transfer |
| V | the value assessment of their GPs | g | The compensation for breach of contract paid by one party to the other after reaching a cooperation agreement between the bank and the third-party intermediary platform |
| a, b | The success rate and return rate of the enterprise projects after obtaining financing loans | D | The loss of goodwill that the enterprise needs to bear if it defaults and cannot repay the principal and interest |
| f | The various fees paid by the enterprise to the third-party platform | S | The credit reward given by the bank when the enterprise repays the principal and interest on time |
| r | The financing interest rate offered by the bank to SMBMEs | θ | The proportion of loss the bank bears if the project investment fails |
| C1 | The cost invested by the bank in the operation of the GPPFG business | m | The cooperation reward paid by the commercial bank to the third-party platform |
| C2 | The cost invested by the third-party intermediary platform during the operation of the GPPFG | n | The change rate of the value of GP |

*3.3. Model Construction*

In this game model, three parties choose corresponding game strategies based on their willingness to cooperate. If the probability of an SMBME repaying on time is x, then the probability of defaulting is $1 - x$; if the probability of a bank cooperating is y, then the probability of not cooperating is $1 - y$; if the probability of a third-party intermediary platform cooperating is z, then the probability of not cooperating is $1 - z$. The strategies available to the SMBME in the game model include timely repayment and defaulting. The cooperative behavior of the bank refers to giving rewards to SMBMEs that repay loans on time and paying the agreed cooperation reward to the third-party intermediary platform that completes the cooperation; otherwise, it is considered non-cooperative. The cooperative strategy of the third-party intermediary platform involves assessing the enterprise and its intellectual property to provide the bank with a more comprehensive credit risk evaluation of the enterprise and imposing certain penalties on SMBMEs that default; otherwise, it is considered a non-cooperative strategy. The game payoff matrix is shown in Table 2.

**Table 2.** The mixed strategy game matrix for SMBMEs, commercial banks, and third-party intermediary platforms.

| | | Commercial Bank | Third-Party Mediation Platform | |
|---|---|---|---|---|
| | | | **Cooperation (z)** | **Non-Cooperation (1 − z)** |
| SMBMEs | On-time repayment (x) | Cooperation (y) | (ab − 1 − r)L + S − f<br>Lr − C1 − S − m<br>f − C2 + m | (ab − 1 − r)L + S<br>Lr − C1 − S + g<br>−C2 − g |
| | | Non-cooperation (1 − y) | (ab − 1 − r)L − f<br>Lr − C1 − g<br>f − C2 + g | (ab − 1 − r)L<br>Lr − C1<br>0 |
| | Overdue default (1 − x) | Cooperation (y) | (1 + ab)L − f − V − D<br>−θL − C1 − m<br>f + m − C2 − (1 − θ)L + nV − C3 | (1 + ab)L − V<br>g − θL − C1<br>−g − (1 − θ)L − C2 |
| | | Non-cooperation (1 − y) | (1 + ab)L − f − V − D<br>−g − θL − C1<br>g + f − C2 − (1 − θ)L | (1 + ab)L − V<br>−L − C1<br>0 |

### 3.4. Model Analysis

#### 3.4.1. Strategic Stability Analysis for SMBMEs

For SMBMEs, the expected benefits of choosing timely repayment versus default, as well as the average expected benefits (E11, E12, E1), are, respectively:

$$\begin{aligned} E11 = yz[(ab - 1 - r)L + S - f] + y(1 - z)[(ab - 1 - r)L + S] \\ + (1 - y)z[(ab - 1 - r)L - f] + (1 - y)(1 - z)(ab - 1 - r)L \end{aligned} \quad (1)$$

$$\begin{aligned} E12 = yz[(1 + ab)L - f - V - D] + y(1 - z)[(1 + ab)L - V] + \\ (1 - y)z[(1 + ab)L - f - V - D] + (1 - y)(1 - z)[(1 + ab)L - V] \end{aligned} \quad (2)$$

$$\overline{E}1 = xE11 + (1 - x)E12 \quad (3)$$

The replicator dynamics equation for strategy choice in SMBMEs is:

$$F(x) = dx/dt = x(E11 - \overline{E}1) = x(1 - x)[Sy + Dz - (2 + r)L + V] \quad (4)$$

The first derivative of x and the function G(y) are defined as follows:

$$d(F(x))/dx = (1 - 2x)[Sy + Dz - (2 + r)L + V] \quad (5)$$

$$G(y) = Sy + Dz - (2 + r)L + V \quad (6)$$

According to the stability theorem of differential equations, for SMBMEs, the condition for the probability of choosing to repay on time to be in a stable state must satisfy: $F(x) = 0$ and $d(F(x))/dx < 0$. Since $G(y)/y > 0$, G(y) is an increasing function of y. Therefore, when $y = [(2 + r)L - V - Dz]/Sy^*$, $G(y) = 0$, and at this point $d(F(x))/dx \mid x = 0 < 0$, implying that $x = 0$ is the evolutionary stable strategy (ESS) for the SMBMEs. On the contrary, when $y > y^*$, $x = 1$ becomes the ESS. The phase diagram of the strategic evolution of SMBMEs can be depicted as Figure 2.

Figure 2 shows that the volume VA represents the probability A that SMBMEs choose to default on payments past the due date. Conversely, the volume VB represents the probability B that they choose to make repayments on time. The calculation goes as follows:

$$VB = \int_0^1 \int_0^{\frac{(2+r)L-V}{D}} \frac{(2 + r)L - V - Dz}{S} dzdx = \frac{2D - [(2 + r)L - V]^2}{2DS} \quad (7)$$

Inference 1: The probability of SMBMEs choosing to make timely repayments is positively correlated with the reputation penalty, reputation rewards, and the assessed value of GP rights, and is negatively correlated with the interest rates on GP loans set by banks.

Inference 1 indicates that the greater the value of an SMBME's own GP rights, the more it enables the company to fulfill its contractual obligations during the process of GPPFG. At the same time, because the reputation penalty for defaulting companies greatly influences the choice of behavioral strategies among SMBMEs, it prompts them to consider the consequences of such actions in the development of the market. In addition, when setting loan interest rates, banks need to consider the bearing capacity of SMBMEs to define reasonable rates and achieve a win–win situation.

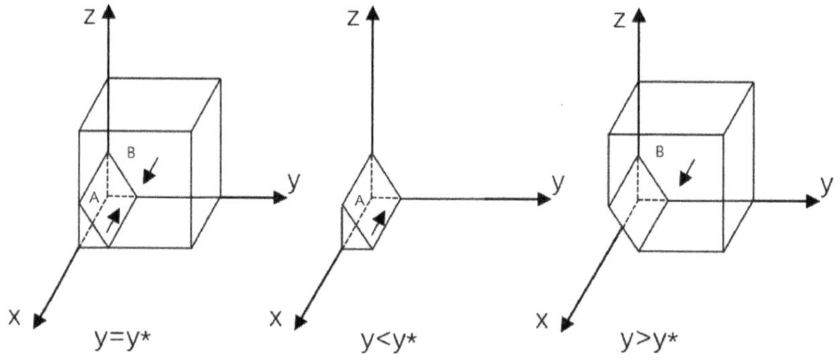


**Figure 2.** Phase diagram of strategic evolution of SMBMEs.

3.4.2. Analysis of Bank Strategy Stability

The expected returns for the bank's choice of cooperation or non-cooperation, as well as the average expected returns (E21, E22, E2), are as follows:

$$E21 = xz(Lr - C1 - S - m) + x(1 - z)(Lr - C1 - S + g) + (1 - x)z(-\theta L - C1 - m) + (1 - x)(1 - z)(g - \theta L - C1) \quad (8)$$

$$E22 = xz(Lr - C1 - g) + x(1 - z)(Lr - C1) + (1 - x)z(-g - \theta L - C1) + (1 - x)(1 - z)(-L - C1) \quad (9)$$

$$\overline{E}2 = yE21 + (1 - y)E22 \quad (10)$$

The replicator dynamics equation for the bank's strategy selection is:

$$F(y) = dy/dt = y(E21 - \overline{E}2) = y(1 - y)[(L - \theta L)xz + x(-S + \theta L - L) + z(-m + \theta L - L) + g - \theta L + L] \quad (11)$$

The first derivative of y and the set J(z) are, respectively:

$$d(F(y))/dy = (1 - 2y)[(L - \theta L)xz + x(-S + \theta L - L) + z(-m + \theta L - L) + g - \theta L + L] \quad (12)$$

$$J(z) = (L - \theta L)xz + x(-S + \theta L - L) + z(-m + \theta L - L) + g - \theta L + L \quad (13)$$

According to the Stability Theorem of Differential Equations, in order to be in a stable state, the bank's probability of choosing to cooperate must meet: $F(y) = 0$ and $d(F(y))/dy < 0$.

Since J(z) is a decreasing function, therefore when $z = [x(L + S - \theta L) - g + \theta L - L]/[x(L - \theta L) + \theta L - L - m] \equiv z^*$, $J(z) = 0$, $d(F(y))/dy \equiv 0$,the bank cannot determine a stable strategy; when $z < z^*$, $J(z) > 0$, $d(F(y))/dy \mid y = 1 < 0$, then $y = 1$ is an ESS; Conversely, $y = 0$ is an ESS. The phase diagram of the bank's strategy evolution is shown in Figure 3.

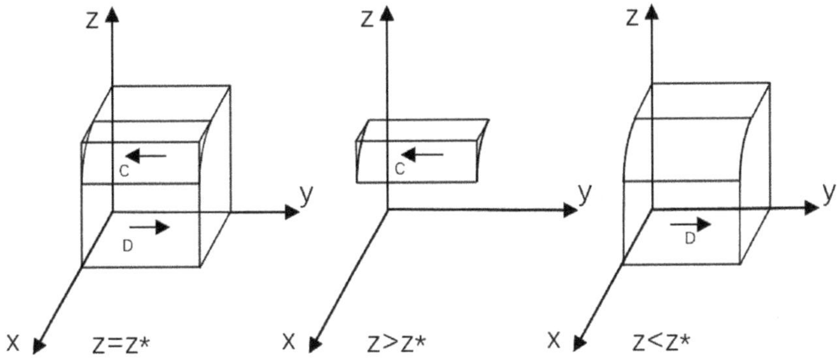


**Figure 3.** Evolutionary phase diagram of bank strategies.

According to Figure 3, the volume of the probability C that banks choose not to cooperate is VC, and the volume for D is VD, yielding the following calculation:

$$\begin{aligned} VC &= \int_0^1 \int_{\frac{g-m}{S}}^1 \frac{x(L+S-\theta L)-g+\theta L-L}{x(L-\theta L)+\theta L-L-m} dxdy \\ &= -\left[\frac{\ln[(g-m)(L-\theta L)/S+(\theta L-L-m)]}{m} + \right. \\ &\left. \ln^2\left(\frac{(g-m)(L-\theta L)+(\theta L-L-m)S}{S}\right)\right] \end{aligned} \tag{14}$$

$$VD = 1 - VC \tag{15}$$

Inference 2: The probability of active cooperation by commercial banks is positively correlated with the default compensation and the pledge value of the GP, and negatively correlated with the credibility reward for timely repayment by SMBMEs.

Inference 2 indicates that when making strategic choices, commercial banks will consider the cost and potential expenses incurred during the process of GPPFG. When the default compensation is too high, the bank may choose to cooperate to avoid the potential expenses, and when the amount of credibility reward is substantial, it may opt not to cooperate to increase profit margins. Moreover, when the value of the pledged GP by the enterprise is high, it indicates a strong market competitiveness and potential for commercialization, which in turn reduces the risk level faced by the banks and increases the probability of cooperation.

#### 3.4.3. Analysis of the Strategic Stability of the Third-Party Intermediary Platform

The expected revenues for the third-party intermediary platform choosing to cooperate or not to cooperate, as well as the average expected revenue (E31, E32, E3), are, respectively:

$$\begin{gathered} E31 = xy(f - C2 + m) + x(1 - y)(f - C2 + g) + (1 - x)y[f + m - C2 - (1 - \theta)L + \\ nV - C3] + (1 - x)(1 - y)[g + f - C2 - (1 - \theta)L \end{gathered} \tag{16}$$

$$E32 = xy(-C2 - g) + (1 - x)y[-g - (1 - \theta)L - C2] \tag{17}$$

$$\overline{E}3 = zE31 + (1 - z)E32 \tag{18}$$

The replicator dynamics equation for the third-party intermediary platform's strategy choice, the first derivative of z, and the set H(y) are, respectively:

$$\begin{gathered} F(z) = z(1 - z)[(C3 - nV - (1 - \theta)L)xy + (1 - \theta)Lx + (m + nV - C3 + (1 - \theta)L \\ + C2)y + g + f - C2 - (1 - \theta)L] \end{gathered} \tag{19}$$

$$\begin{gathered} d(F(z))/dz = (1 - 2z)[(C3 - nV - (1 - \theta)L)xy + (1 - \theta)Lx + (m + nV - C3 \\ +(1 - \theta)L + C2)y + g + f - C2 - (1 - \theta)L] \end{gathered} \tag{20}$$

$$H(y) = (C3 - nV - (1 - \theta)L)xy + (1 - \theta)Lx + (m + nV - C3 + (1 - \theta)L + C2)y + g + f - C2 - (1 - \theta)L \quad (21)$$

To attain a stable state of cooperation, the third-party intermediary platform must satisfy: $F(z) = 0$ and $d(F(z))/dz < 0$. Given that $H(y)/y > 0$, $H(y)$ is an increasing function of y. Therefore, when $y = [(1 - \theta)Lx + g + f - C2 - (1 - \theta)L]/[x(nV - C3 + (1 - \theta)L) - m - nV + C3 - (1 - \theta)L - C2] \equiv y^*$, $H(y) = 0$, $d(F(z))/dz \equiv 0$, the stable strategy cannot be determined; when $y < y^*$, $H(y) < 0$, $d(F(z))/dz \mid z = 0 < 0$, $z = 0$ is an ESS; otherwise, $z = 1$ is an ESS. The phase portrait of the third-party intermediary platform is as illustrated in Figure 4.

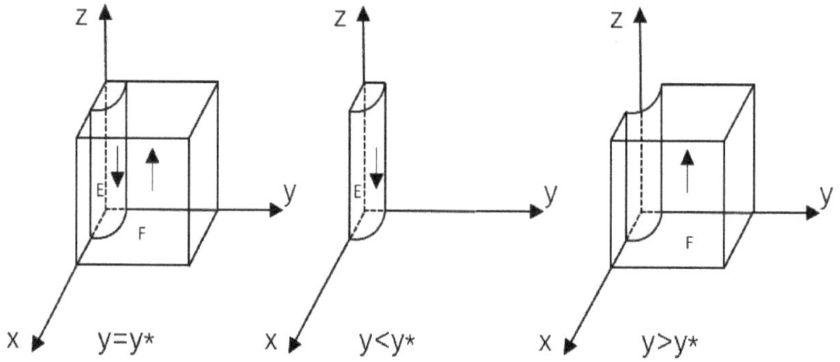


**Figure 4.** The evolutionary phase diagram of third-party intermediary platform strategies.

According to Figure 4, if the probability that the third-party intermediary platform chooses not to cooperate is E and its volume is VE, then the volume of the probability that it chooses to cooperate is VF. The calculation is as follows:

$$\begin{aligned} VE &= \int_0^1 \int_0^{\frac{C2+(1-\theta)L-g-f}{(1-\theta)L}} \frac{(1-\theta)Lx+g+f-C2-(1-\theta)L}{x(nV-C3+(1-\theta)L)-m-nV+C3-(1-\theta)L-C2} dxdz \\ &= \frac{(m+nV-C3+(1-\theta)LnV(C2+(1-\theta)L-g-f)-(m+nV-C3+(1-\theta)L-C2)(1-\theta)Lm(C2+(1-\theta)L-g-f)^2}{(1-\theta)L(m+nV-C3+(1-\theta)L-C2)^2} \Bigg] \end{aligned} \quad (22)$$

$$VF = 1 - VE \quad (23)$$

Inference 3: The probability of a third-party intermediary platform choosing to cooperate is positively correlated with the default compensation, the sum of various fees paid by SMBMEs, the proceeds from the disposition of pledged GP in breach of contract, and the cooperation remuneration given by commercial banks. It is negatively correlated with the input cost in the process of GPPFG cooperation, as well as the disposal cost of the default GP.

Inference 3 indicates that before adopting a behavioral strategy, a third-party intermediary platform considers its own earnings against the cost of investment. When the cost of investment is higher than expected, the probability of choosing not to cooperate increases. Conversely, when the income derived from transaction fees paid by SMBMEs and the cooperation remuneration given by banks is higher than expected, the third-party intermediary platform is likely to adopt a proactive cooperative behavioral strategy.

### 3.4.4. Stability Analysis of Equilibrium Points in the Evolutionary Game System

From $F(x) = F(y) = F(z) = 0$, the following nine local equilibrium points of this game model can be obtained, which are (0, 0, 0), (0, 1, 0), (0, 0, 1), (0, 1, 1), (1, 0, 1), (1, 1, 0), (1, 0, 0), (1, 1, 1), $(x\hat{}, y\hat{}, z\hat{})$. $(x\hat{}, y\hat{}, z\hat{})$ is the mixed strategy of evolutionary games, whereas in multi-party evolutionary game systems, stable strategies only appear in pure strategies, hence the subsequent game analysis does not discuss mixed strategies. The Jacobian matrix of the three-party evolutionary game system is as follows:

$$J = \begin{bmatrix} (1-2x)\binom{Sy+Dz-}{(2+r)L+V} & x(1-x)S & x(1-x)D \\ y(1-y)\binom{(L-\theta L)z+}{\theta L-S-L} & (1-2y)\binom{(L-\theta L)xz+x(\theta L-S-L)}{+z(\theta L-m-L)+g-\theta L+L} & y(1-y)\binom{(L-\theta L)x}{\theta L-L-m} \\ z(1-z)\binom{(C3-(1-\theta)L)y}{+(1-\theta)L-nV} & z(1-z)\binom{(C3-nV-(1-\theta)L)x+m}{+nV-C3+(1-\theta)L+C2} & (1-2z)\binom{(C3-nV-(1-\theta)L)xy+(1-\theta)Lx++g+f-C2}{+(m+nV-C3+(1-\theta)L+C2)y-(1-\theta)L} \end{bmatrix} \quad (24)$$

According to the method proposed by Lyapunov, the sufficient and necessary conditions for an evolutionary game to be asymptotically stable is that all eigenvalues of the Jacobian matrix have negative real parts. Thus, we analyze the stability of each equilibrium point, as shown in Table 3.

**Table 3.** Analysis of equilibrium point stability.

| Equilibrium Point | Eigenvalues of the Jacobian Matrix | | Stability Conclusion |
|---|---|---|---|
| | λ1, λ2, λ3 | Sign of the Real Part | |
| (0, 0, 0) | L + g − θL, V − 2L − Lr, f − L − C2 + g + θL | (+, +, +/−) | Unstable Point |
| (0, 1, 0) | θL − g − L, S − 2L + V − Lr, f − C3 + g + m + Vn | (−, +, +) | Unstable Point |
| (0, 0, 1) | g − m, D − 2L + V − Lr, C2 + L − f − g − θL | (+, +, +/−) | Unstable Point |
| (1, 0, 1) | C2 − f − g, g − S − m, 2L − D − V + Lr | (−, +, −) | Unstable Point |
| (0, 1, 1) | m − g, D − 2L + S + V − Lr, C3 − f − g − m − Vn | (−, +, −) | Unstable Point |
| (1, 1, 0) | S − g, f + g + m, 2L − S − V + Lr | (−, +, −) | Unstable Point |
| (1, 0, 0) | g − S, f − C2 + g, 2L − V + Lr | (+, +, −) | Unstable Point |
| (1, 1, 1) | −f − g − m, S − g + m, 2L − D − S − V + Lr | (−, −, −) | ESS |

According to the stability judgment, it is known that the system has an evolutionary game equilibrium point (1, 1, 1), which eventually converges to SMBMEs repaying on time, and commercial banks and third-party intermediary platforms are willing to cooperate with each other, fulfill their promises, and provide high-quality services.

## 4. Simulation and Result Analysis

Based on the results of the evolutionary game above, to verify the effectiveness of the evolutionary stability analysis, we used MATLAB R2022b to simulate the dynamic evolution of the strategies of SMBMEs, banks, and third-party intermediary platforms under different parameter levels. This further proves the effectiveness of the evolutionary game model and shows that all different initial points will evolve towards the equilibrium point (1, 1, 1). Chen et al. (2022), Huang et al. (2018), and Xu et al. (2019) have summarized the parameter value range for the evolutionary game model simulation parameters [49–51]. The parameter values are adjusted according to the inequality relationship satisfied by the parameters. The parameter values of the model are shown in Table 4.

**Table 4.** Parameter value table.

| Parameter | V | L | f | r | C1 | C2 | C3 | g | D | S | θ | m | n |
|---|---|---|---|---|---|---|---|---|---|---|---|---|---|
| Value | 20 | 8 | 4 | 0.05 | 2 | 2 | 1 | 3 | 13 | 0.1 | 0.1 | 2.5 | 0.7 |

Figure 5 shows that there exists an evolutionary stable point (1, 1, 1), that is, the strategy combination of SMBMEs, banks, and third-party intermediary platforms of (On-time repayment, Cooperation, Cooperation) is the ESS combination. Therefore, this evolutionary result shows that under the participation of a third-party intermediary platform as a cooperative guarantee, SMBMEs repay bank loans on time, banks actively cooperate to issue loans, and the three parties jointly cooperate in the operation of the GPPFG business. It can be seen that the simulation analysis is consistent with the conclusion of the stability analysis of each party's strategy and has effectiveness, which is of practical guiding significance for the promotion of the GPPFG business.

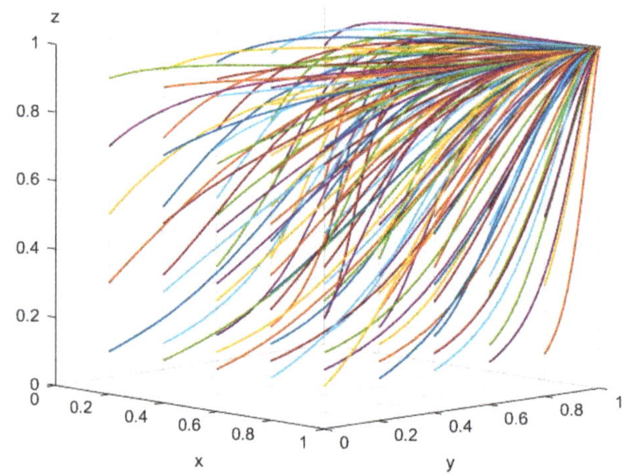


**Figure 5.** Results of 50 system evolutions.

### 4.1. *The Impact of Corporate Goodwill Loss on GPPFG Cooperation*

As seen from Figure 6, as the reputation penalty D that SMBMEs need to bear for overdue defaults gradually increases, it can accelerate the evolutionary speed of SMBMEs choosing to repay on time. With the increase of D, the probability of SMBMEs repaying on time increases, and the probability of active cooperation by third-party intermediary platforms rises. This indicates that the loss of goodwill concerns the future development prospects of the enterprise. A higher possibility of goodwill loss forces SMBMEs to diligently build projects to repay loans on time, reducing their own risks, thereby prompting the participation of third-party intermediary platforms. Therefore, third-party intermediary platforms and banks should have an open and transparent data platform to timely feedback the operation of enterprise projects and loan repayments, allowing the market and the public to form better transparent supervision of such SMBMEs.

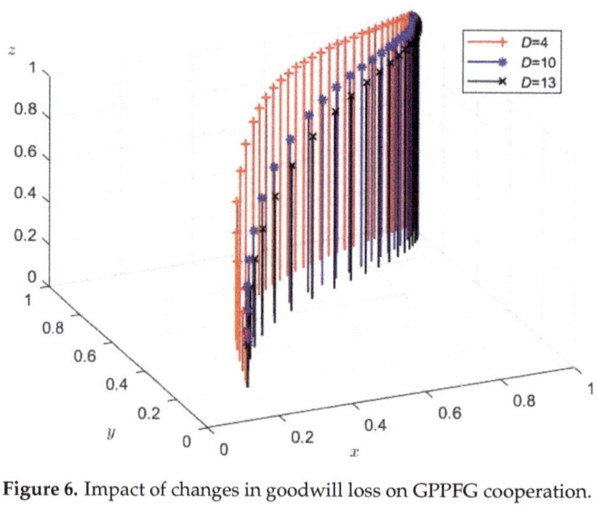


**Figure 6.** Impact of changes in goodwill loss on GPPFG cooperation.

### 4.2. *The Impact of Changes in Default Compensation between Banks and Third-Party Intermediary Platforms on GPPFG Cooperation*

As seen from Figure 7, during the evolution process, as the default compensation g increases, the likelihood of default by a third-party intermediary platform in the process of promoting GPPFG business decreases. At the same time, the probability of commercial banks choosing to default decreases. Therefore, when banks or third-party intermediary platforms violate their prior contract stipulations and need to pay a certain default compensation, such a reward and punishment system can better compel third-party intermediary platforms to actively provide quality services and actively participate in the GPPFG business in cooperation with banks.

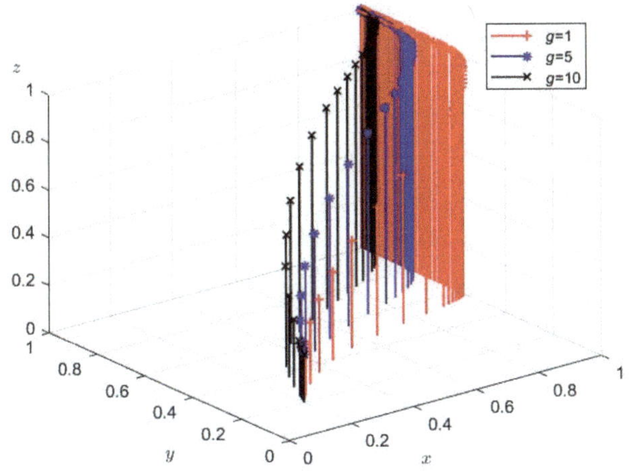


**Figure 7.** The impact of changes in default compensation between banks and third-party intermediary platforms on GPPFG cooperation.

### *4.3. The Impact of Changes in the Value of GP on GPPFG Cooperation*

Figure 8 analyzes the impact on the three-party strategy under different levels of GP value change rates, respectively setting the change rate of the GP of SMBMEs in the trading market as n = −0.15, 0.5, 0.7. Figure 8 shows that as the rate of change of the GP pledged by the enterprise turns from negative to positive and gradually increases, the willingness of third-party intermediary platforms to choose to provide services for SMBMEs and actively cooperate is higher, and they will reach a stable state of cooperation in a shorter time. This shows that the value changes of the SMBME's GPs are related to the economic losses that the third-party intermediary platform may have to share for the company in the future, as well as the high and low of the GP market transaction amount. At the same time, it is also closely related to the enterprise risk considered by the bank. Therefore, the third-party intermediary platform should establish a complete and unified database, set up standard scientific evaluation criteria, and do a good job in the preliminary scientific and professional patent value evaluation, to reduce the depreciation of GP value in the financing guarantee process caused by evaluation problems.

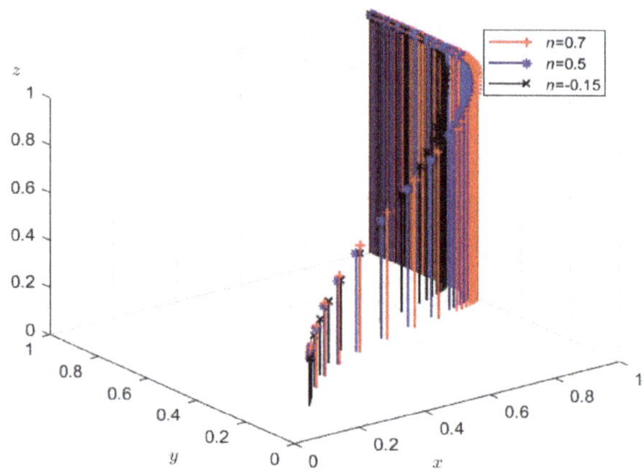


**Figure 8.** The impact of changes in the value of GP on GPPFG cooperation.

### 4.4. Impact of Changes in Bank Loan Interest Rates on GPPFG Cooperation

An analysis is conducted on the strategy selection willingness at different levels of bank GPPFG interest rates. Figure 9 shows that prior to the three-party strategy becoming stable at 1, a higher loan interest rate will increase the bank's willingness to participate in cooperation. However, for SMBMEs, the likelihood of choosing cooperation is relatively lower under high interest rate requirements. This requires a coordination of the interests between banks and SMBMEs, setting a reasonable interest rate for the GPPFG, ensuring that banks obtain reasonable and appropriate interest income while considering the debt-paying ability of the SMBMEs.

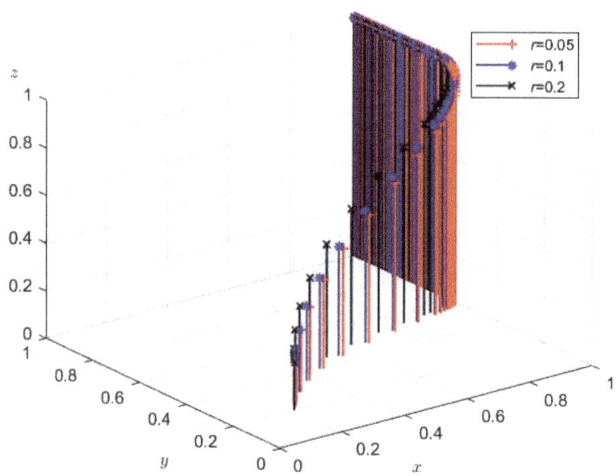


**Figure 9.** The impact of changes in bank loan interest rates on GPPFG cooperation.

## 5. Discussion

The intellectual property pledge financing may involve multiple entities such as borrowers, lenders, and third-party intermediary organizations (or platforms). Studying the behavioral interactions among these entities and their influencing factors is of great significance for promoting stable and sustainable cooperation among the main bodies. Currently, scholars' research on the game among multiple entities in intellectual property pledge financing mainly focuses on the game between financial institutions and enterprises, and the game between enterprises and third-party intermediary organizations (or platforms). However, the research on the game among financial institutions, banks, and third-party intermediary organizations (or platforms) is relatively weak [52]. Existing studies have shown that analyzing the game among multiple entities helps to achieve market equilibrium and improve market efficiency [53].

Third-party intermediary organizations provide professional services such as green patent evaluation, licensing, transfer, and guarantee, helping to solve the problems of green patent pledge financing. This paper contributes theoretically by exploring the financing challenges faced by SMBMEs during their green transformation process and proposes an innovative solution of incorporating third-party intermediary services to establish an efficient GPPFG cooperation mechanism. By constructing a tripartite evolutionary game model and conducting stability and numerical simulation analyses on this model, this paper advances theoretical innovation in the field of GPPFG. It compensates for the previous research's insufficient attention to the involvement of third-party intermediary organizations in the cooperation of GPPFG for SMBMEs. Although some studies have proposed the important role of third-party intermediary service platforms in improving the enthusiasm of banks for lending and reducing financing risks [54], and have pointed out that the difficulty in evaluating and transferring intellectual property rights is a challenging risk issue that restricts intellectual property pledge financing [55], they have not proposed corresponding solutions, especially incorporating third-party intellectual property service platforms into the financing guarantee system, exploring how to promote the optimal cooperation of financing guarantee parties, and reducing the risks of intellectual property pledge financing [56].

In summary, this paper's evolutionary game analysis of the GPPFG model system verifies the effectiveness of the overall model and provides references for effectively promoting risk diversification and upgrading the accessibility of enterprise GPPFG. By advocating for the strategy of encouraging multi-party cooperative efforts, this study aims to reduce financing risks and promote effective cooperation among banks, third-party intermediary platforms, and SMBMEs, thereby providing theoretical support for achieving a win–win situation. The in-depth analysis of the GPPFG cooperation mechanism in this paper also offers theoretical insights for similar financing issues in other industries, contributing to the broader promotion of green transformation and innovative development across industries.

Specifically, this paper starts from the third-party intermediary organizations assisting SMBMEs in the GPPFG, establishes a tripartite evolutionary game model, and explores the key influencing factors and mechanisms affecting the sustainable cooperation among enterprises, banks, and third-party intermediary organizations. Key influencing factors such as SMBMEs' reputation loss, the penalty between banks and third-party intermediary organizations, the patent value fluctuation rate, and the impact of bank loan interest rates on GPPFG cooperation are discussed. Since the reputation loss of SMBMEs is positively correlated with the probability of SMBMEs repaying on time and the probability of third-party intermediary platforms actively cooperating, it is recommended to establish a transparent information platform to increase public and social supervision of enterprises, forcing enterprises to repay on time. Although previous studies have not mentioned the specific impact of enterprise reputation loss on intellectual property pledge financing cooperation, they have emphasized the importance of information transparency for cooperation, indirectly indicating the importance of enterprise reputation protection [23]. Different from existing studies that propose that banks establishing a reward and punishment mechanism helps to prevent enterprise default risks [57], this paper also proposes that through the contractual constraints between banks and third-party intermediary organizations, setting reasonable compensation for breach of contract can enhance the enthusiasm of banks for cooperation and the quality of green patent evaluation and transfer. In addition, considering the impact of the patent value fluctuation rate on GPPFG cooperation, it intuitively shows the adverse impact of patent devaluation on cooperation, highlighting the importance of accurate patent evaluation. This research result is consistent with previous studies, as previous evolutionary game studies have shown that the issue of patent devaluation has a negative impact on the decision-making of banks and enterprises in intellectual property pledge financing [58]. However, it should not be overlooked that we have creatively numerically simulated the positive impact of patent value enhancement on GPPFG cooperation. Finally, we propose that too high or too low loan interest rates are not conducive to GPPFG cooperation because, although higher loan interest rates will increase the willingness of banks to cooperate, they will inhibit the enthusiasm of SMBMEs.

The practical contribution of this article lies in proposing the introduction of third-party GP intermediary services to construct an efficient GPPFG cooperation mechanism. This mechanism helps alleviate the financial pressure on SMBMEs and provides robust financial support for their green transformation and sustainable development. Through in-depth analysis and recommendations, this article facilitates the formulation and implementation of related policies, thereby contributing to sustainable development. Furthermore, this paper verifies the effectiveness of the financing guarantee business model centered on third-party intermediary service platforms through simulation analysis, offering practical guidance for promoting risk diversification and enhancing the financing capacity of enterprises. It also builds a bridge between financial institutions and policymakers for SMBMEs, offering strong support in aspects such as fund mobilization, technology research and development, and market promotion. By establishing a comprehensive risk-sharing mechanism and a reward and penalty system, this paper aids in creating a stable and reliable financing environment, encouraging and promoting the innovation and development of small and medium enterprises. Ultimately, the recommendations and implementation strategies proposed in this paper provide a viable financing pathway for SMBMEs, opening new possibilities for their green transformation and contributing to the green and sustainable development of the entire construction industry and beyond.

This study's results can provide management and policy improvement suggestions for the GPPFG of SMBMEs. Firstly, multiple participating entities should strengthen risk sharing and improve the risk system of the GPPFG, promote the establishment of a GPPFG model, and enhance the confidence of SMBMEs in applying for the GPPFG and the confidence of banks in issuing loans. The financing guarantee operations of government guarantee agencies should act as a catalyst to promote more banks and private guarantee agencies to participate in the GPPFG. Third-party intermediary service platforms should play a role in resource integration and diligently fulfill their evaluation, operation, and guarantee responsibilities, reducing the risk of GP and enterprises as well as financing cost issues, through bilateral market attributes, solving the dilemma of GP breach disposal and promoting the smooth development of innovation and entrepreneurship activities with the help of social capital.

Secondly, a sound reward and punishment mechanism for participants in the GPPFG should be established. Research results show that if banks and third-party intermediary platforms violate the previous contract provisions, they should pay reasonable compensation for breach of contract. In this way, it can encourage participating entities to actively cooperate. Thirdly, a GPPFG cooperation mechanism around the third-party intermediary service platform should be established. Third-party intermediary platforms should adapt to the operating conditions, profitability, and legal status, use status, and service life of the GP of financing SMBMEs, and develop appropriate financing guarantee models according to the financing conditions of different SMBMEs. Third-party intermediary platforms should establish a cooperative relationship of responsibility and profit sharing between guarantee companies, evaluation institutions, and operation institutions. Due to the large fluctuation in the value of GPs, to encourage evaluation institutions and operation institutions to cooperate with guarantee companies, an income sharing distribution method should be implemented. Through multi-entity cooperation within the platform, reduce the business operation costs of all parties through information and channel sharing. In solving the valuation difficulties, a scientific and professional GP evaluation talent team should be established, widely understand finance and law, and set up lectures with the help of relevant experts from universities, governments, and other social institutions. GP personnel should be regularly trained to improve the professional qualifications of evaluators. Cooperation with government intellectual property-related management departments, law firms, and other institutions should be strengthened to keep up with the latest changes in the field of GP valuation.

However, this paper also has some research limitations. For instance, while it has contributed through the construction of theoretical models and simulation analyses, it lacks empirical research or case analyses to validate the practical application effects of the model. The implementation details and effects of the GPPFG mechanism may vary across different regions and industries, and this study has not fully explored the impact of these differences on the execution of the mechanism. Additionally, although this paper underscores the importance of third-party GP intermediary service organizations and constructs a model to analyze their role, the research on critical issues such as how to enhance the operational efficiency of these platforms and mitigate their operational risks is not thorough enough. Future research should delve into the operational models and risk control mechanisms of third-party intermediary service platforms and how to leverage technological innovation (such as blockchain technology) to improve service efficiency and reduce costs. It should explore how to construct a more flexible and efficient intermediary service system to support the green transformation and sustainable development of SMBMEs.

## 6. Conclusions

In view of the increasing research on financial technology and intellectual property pledge financing, and the independent third-party intermediary service platform that fits the background of the times in the market operation process, this paper constructs a tripartite evolutionary game model between SMBMEs, banks, and third-party intermediary platforms. It analyzes the stability of the strategic choices of all parties, the stability of the equilibrium strategic combination of the game system, and the impact of key elements on financing guarantees. Through simulation analysis, it verifies the effectiveness of the analysis conclusions and concludes that the third-party intermediary service platform, as a breakthrough, deeply analyzes the conditions of the new model of intellectual property pledge financing guarantees, and puts forward relevant countermeasures and suggestions for the influence relationship of key factors and stability conditions.

This research work shows that the reputation penalty faced by SMBMEs and the increase in their own intellectual property valuation both promote the on-time repayment behavior in the financing guarantee process. However, an increase in the loan interest rate set by banks for intellectual property pledge financing is not conducive to their choice of this innovative financing method. In addition, banks are greatly influenced by the market value of the intellectual property pledged by enterprises when deciding whether to lend, tend to avoid risks and reduce costs when carrying out intellectual property pledge financing guarantee operations, and are more willing to carry out green patent pledge financing guarantee business when the loan interest rate is relatively high and there is third-party intermediary platform participation in the financing process. It is worth noting that SMBMEs can use the guarantee and insurance services of third-party intermediary platforms to reduce their own risks, increase the possibility of bank loans, and reduce the risk losses of all parties in the financing guarantee process [59]. This is because the third-party intermediary platform has a bilateral market attribute, which can reduce the evaluation risk and cost issues in the business development process, and solve the market transaction dilemma of breach of contract quality disposal. Interestingly, when it faces a penalty higher than its income in the financing guarantee business, it will play a more active role as an intermediary, promoting cooperation between commercial banks and small and medium building materials enterprises.

**Author Contributions:** Conceptualization, Z.M.; methodology, Q.Z.; software, J.M.; validation, J.Z.; formal analysis, Q.Z.; investigation, Z.M.; resources, J.Z.; data curation, J.M.; writing—original draft preparation, Q.Z.; writing—review and editing, Q.Z.; visualization, J.M.; supervision, J.Z.; project administration, Z.M.; funding acquisition, J.Z. All authors have read and agreed to the published version of the manuscript.

**Funding:** This work was supported by the National Natural Science Foundation of China [grant number 72004081], the Key Project of Philosophy and Social Science Research in Colleges, Universities in Jiangsu Province [grant number 2022SJZD017], and the Pre-Research Project of the Intellectual Property College at Jiangsu University [grant number ZY202303].

**Data Availability Statement:** The authors confirm that the data supporting the findings of this study are available within the article.

**Conflicts of Interest:** The authors declare no conflicts of interest.

## References

1. Feng, B.; Ying, X.; Feng, J. Study on the Value Evaluation of Intellectual Property Pledge Financing. *Can. Soc. Sci.* **2021**, *17*, 113–122. [CrossRef]
2. Su, R.; Li, M.; Fang, Y.; Yang, C. Valuation Method of Intellectual Property Pledge Financing Based on Income Interval Analysis and Risk Adjustment Coefficient. *Humanit. Soc. Sci. Commun.* **2023**, *10*, 501. [CrossRef]
3. Panda, B.; Joy, S. Intellectual Property Rights-Based Debt Financing to Startups: Need for a Changing Role of Indian Banks. *Vikalpa* **2021**, *46*, 143–152. [CrossRef]
4. Owen, R.; Botelho, T.; Hussain, J.; Anwar, O. Solving the SME Finance Puzzle: An Examination of Demand and Supply Failure in the UK. *Ventur. Cap.* **2023**, *25*, 31–63. [CrossRef]
5. Fischer, T.; Ringler, P. What Patents Are Used as Collateral?—An Empirical Analysis of Patent Reassignment Data. *J. Bus. Ventur.* **2014**, *29*, 633–650. [CrossRef]
6. Wan, X.; Wang, K.; Fan, X.; Xu, W. Can Government Subsidy and Marine Carbon Capture, Utilization and Storage Technology Improve the Decision Level of Supply Chain Intellectual Property Pledge Financing? *Ocean Coast. Manag.* **2023**, *240*, 106663. [CrossRef]
7. Abdin, J.; Sharma, A.; Trivedi, R.; Wang, C. Financing Constraints, Intellectual Property Rights Protection and Incremental Innovation: Evidence from Transition Economy Firms. *Technol. Forecast. Soc. Chang.* **2024**, *198*, 122982. [CrossRef]
8. Bao, X.; Qu, Q.; Dong, Y.; Wang, Y.; Sheng, X. Risk Evaluation of Value Assessment in IPR Pledge Financing Based on Interval Value TOPSIS Method. In Proceedings of the 2015 International Conference on Logistics, Informatics and Service Sciences (LISS), Sydney, Australia, 24–27 July 2015; pp. 1–7.
9. Xie, L.; Liu, G.; Liu, B. Patent Pledge Policy and Stock Price Crash Risk: Evidence from China. *Res. Int. Bus. Financ.* **2023**, *65*, 101967. [CrossRef]
10. Ge, X.; Jin, Y.; Li, Q.; Yao, X.; Liu, S. Research on a Dynamic Evaluation of Financial Risks of Intellectual Property Pledge Financing of High-Tech Enterprises. *Discret. Dyn. Nat. Soc.* **2021**, *2021*, 2048811. [CrossRef]
11. Nikitenko, S.M.; Mesyats, M.A.; Rozhkova, O.V. Intellectual Property as an Instrument of Interaction between Government, Business, Science and Society. *IOP Conf. Ser. Earth Environ. Sci.* **2017**, *84*, 012015. [CrossRef]
12. Qian, K.; Shen, H.; Huang, Z. The Research Status and Development Trend of Intellectual Property Pledge Financing. *Technol. Econ.* **2013**, *26*, 51–55.
13. Yu, M.; Li, B. On the Status and Scale Expansion of Intellectual Property Rights Pledge Financing of Enterprises in Our Country—In Perspective of High and New Technology Enterprise. *J. Heilongjiang Adm. Cadre Coll. Politics Law* **2013**, *6*, 61–63.
14. Njatrijani, R.; Rahmanda, B.; Prananda, R.R.; Zaman, M.N.U. Business Development of Copyright and Fiduciary Guarantee in Indonesia. *Int. J. Econ. Bus. Adm.* **2020**, *8*, 278–284. [CrossRef] [PubMed]
15. Yang, Z.; Shi, Y.; Li, Y. Analysis of Intellectual Property Cooperation Behavior and Its Simulation under Two Types of Scenarios Using Evolutionary Game Theory. *Comput. Ind. Eng.* **2018**, *125*, 739–750. [CrossRef]
16. Qian, K.; Shen, H.; Huang, Z. Research on Evolutionary Game Analysis on the Intellectual Property Backed Loans of High-Tech SMEs. In Proceedings of the ICSSSM11, Tianjin, China, 25–27 June 2011; pp. 1–5.
17. Dong, X. Types and Methods of Assessing the Value of Intellectual Property Pledge Financing. *Acad. J. Manag. Soc. Sci.* **2023**, *5*, 94–98. [CrossRef]
18. Chen, B.; Wang, Y. High-Tech Small and Medium-Sized Enterprises Pledge Loans of Intellectual Property Rights Should Go Where. *China Soft Sci.* **2009**, *S1*, 118–120.
19. Santoso, M.G.; Dachyar, M. Modeling the Barriers and Enablers of Intellectual Property-Based Financing Scheme for Developing Economies. In Proceedings of the 11th Annual International Conference on Industrial Engineering and Operations Management, Singapore, 7–11 March 2021; pp. 1559–1571.

20. Xin, W. Problems and Solutions in the Small and Medium-Sized Enterprise Intellectual Property Pledge Financing. In Proceedings of the 2nd International Conference on Information Science and Engineering, Hangzhou, China, 4–6 December 2010; pp. 2982–2985.
21. Raymond, A. Intellectual Property as Collateral in Secured Transactions: Collision of Divergent Approaches. *Bus. Law Int.* **2009**, *10*, 27.
22. Panda, B.; Joy, S. Intellectual Property-Based Debt Financing by Indian Banks: Scope and Challenges. *IUP J. Bank Manag.* **2019**, *18*, 32–44.
23. Yan, H.; Wang, J.; Cui, J.; Yang, Z.; Ren, M. The Influencing Factors of Intellectual Property Pledge Loan Trust: A Social Network Theory-Based Experimental Study. *Discret. Dyn. Nat. Soc.* **2021**, *2021*, 4441406. [CrossRef]
24. Chen, Z.; Zhou, Z. Problems and Suggestions on the Mode of Intellectual Property Financing of Small and Medium-Sized Technological Enterprises. *DEStech Trans. Soc. Sci. Educ. Hum. Sci.* **2017**. [CrossRef]
25. Zhao, X.; Liu, Y. Empirical Prediction of Patent Pledge Financing of Pharmaceutical Enterprises—A Case Study in Jiangsu China. *PLoS ONE* **2020**, *15*, e0233601. [CrossRef] [PubMed]
26. Feng, B.; Li, H.; Hui, Z.; Feng, J. Intellectual Property Pledge Value Evaluation for Listed Companies: A Case Study of Yunnan Baiyao Enterprise. *High. Educ. Soc. Sci.* **2021**, *30*, 10–20.
27. Wang, S.; Luo, Y.; Luo, L. Study on Risk Evaluation of Intellectual Property Pledge Loan Based on Fuzzy Analytic Hierarchy Process. *J. Phys. Conf. Ser.* **2020**, *1486*, 042039. [CrossRef]
28. Zhang, J. Risk Management on Intellectual Property Pledge Financing by Small and Medium S&T Enterprises—From the Aspect of Bank. *Sci. Manag. Res.* **2013**, *31*, 98–101.
29. Zhang, J. The Study of Intellectual Property Pledge Financing Risk Evaluation Index. *DEStech Trans. Soc. Sci. Educ. Hum. Sci.* **2017**. [CrossRef] [PubMed]
30. Su, R. New Model of Intellectual Property Evaluation Based on Game Theory and Creditrisk+ Pledge Model. *J. Hebei North Univ.* **2013**, *29*, 49–58.
31. Ren, Y. Study on Financial Model of Intellectual Property Rights Pledge of Technology-Based SMEs-Based on Shaanxi. *Sci. Manag. Res.* **2012**, *30*, 105–108. [CrossRef]
32. Cheng, S.; Zhou, R. Research on Policy Tools and Modes in Intellectual Property Pledge Financing. *East China Econ. Manag.* **2013**, *27*, 159–166.
33. Feng, X. Pledge Financing of Enterprise Intellectual Property in China and Corresponding Institutional Improvement. *Hebei Law Sci.* **2012**, *30*, 39–46.
34. Wu, X.; Zhao, X.; Zhang, F.; Yang, W.; Cui, Y. Small and Mid-Sized Enterprise Intellectual Property Rights Financing Mode Analysis. In Proceedings of the 2012 International Symposium on Management of Technology (ISMOT), Hangzhou, China, 8–9 November 2012; pp. 421–425.
35. Li, Y. Probe into the Intellectual Property Rights Pledge Financing and the Construction of Its Operating Environment. *Sci-Tech Inf. Dev. Econ.* **2012**, *22*, 104–106.
36. Chen, H.; Zhang, M. *Simulation Research of Evolutionary Game to Bank and Technological SME under the Pledge Financing Mode*; Atlantis Press: Amsterdam, The Netherlands, 2016; pp. 928–933.
37. Chen, C.; Yin, W. Research on Intellectual Property Financing of Small and Medium-Sized Enterprises. In Proceedings of the 2011 International Conference on Management and Service Science, Wuhan, China, 12–14 August 2011; pp. 1–4.
38. Zhou, W.; Ouyang, W. Analysis of Industrial Intellectual Property Operation Service Platform Construction Based on the DART Model. *Appl. Math. Nonlinear Sci.* **2023**, *8*, 1951–1960. [CrossRef]
39. Long, L. Japan's Evaluation and Pledge of Intellectual Property Financing. *J. East China Univ. Sci. Technol. (Soc. Sci. Ed.)* **2009**, *24*, 79–85+99.
40. Li, L.; Shi, J. The Risk Compensation Mechanism of Intellectual Property Pledge Financing: American Experience and Enlightenment. *Proc. Int. J. Soc. Sci. Educ. Res.* **2020**, *9*, 47–55.
41. Marcello, G.; Mastiyanto, R.R. Intellectual Property as an Object of Banks Collateral in Startup Development in Indonesia (Comparation Study of Singapore and Malaysia). *J. Priv. Commer. Law* **2023**, *7*. [CrossRef]
42. Xie, L. Financing Mechanism of Intellectual Property in the U.S. and Its Implication. *J. Soc. Sci. Harbin Norm. Univ.* **2010**, *1*, 138–142.
43. Cheng, C.-H.; Tang, B.-J.; Cheng, Y.-R. Strategies and Tools for Small- and Medium-Sized Enterprises (SMEs) to Move toward Green Operations: The Case of the Taiwan Metal Industry. *Sustainability* **2024**, *16*, 4705. [CrossRef]
44. Mazur, E. *Green Transformation of Small Businesses: Achieving and Going beyond Environmental Requirements*; OECD: Paris, France, 2012.
45. Baranova, P.; Paterson, F. Environmental Capabilities of Small and Medium Sized Enterprises: Towards Transition to a Low Carbon Economy in the East Midlands. *Local Econ.* **2017**, *32*, 835–853. [CrossRef]
46. Jarboe, K.P.; Ellis, I. Intangible Assets: Innovative Financing for Innovation. *Issues Sci. Technol.* **2010**, *26*, 75–80.
47. Mann, W. Creditor Rights and Innovation: Evidence from Patent Collateral. *J. Financ. Econ.* **2018**, *130*, 25–47. [CrossRef]
48. Yang, X.; Gu, X.; Yang, X. Firm Age and Loan Financing with Patents as Collateral of Chinese Startups: The Roles of Innovations and Experience. *Econ. Innov. New Technol.* **2023**, *32*, 343–369. [CrossRef]

49. Zhang, Z.; Yang, X.; Yang, X.; Gu, X. The Effects of Knowledge Base Characteristics on Patent Pledge Financing: The Moderating Effects of Technological Munificence and Dynamism. *J. Knowl. Manag.* **2024**, *28*, 1228–1248. [CrossRef]
50. Rassenfosse, G.; Palangkaraya, A. Do Patent Pledges Accelerate Innovation? *Res. Policy* **2023**, *52*, 104745. [CrossRef]
51. Jiang, C.; Zhou, Y.; Chen, B. Mining Semantic Features in Patent Text for Financial Distress Prediction. *Technol. Forecast. Soc. Chang.* **2023**, *190*, 122450. [CrossRef]
52. Guo, C. A Review of Multi-Agent Game in Intellectual Property Pledge Financing. *J. Risk Anal. Crisis Response* **2024**, *14*, 228–239. [CrossRef]
53. Wang, R.; Sun, B. Risk Allocation of Intellectual Property Pledge Financing Based on Blockchain Technology: An Evolutionary Strategy Study. In Proceedings of the 2023 6th International Conference on Information Management and Management Science, Chengdu China, 25–27 August 2023; Association for Computing Machinery: New York, NY, USA, 2023; pp. 301–307.
54. Berlinger, E.; Gosztonyi, M.; Havran, D.; Pollak, Z. Interpersonal versus Interbank Lending Networks: The Role of Intermediation in Risk-Sharing. *Emerg. Mark. Rev.* **2023**, *54*, 100989. [CrossRef]
55. Kang, Y. Research on Intellectual Property Financing Mode of Technology-Based Small and Medium-Sized Enterprises. *Financ. Eng. Risk Manag.* **2023**, *6*, 34–42. [CrossRef]
56. Talisa, R.; Badriyah, S.M. The Implications of Intellectual Property as Objects of Fiduciary Guarantee for Creative Economy Entrepreneurs. *J. Daulat Huk.* **2023**, *6*, 1–13. [CrossRef]
57. Choi, J.; Choi, P.M. Do Bank Loans Curb Corporate Moral Hazard? *J. Appl. Bus. Res.* **2017**, *33*, 115–122. [CrossRef]
58. Qian, K.; Pan, Y.; Huang, Z.Q. Game Analysis of P2P Lending Signal Based on Patent Pledge. *Soft Sci.* **2018**, *32*, 108–112+118. [CrossRef]
59. Yu, X.; Chen, H.; Xiang, K.; Wan, Z. Supply Chain Financing Mechanism with Guarantee Insurance. *Manag. Decis. Econ.* **2021**, *42*, 308–318. [CrossRef]

buildings

MDPI

*Article*

# Adoption of Innovative Energy Facilities in the Tertiary Sector Buildings: Exploring Interdependencies and Key Drivers

**Ruijin Huang [1] and Gaofeng Gu [2,*]**

[1] School of Information and Business Management, Dalian Neusoft University of Information, Dalian 116023, China; huangruijin@neusoft.edu.cn
[2] College of Transportation Engineering, Dalian Maritime University, Dalian 116026, China
* Correspondence: ggu@dlmu.edu.cn

**Abstract:** Innovative energy facilities, such as solar panels, heat pumps, and smart control ventilation, offer substantial opportunities to improve energy efficiency and environmental performance in the tertiary sector, aligning with green building objectives. This study aims to identify the key factors influencing the adoption of these facilities by small and medium-sized enterprise owners in the tertiary sector and to explore the interdependencies among them. To achieve this, we employed a stated choice experiment to assess preferences and decision-making by presenting respondents with multiple hypothetical scenarios, each containing alternatives described by varying attributes. A simultaneous equation model was used to analyze the key drivers of adoption and the interrelationships among these facilities. The results reveal that cost-related attributes and government incentives significantly impact the acceptance of energy facilities. Notably, while environmental responsibility is slightly associated with solar panel adoption, it shows no significant link with heat pumps or ventilation systems. Furthermore, we identified a bi-directional relationship between the adoption of solar panels and heat pumps, suggesting that acceptance of one positively influences the other. Conversely, a unidirectional relationship exists between ventilation and solar panels, where the adoption of ventilation positively influences solar panel adoption, but not the other way around. These findings contribute to a deeper understanding of decision-making processes in green building projects and provide valuable insights for policymakers and industry stakeholders aiming to promote sustainable energy solutions in the tertiary sector.

**Keywords:** building energy; sustainable construction; adoption intentions; energy facilities; simultaneous equation model

## 1. Introduction

Accelerating electrification and promoting renewable energy are crucial strategies for mitigating global warming, aligning with the European Green Deal's target to cut emissions by 55% from 1990 levels by 2030. The building sector represents a significant component of energy consumption, accounting for nearly 40% of total energy use in the EU [1]. Commercial buildings, in particular, contribute substantially, making up 12% of end-use energy consumption in the US [2]. Yet, fewer than 30% of companies in Europe have implemented energy conservation measures, indicating a significant potential for energy savings through effective strategies.

Innovative energy facilities not only offer higher energy efficiency but also provide flexibility, automation, and various economic, social, and environmental benefits in the long run. The term "energy facilities" in this study means equipment or facilities that can be used to generate, storage, transfer, or save energy sources. The "innovative" refers to recently developed renewable energy facilities and energy-efficient systems that have not yet been widely adopted. We specifically focus on technologies such as solar panels, air source heat pumps, and smart control ventilation systems, which represent advancements designed to enhance energy efficiency and sustainability. These innovations have significant potential for improving energy performance and reducing carbon emissions, particularly for small and medium-sized enterprises (SMEs) in the tertiary sector. Solar photovoltaic panels and heat pumps are among the key renewable energy technologies expected to play pivotal roles in achieving the EU 2030 and 2050 energy and climate goals. Traditional ventilation systems also contribute significantly to energy consumption, with heating, ventilation, and air conditioning systems accounting for about 40% of energy use in office buildings [3]. The integration of smart control systems, such as advanced mechanical ventilation with heat recovery, can lead to energy consumption reductions of approximately 30% while maintaining thermal comfort levels [4].

Despite the advantages, the market penetration of renewable energy technologies remains slow in commercial buildings [5,6]. Factors hindering adoption include relatively low gas prices compared with electricity, the need for customer awareness regarding the benefits and features of innovative facilities, and the alignment of facility characteristics with specific business needs. For instance, restaurants prioritize temperature and air quality, while butcher shops focus more on cooling requirements. In addition, the functional integration of multiple energy facilities often brings more possibility, e.g., heat pump plus ventilation may bring not only good air quality but also comfortable temperature; however, such a combination often comes with a price, especially at the initial stage, which might hinder the adoption process.

Thus, it is also critical whether companies would want to accept the new energy facilities because the market potential of energy facilities depends on the adoption behavior of customers. It is necessary and timely to understand the decision-making mechanism of companies in the adoption of different innovative energy facilities. This research specifically focuses on the willingness of small and medium-sized enterprises (SMEs) owners in the service industry to adopt these innovative energy facilities. The tertiary sector includes SMEs such as butchers, barbers, small food and non-food shops, restaurants, and pubs.

This study aims to investigate the adoption of energy facilities, paying particular attention to the mutual dependencies between different technologies. To achieve this, we designed a dedicated stated choice experiment to collect data from targeted companies, enabling us to capture their adoption behaviors regarding a set of innovative energy facilities characterized by specific attributes and choice contexts. This methodological approach is especially useful in situations where historical data and variation are insufficient.

Furthermore, we estimated a simultaneous equation model, which is particularly suited for examining potential interdependencies and interactions among multiple alternatives. By enhancing our understanding of the decision-making mechanisms influencing companies' adoption of energy facilities, this research aims to inform policymakers and technology developers about effectively promoting the use of green energy solutions. Former studies largely emphasize the residential areas [7–11], while the decision-making mechanism of companies could be unique or dramatically different. To the best of our knowledge, this is perhaps the first attempt to examine companies' adoption of innovative energy facilities systematically.

The remainder of this paper is organized as follows: a literature review will precede a discussion of data collection and the design of the stated choice experiment, followed by a detailed analysis. We will then present the results from the simultaneous equation model estimation, concluding with future considerations.

## 2. Literature Review

The adoption of renewable energy facilities was influenced by a variety of factors, including financial, technological, policy-related, and environmental considerations, as shown in Table 1. While these factors have been widely recognized in the decision-making process of households [12,13], their relative importance and interplay remain underexplored in the context of SME adoption.

**Table 1.** Variables influencing preferences for energy facilities.

| | Variables | References |
|---|---|---|
| Finance-related variables | purchase price and maintenance costs | [14–18] |
| | payback period | [19,20] |
| Technical performance-related variables | off-grid capability | [21] |
| | cooling capability | [22] |
| | energy recovery performance | [23] |
| | air filtration | [24] |
| Policy-related variables | policy incentives and tax refunds | [25] |
| environmental considerations-related variables | environmental concerns | [15,21] |

### 2.1. Financial Barriers and Constraints

A consistent theme in the literature is the financial barriers that customers face when considering the adoption of energy-efficient technologies. The high initial investment costs and ongoing maintenance expenses are often cited as major deterrents [14]. For instance, access to capital is frequently identified as a significant constraint for companies, with studies indicating that many businesses lack the necessary financial resources to invest in energy-efficient solutions. This issue is particularly pronounced in SMEs, which generally face more stringent budgetary limitations compared with larger enterprises [15]. While financial barriers such as these are well-documented in residential energy adoption research, SMEs face additional complexities, such as concerns about return on investment and the long payback periods associated with renewable energy systems [16]. Research has highlighted that these concerns often lead to delayed or avoided investment despite the long-term benefits of renewable technologies [17].

### 2.2. Technological Considerations and Performance

The technical performance of energy facilities plays a critical role in adoption decisions. Businesses often prioritize technologies that offer tangible benefits, such as cost savings or increased energy efficiency. For example, the self-sufficiency and off-grid capabilities of solar panels have made them a popular choice, particularly in regions where energy security is a concern [18]. Similarly, heat pumps are attractive because of their energy-saving potential, especially for businesses seeking to reduce heating and cooling costs [19]. The energy recovery performance of ventilation systems is a significant concern for small businesses, with advanced features [20] such as demand control and air filtration being critical for energy conservation [21].

### 2.3. Policy and Incentive Structures

Government policies, particularly financial incentives such as subsidies, tax breaks, and rebates, have been widely recognized as important drivers of adoption. In many countries, these policies have successfully boosted the adoption of renewable energy technologies. For example, tax incentives for solar panels have been shown to increase adoption rates in the residential sector significantly [22]; however, the impact of policy incentives on SMEs has not been analyzed extensively. While some studies indicate a positive relationship between government incentives and adoption rates, it is important to note that the effectiveness of such policies may vary across different types of businesses, depending on their specific needs, financial situations, and energy consumption patterns. Furthermore, the influence of policies on SMEs may be overshadowed by other constraints, such as a lack of concern about the long-term viability of the technologies.

### 2.4. Environmental Considerations and Corporate Social Responsibility

Environmental concerns have also been cited as motivators for renewable energy adoption, though their influence on SMEs is less clear. On the one hand, SMEs in certain industries are increasingly recognizing the importance of corporate social responsibility (CSR) and the role they can play in reducing carbon emissions; however, there is evidence that environmental motivations do not always translate into a willingness to invest in renewable energy technologies. While some studies have found that environmental concerns significantly influence the adoption of energy-efficient solutions [23], others suggest that these motivations are less important for SMEs [15], which often prioritize financial and operational considerations over environmental benefits. This is particularly true for businesses in the tertiary sector, where the adoption of green technologies may be seen as a way to enhance their social image and meet consumer demand for sustainability rather than being driven by intrinsic environmental values.

### *2.5. Research Gaps*

In summary, while the existing literature has provided valuable insights into the factors influencing renewable energy adoption [24], there remain significant gaps, particularly in understanding the complex decision-making processes of SMEs. Companies may have a more significant role in energy-saving because of their responsibility for carbon reduction [25]; however, balancing this responsibility with innovation efforts can be challenging because of various constraints.

Additionally, the integration of multiple energy technologies and the synergies between them have been largely overlooked. Existing studies have primarily focused on the adoption of individual renewable energy technologies [26], with limited attention given to the potential benefits of integrating multiple systems. For example, the combination of solar panels and heat pumps has been shown to reduce energy consumption. The literature largely overlooks the synergies that may exist between different energy technologies, which could influence adoption decisions [27]. This lack of focus on the integration of multiple technologies in SME settings represents a critical gap in the literature. This study aims to address these gaps by examining the adoption behavior of SMEs in the tertiary sector, with a focus on the interdependencies between solar panels, heat pumps, and ventilation systems.

## 3. Materials and Methods

In order to collect data for this analysis, a dedicated experiment needs to be designed. For this purpose, we designed a stated choice experiment and implemented a survey to obtain data from SMEs in the tertiary sector. The stated choice experiment methodology for this study is due to its effectiveness in capturing the nuanced decision-making processes. Given limited historical data on these specific technologies in the tertiary sector, stated choice experiments offer a robust framework for evaluating preferences based on hypothetical scenarios. By the stated choice experiment methodology, we aim to uncover the key factors influencing adoption intentions, which is essential for informing policy and practice in promoting renewable energy solutions. This section will describe in detail the experimental design and the survey management.

### *3.1. Experiment Design*

The purpose of designing a choice experiment is to capture the adoption behavior of companies given a set of innovative energy facilities represented by the characteristics of the alternatives and the choice contexts. The stated choice experiment is a research method used to understand preferences and decision-making processes by presenting respondents with hypothetical scenarios. This approach is especially useful when no sufficient historical data and variation exist. Through the definition of attributes and the associated levels, one can create a hypothetical situation in which respondents can make a choice on the provided alternatives. Normally, respondents are presented with a series of randomized hypothetical choice sets in which the attribute values of the alternatives are varied across choice situations. Thus, defining the attributes and attribute levels is critical to warrant reliable observations. Since companies' adoption of innovative energy facilities has not been addressed systematically in the literature, we borrow the knowledge of relevant existing works to define the attributes and attribute levels for each energy facility.

3.1.1. Attributes of Solar Panel

As presented in the literature review, cost, policy, and technology performance are generally found to affect the adoption of solar panels. The cost in practice involves many types, including installation, insurance, number of solar panels, brand, etc. The policy could refer to the incentive programs operated at the municipality or national level, e.g., tax reduction for purchasing. The technology performance normally means the efficiency of the solar panel, which influences the total payback years compared with the initial investment for installation. These different factors play a different role in affecting the adoption of solar panels. Furthermore, solar panel as a renewable energy source also helps companies to reduce $CO_2$ emissions; therefore, in the choice experiment, we consider $CO_2$ emission as a factor that might indirectly influence the choice decisions bounded by social responsibility. In the end, the alternative of solar panels involves five attributes: total installation price, municipal incentive, annual reduction in energy charges, payback years, and the percentage of $CO_2$ reduction.

To define proper levels for each attribute, company profiles need to be considered because of the large variation in types and scales. For instance, the installation price may be associated with the company size because larger companies may demand more solar panels due to bigger energy demand. Respondents were informed that the prices included acquisition, installation, and maintenance costs before conducting the experiment. Considering the market situation in the Netherlands, the range of price is set from €2500 to €10,000 if the size of the company (building square meters) is smaller than 70 $m^2$, from €9000 to €18,000 if the company size is between 70 $m^2$ and 140 $m^2$, and from €12,000 to €24,000 if company size is larger than 140 $m^2$.

The range of payback years is another important factor. The payback period is calculated based on the rate of return, which depends on the price of solar panels, the price of electricity, and the annual power yield. In Belgium and the Netherlands, the maximum annual power generation is approximately 850–900 h. Considering the local solar power efficiency and dynamic electricity rates, €10,000 can purchase 4700 Watt Peak solar panels, with an annual power generation value ranging from €750 to €1800. To cover the actual situation as much as possible, the payback year of solar panels in this study is varied from 5 years to 20 years.

Apart from that, we define policy incentives as the incentives from municipalities. This is also the common practice in the Netherlands where a certain amount of tax refund is offered for purchasing solar panels. In our experiment, the amount of the incentives is a percentage that is relative to the purchase price. We consider the values from the extreme case of no incentives (i.e., when solar panels are popular) to 30% of the purchase price as the maximum tax refund.

In spite of the effects of saved energy charge can be partly reflected by the payback year, this is an indicator related to the energy use behavior, e.g., energy-efficient behavior will save more energy cost; however, the magnitude of the cost savings is also closely related to the number of solar panels. In general, the more solar panels installed, the larger energy savings can be achieved. A small number of solar panels can support basic energy use partly or fully, while a large number of solar panels will generate more profit. Thus, we set the range of potential cost saving into four levels: 20%, 40%, 60%, and 80% of the current energy cost.

Apart from these cost-related factors, we also include social responsibility in reducing carbon emissions. By informing users of a certain amount of carbon reduction attached to solar panels, we can observe their responses to solar panels respective to the carbon reduction. Here, we set the percentage of emission reduction as 10%, 20%, 30%, and 40%, respectively.

3.1.2. Attributes of Heat Pump

Similar to solar panels, cost-related factors are the primary concerns in the acceptance of heat pumps, such as installation price, policy incentives, and performance in carbon emission reduction. In addition, differing from solar panels, which generate electricity, heat pumps consume electricity and are mainly featured with functions of cooling and/or warming; therefore, a binary attribute related to the function, varied with cooling and heating, was specified for heat pumps.

The installation price was defined according to the size of the company, ranging from €5000 to €8000 if the company size is smaller than 70 $m^2$, from €8000 to €10,000 if the company size is between 70 and 140 $m^2$, and from €10,000 to €16,000 if company size is larger than 140 $m^2$. Note that the price here is approximated to the average of a standard heat pump without consideration of luxury additions.

The payback period of a heat pump is influenced by several key factors, including the initial costs of the heat pump itself, the energy consumption it incurs, the prevailing energy prices, and any available tax refunds or grants. In the Netherlands, the price of natural gas for small-sized enterprises ranges from €1.0 to €1.5 per cubic meter. For instance, a heat pump priced at €5000 can lead to substantial savings by reducing natural gas consumption by approximately 500 $m^3$ annually; however, it is important to note that this system also requires electricity, consuming between 600 and 1200 kWh each year. Given the average electricity price of €0.40 per kWh, the annual savings generated from the heat pump can be estimated to fall between €200 and €500. Furthermore, considering the availability of a 30% grant, the overall financial landscape becomes more favorable. With this grant taken into account, we project that the payback period for the investment in a heat pump could extend from 5 to 20 years, depending on the specific operational parameters and energy usage patterns of the enterprise. Similar to solar panels, the tax refund is set to range from no incentives to a maximum of 30%, considering the realistic policy situation in the Netherlands.

The carbon emissions of heat pumps are correspondingly high when the supplied electricity is produced by high-emission energy resources and relatively low in the case of renewable energy-generated electricity. Overall, the performance of carbon emission of heat pumps is lower than that of solar panels; therefore, the proportion of carbon emission for heat pumps is set from 10% to 40%.

3.1.3. Attributes of Ventilation

The ventilation was characterized by installation price, incentive, carbon emission reduction, and the function of demand control, pollutant filter, and heat recovery. According to the market situation of ventilation in the Netherlands, prices were set from €1000 to €10,000. The possible value of the tax incentive in the stated choice experiment was set as 0, 10%, 20%, and 30%, considering the realistic policy situation. In addition to the price, incentive, environmental benefits, and the function of demand control were considered since it can provide an optimized ventilation scheme and has higher energy efficiency. According to the literature review section, the function of recovering heat can reduce energy consumption for heating. According to the previous study [28], ventilation with demand response strategies or heat recovery can reduce around 20% of emissions. The filter-out pollutants function can provide the promotion of comfort and a healthy environment. The ranges of attributes for this experiment are shown in Table 2.

**Table 2.** Attributes of renewable energy facilities.

| Attribute | Solar Panel | Heat Pump | Ventilation |
|---|---|---|---|
| Price (€) (<70 $m^2$) | 2500; 5000; 7500; 10,000 | 5000; 6000; 7000; 8000 | 1000; 3000; 5000; 10,000 |
| Price (€) (70–140 $m^2$) | 9000; 12,000; 15,000; 18,000 | 8000; 10,000; 2000; 14,000 | 1000; 3000; 5000; 10,000 |

**Table 2.** *Cont.*

| Attribute | Solar Panel | Heat Pump | Ventilation |
|---|---|---|---|
| Price (€) (>140 m$^2$) | 12,000; 16,000; 20,000; 24,000 | 10,000; 12,000; 4000; 16,000 | 1000; 3000; 5000; 10,000 |
| Incentive from municipality | 0; 10%; 20%; 30% | 0; 10%; 20%; 30% | 0; 10%; 20%; 30% |
| Annual reduction in energy charges | 20%; 40%; 60%; 80% | 20%; 40%; 60%; 80% | - |
| $CO_2$ reduction | 10%; 20%; 30%; 40% | 10%; 20%; 30%; 40% | - |
| Payback years | 5; 10; 15; 20 | 5; 10; 15; 20 | - |
| Function | - | Cooling + Heating; Heating only | - |
| Have demand control? | - | - | Yes/No |
| Can recover heat? | - | - | Yes/No |
| Can filter out pollutants? | - | - | Yes/No |

#### 3.1.4. Choice Set Generation and Implementation

A detailed list of attributes and attribute levels is presented in Table 2, where 2-levels and 4-levels are adopted according to each alternative. To generate the choice sets, the feasible number of combinations between attributes and levels needs to be identified. Because, in total, we have twenty 4-level attributes and three 2-level attributes, a full factorial design will result in $4^{20} \times 2^3$ combinations, which makes the actual implementation unrealistic. Therefore, we use an orthogonal fractional factorial design which still warrants the orthogonality and the observation of the main effects.

The fractional factorial design includes 64 combined profiles varied by the levels of attributes. The choice sets were then randomly assigned to the respondents. Each respondent was given eight different choice situations, which means every participant answered eight times the choice questions, and each question included different combinations of attribute levels. Figure 1 shows an example of the choice task, which includes three alternatives: solar panel, heat pump, and ventilation.

In a standard stated choice experiment, each respondent is often asked to choose the option they prefer the most, leaving other alternatives not chosen. This assumes that the choice alternatives are independent. As noted in the first section, the interrelationship between different energy facilities should be captured. This means the synergy effect between alternatives will not be observed if only one alternative can be selected. For example, if one respondent chooses the solar panel as the preferred option, then how the respondent values the integration between solar panels and other options becomes unknown, losing the information on the correlation between alternatives.

Therefore, we propose a different approach. Instead of choosing only one of the options, respondents were asked to report the intention to adopt each of the provided options at a level of scale. More specifically, respondents were asked to indicate their intention to install each type of energy facility using the options including "definitely not", "unlikely", "neutral", "likely", and "definitely yes". The intention to install each innovative energy facility is then recorded. In this way, the potential mutual dependency underlying individuals' decision-making behavior is captured. Table 3 presents an example of the choice scenario provided to respondents.

**Figure 1.** Distribution of companies who provided valid answers.

**Table 3.** Example of the choice scenario.

| | 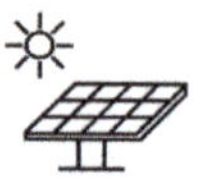 **Solar Panel** |  **Heat Pump** |  **Ventilation** |
|---|---|---|---|
| Price (€) | 4500 | 8000 | 1000 |
| Incentive from municipality | No incentive | 10% | 20% |
| Annual reduction in energy charges | 80% | 20% | |
| $CO_2$ reduction | 20% | 10% | |
| Payback years | 20 | 5 | |
| Function | | Cooling + Heating; | |
| Demand control | | | Yes |
| Can recover heat? | | | No |
| Can filter out pollutants? | | | No |
| Your choice | □ Definitely NOT<br>□Unlikely<br>□Neutral<br>□Likely<br>□Definitely YES | □Definitely NOT<br>□Unlikely<br>□Neutral<br>□Likely<br>□Definitely YES | □Definitely NOT<br>□Unlikely<br>□Neutral<br>□Likely<br>□Definitely YES |

*3.2. Survey Management and Descriptive Statistics*

The data collection for this study was conducted in the Netherlands in October 2020. The questionnaire started with a general introduction to the research objective. Then, respondents were asked to fill in the general profiles of the company, e.g., location, business scope, size, investment plan, and energy consumption details of their companies, and then the choices of the experiment. Because managers may not be familiar with the innovative energy facilities, i.e., features and price, some screening information about these facilities was explained before the choice experiment. Completing the survey took about 15–20 min.

A marketing company was engaged to recruit respondents from various cities, ensuring diverse representation from the tertiary sector. Given the focus on the purchasing intentions of small and medium-sized enterprises (SMEs) regarding energy facilities, the respondents were typically SME owners or individuals in managerial roles. Figure 2 shows the spatial distribution of the companies (several are located within Belgium). Additionally, we required respondents to have a recent investment intention, which led to the relatively low response rate of 15.6%, with 126 out of 810 contacted individuals providing valid data. While this response rate may seem modest, it reflects the challenges inherent in engaging this specific demographic. To enhance the sample size, we employed the stated choice method, where each respondent answered eight choice scenarios, each varying in context, resulting in a total of 1008 choice observations for analysis. This approach not only increases the robustness of the dataset but also strengthens the validity and reliability of the findings presented in this study.

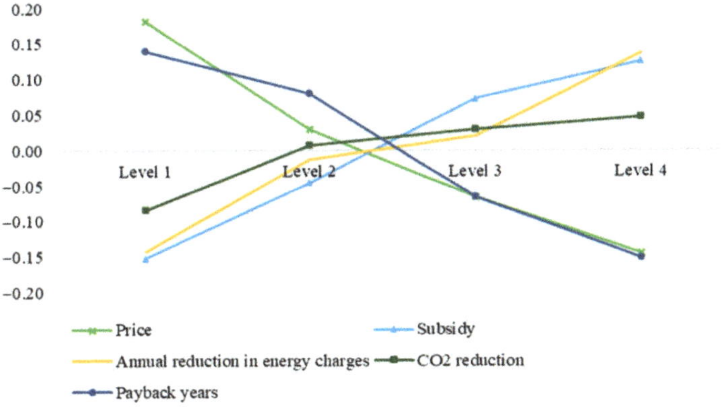


**Figure 2.** Marginal effects of solar panel attributes.

Table 4 presents the descriptive statistics of the company characteristics. The location represents the built environment around the company. It is related to the building density and available space for the installation of the energy facilities. Companies are mostly located in the suburbs (42.9%) and rural areas (37.3%) and less in the city centers (19.8%). Company size is another important indicator as it approximates the potential energy demand; the size of companies is almost equally distributed, and each takes about one-third.

Building ownership is linked to influencing the authority to install a facility and uncertainty about the future. Obviously, those who own the building are more flexible in the adoption of energy facilities. The results show a slightly larger number of companies rented the building (53.2%) instead of ownership. In addition, the business scope of companies can significantly impact the structure of energy demand and the facilities. Here, we defined detailed categories in the survey; however, considering the high level of variations, we recategorized the type into food-related (butcher shops, bakery, food shops, restaurants, bars, and hotels), non-food related (retailing, haircut, and offices), and others.

Moreover, whether a company owner has an investment plan could be related to the reserved budget to invest in energy facilities. Interestingly, these data show more than 60%

of the companies have a plan in the future to install innovative energy facilities if possible. This means that keeping everything else constant, the intention to upgrade companies' current facilities is relatively big.

**Table 4.** Characteristics of the companies.

| Variable | Classification | # of Cases | Percentage |
|---|---|---|---|
| Location | Center | 25 | 19.8% |
| | Suburb | 54 | 42.9% |
| | Rural | 47 | 37.3% |
| Number of employees | Less than 5 | 45 | 35.7% |
| | From 6 to 20 | 40 | 31.8% |
| | More than 20 | 41 | 32.5% |
| Business scope | No Food-related | 51 | 40.5% |
| | Food-related | 43 | 34.1% |
| | Others | 32 | 25.4% |
| Plan to install innovative energy facilities | Yes | 77 | 61.1% |
| | No | 49 | 38.9% |
| Building ownership | Own | 59 | 46.8% |
| | Rent | 67 | 53.2% |

### 3.3. Methodology

In this study, the adoption intentions of company owners toward innovative energy facilities are examined using a simultaneous equation model. This methodological approach is particularly suitable for capturing the potential interdependencies and interactions among multiple alternatives. A simultaneous equation model is a robust econometric tool that can simultaneously deal with two or more equations that interact with each other. It is specifically designed to address situations where the observational phenomena are assumed to be reciprocally causal, thereby enabling a more comprehensive understanding of the dynamics at play. By modeling such interdependencies, the simultaneous equation model can effectively manage issues of endogeneity, which often arise when there are mutual causal relationships between variables. This capacity to disentangle and analyze the relationships between multiple endogenous variables is a distinguishing feature of the simultaneous equation model, making it an invaluable tool for exploring complex decision-making processes, such as those involving the adoption of innovative technologies. Through this approach, this study aims to provide insights into how different factors jointly influence the adoption of energy innovations, offering a more nuanced perspective on corporate decision-making behavior in this domain. Here, we consider the attributes of energy facilities and company characteristics as the exogenous variables and the intention to purchase innovative energy facilities as endogenous variables in a simultaneous equation model. In specific, the model can be written as follows:

$$y_{sp} = c_{sp} + \alpha_{sp}z + \beta_{sp}x_{sp} + \gamma_{sp}\mathbf{y}_{h,v} + \varepsilon_{sp} \quad (1)$$

$$y_{hp} = c_{hp} + \alpha_{hp}z + \beta_{hp}x_{hp} + \gamma_{hp}\mathbf{y}_{s,v} + \varepsilon_{hp} \quad (2)$$

$$y_{vt} = c_{vt} + \alpha_{vt}z + \beta_{vt}x_{vt} + \gamma_{vt}\mathbf{y}_{s,h} + \varepsilon_{vt} \quad (3)$$

where $y_{sp}$, $y_{hp}$, and $y_{vt}$ are dependent variables, representing the level of acceptance of solar panel (*sp*), heat pump (*hp*), and ventilation (*vt*), respectively.$c_{sp}$, $c_{hp}$, and $c_{vt}$ are the intercepts for the three alternatives. $\mathbf{z}$ denotes the vector of company characteristics that are invariant across alternatives and refers to location, size, building ownership, business scope, and investment plan. $\mathbf{x_{sp}}$, $\mathbf{x_{hp}}$, and $\mathbf{x_{vt}}$ are the vectors of the main attributes included in the stated choice experiment with respect to solar panels, heat pumps, and ventilation, respectively. $\mathbf{x_{sp}}$ includes the price, the incentive from the municipality, the annual reduction in energy charges, $CO_2$ reduction, and payback years. $\mathbf{x_{hp}}$ includes price, the incentive from the municipality, the annual reduction in energy charges, $CO_2$ reduction, payback years, and cooling function. $x_{vt}$ includes price, the incentive from the municipality, the annual reduction in energy charges, payback years and the function of demand control, heat recovery, and pollutants filtering. **y** is a vector to address the endogeneity, representing the acceptance of other energy facilities. In concrete, $\mathbf{y}_{h,v}$ represents the effects of the choices of heat pump and ventilation, $\mathbf{y}_{s,v}$ represents the effects of the choices of solar panels and ventilation, and $\mathbf{y}_{s,h}$ represents the effects of the choice of solar panel and heat pump. $\alpha_{sp}, \alpha_{hp}, \alpha_{vt}, \beta_{sp}, \beta_{hp}, \beta_{vt}, \gamma_{sp}, \gamma_{hp},$ and $\gamma_{vt}$ are the vectors of the parameters to be estimated. $\varepsilon_{sp}, \varepsilon_{hp}$ and $\varepsilon_{vt}$ denote the error terms, which are normally distributed with 0 mean and a variance-covariance matrix $V$ ($\varepsilon_{sp}$,$\varepsilon_{hp}$, $\varepsilon_{vt}$). The variance covariance matrix is given by

$$V = \begin{pmatrix} 1 & \rho_{sp,hp} & \rho_{sp,vt} \\ \rho_{hp,sp} & 1 & \rho_{hp,vt} \\ \rho_{vt,sp} & \rho_{vt,hp} & 1 \end{pmatrix} \quad (4)$$

where the off-diagonal elements in the covariate matrix, $\rho$, represent the correlation between the endogenous components of three alternatives. To deal with the endogeneity in the simultaneous equation model, the error terms are correlated with each other. This means the traditional ordinary least square (LS) estimation, which deals with a single equation standalone, does not apply. In addition, because of the ignorance of the simultaneity in the equation system, LS normally results in simultaneity bias and inconsistent estimation [29].

Therefore, this study employs the Three-Stage Least Squares (3 SLS) estimation method, a widely recognized approach for addressing both contemporaneous residual correlation and predictor endogeneity. The 3 SLS method extends the capabilities of the Two-Stage Least Squares (2 SLS) approach by not only accounting for the potential endogeneity of explanatory variables but also allowing for correlations between the error terms of different equations within the system. This is particularly relevant when multiple equations are involved, as the unobserved disturbances across these equations may be correlated, which can bias the estimates if not properly accounted for. It is more efficient compared with 2 SLS and allows correlations between unobserved disturbances across various equations. In this study, the exogenous variables uncorrelated with the error terms were selected as the instrument variables which are used to regress the endogenous predictor. More specifically, the variables that are independent of all the other variables (including error terms) but not independent of the endogenous variables were selected.

## 4. Results

An essential test was conducted before estimating the simultaneous equation model with 3 SLS. The Hausman test was applied to test the endogeneity of the equations. The null hypothesis of the endogeneity is that the ordinary least squares (OLS) estimator would not be significantly impacted by the endogeneity problem. The OLS estimators provide consistent estimate results. A rejection of the null hypothesis shows instrument variables are required in the estimate process. The Haussmann statistics provide evidence of acceptance or rejection of the null hypothesis. The test statistic value is 21.429 (df = 3), which is insignificant ($p > 0.99$) such that the null hypothesis can be rejected. We conclude that OLS is inconsistent. The McElroy R-squared value of the simultaneous model with 3 SLS is 0.738, which is satisfactory. The detailed estimation results are summarized in Tables 5–7.

### *4.1. Results of Solar Panel*

As reported in Table 5, most of the estimated parameters are statistically significant. The installation price of €2500 (0.181) has a significant and positive effect on the adoption of solar panels, while the adoption intention is negatively related to the installation price when the price is higher than €7500. This means the acceptance decreases with the increasing installation price. The difference between the effects of price at €2500 and €10,000 is 0.326. These results are broadly in line with the findings of existing studies [30].

**Table 5.** Results of solar panel.

| | | Estimate | | t-Value | *p*-Value |
|---|---|---|---|---|---|
| | Intercept | 1.495 | *** | 6.687 | 0.000 |
| Price | €2500 | 0.181 | *** | 7.491 | 0.000 |
| | €5000 | 0.030 | | 1.263 | 0.207 |
| | €7500 | −0.066 | *** | −2.767 | 0.006 |
| | €10,000 | −0.145 | | | |
| Incentive | 0% | −0.152 | *** | −6.438 | 0.000 |
| | 10% | −0.046 | ** | −1.934 | 0.053 |
| | 20% | 0.073 | *** | 3.110 | 0.002 |
| | 30% | 0.125 | | | |
| Annual reduction in energy charges | 20% | −0.143 | *** | −6.020 | 0.000 |
| | 40% | −0.013 | | −0.561 | 0.575 |
| | 60% | 0.020 | | 0.853 | 0.394 |
| | 80% | 0.136 | | | |
| $CO_2$ reduction | 10% | −0.084 | *** | −3.558 | 0.000 |
| | 20% | 0.007 | | 0.276 | 0.783 |
| | 30% | 0.030 | | 1.287 | 0.198 |
| | 40% | 0.047 | | | |
| Payback years | 5 | 0.139 | *** | 5.835 | 0.000 |
| | 10 | 0.079 | *** | 3.322 | 0.001 |
| | 15 | −0.066 | *** | −2.787 | 0.005 |
| | 20 | −0.152 | | | |
| Location | City center | −0.116 | *** | −1.967 | 0.049 |
| | Suburb | −0.044 | *** | −2.261 | 0.024 |
| | Rural | 0.160 | | | |
| Have an investment plan | Yes | 0.299 | *** | 10.761 | 0.000 |
| | No | −0.299 | *** | −10.761 | 0.000 |
| Building ownership | Own | −0.027 | | −1.074 | 0.283 |
| | Rent | 0.027 | | 1.074 | 0.283 |
| Number of employees | ≤5 | 0.002 | | 0.043 | 0.966 |
| | 5–20 | −0.116 | *** | −5.011 | 0.000 |
| | ≥20 | 0.114 | | | |
| Business scope | Food | 0.228 | *** | 11.192 | 0.000 |
| | Non-food related | −0.391 | *** | −19.847 | 0.000 |
| | Other | 0.163 | | | |
| Endogenous effects | Heat pump | 0.450 | *** | 5.652 | 0.000 |
| | Ventilation | 0.152 | *** | 2.571 | 0.010 |
| R-Squared | | 0.730 | | | |

Notes: "**", "***" indicate significance at the 90%, 95% and 99% confidence level, respectively.

The reduction in annual energy charges is an indicator of the economic benefit for companies. The parameter is found negative at the level of 20% reduction and positive at the level of 80% reduction, indicating that a larger reduction in energy charges brought by the innovative energy facilities does increase the intention to adopt innovative energy facilities.

The payback period has been considered a highly relevant indicator to invest in solar panels. Results show a five-year payback period has a positive impact on the adoption, while such an effect decreases sharply when the payback period is longer than 10 years. This is understandable in that a long payback period for solar panels, in general, hinders the adoption behavior because of the unpredictable future. These results are also consistent with the findings by [31]. Furthermore, in the case of household decisions, the initial cost was recognized to have a higher impact than the payback period on the choice of solar panels; however, the results in this study indicate that the payback period has a higher impact on the acceptance of solar panel [30]. This is perhaps triggered by the difference between companies and households, where companies could be more cautious when facing an uncertain future [32].

Incentive is an important policy measure to promote the adoption of innovative energy facilities. Results show a significant effect of government incentives on companies' acceptance of solar panels. The coefficient of no government incentive is negative, indicating no interest, which is consistent with the recent studies about the effects of the incentive on attitudes toward solar panels [33]. When the incentive increases from 10% (−0.046) to 20% (0.073), the effect increases by 0.119, while when the incentive increases from 20% to 30%, the effect increases by only 0.052 (Figure 3). These results indicate that the marginal effect of a 10% incentive to a 20% incentive is relatively high. In other words, incentives between 10% and 20% are more cost effective to increase the acceptance of solar panels.

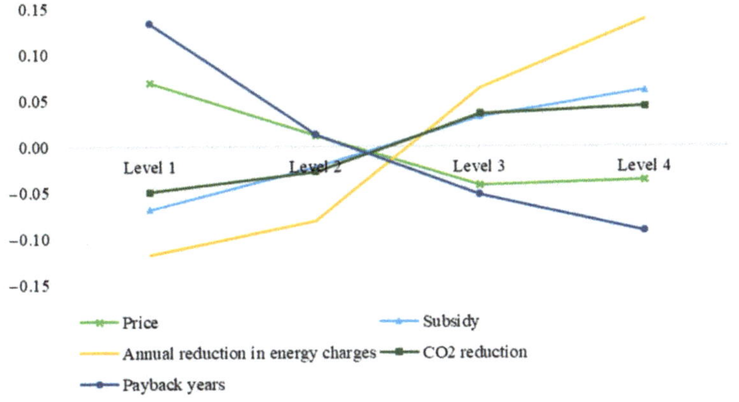


**Figure 3.** Marginal effects of heat pump attributes.

As illustrated before, the $CO_2$ reduction was included as an indicator of environmental benefits related to solar panels. The intention is to examine, by showing the percentage of $CO_2$ reduction, if this environmental feature has an impact on the adoption decision. The results show that $CO_2$ reduction has a positive impact on the acceptance of solar panels in terms of the signs of parameters, although the effects are not fully significant.

Regarding company characteristics, the intention to adopt solar panels for the companies located in city centers is lower than the companies in suburb and rural areas. This may be because of the limited space in central urban areas. The barrier caused by constraint has also been reported by [32].

Moreover, the company size at the level of "more than 20 employees" has a positive sign, which means the acceptance of solar panels increases with the number of employees. Companies with more employees may be more flexible in financial situations, while smaller companies may have to deal with the initial capital. In addition, the companies who have an investment plan are found to have a higher intention to adopt solar panels. This indicates promoting solar panels would be easier for those having investment plans.

The adoption behavior of companies in solar panels also differs according to their business. We found that food-related companies such as restaurants and hotels have a higher intention to install solar panels than other types of companies, which is probably because the electricity demand of food-related businesses is relatively high, and solar panel can significantly reduce their energy cost.

### 4.2. Results of Heat Pump

As presented in Table 6, most parameters are significant. Similar to the heat pump, an installation price of €5000 level has a positive and statistically significant effect on the acceptance of heat pumps. The effect decreases linearly when the price of the heat pump increases from €5000 to €7000; however, the effects do not change much when the price increases from €7000 to €8000.

**Table 6.** Results for heat pumps.

| | | Estimate | | t-Value | *p*-Value |
|---|---|---|---|---|---|
| | Intercept | 2.025 | *** | 12.752 | 0.000 |
| Price | €5000 | 0.069 | *** | 3.008 | 0.003 |
| | €6000 | 0.011 | | 0.502 | 0.615 |
| | €7000 | −0.043 | ** | −1.898 | 0.058 |
| | €8000 | −0.037 | | | |
| Incentive | 0% | −0.069 | *** | −2.979 | 0.003 |
| | 10% | −0.023 | | −1.023 | 0.307 |
| | 20% | 0.032 | | 1.391 | 0.164 |
| | 30% | 0.060 | | | |
| Annual reduction in energy charges | 20% | −0.118 | *** | −5.057 | 0.000 |
| | 40% | −0.081 | *** | −3.564 | 0.000 |
| | 60% | 0.062 | *** | 2.709 | 0.007 |
| | 80% | 0.137 | | | |
| $CO_2$ reduction | 10% | −0.044 | ** | −1.934 | 0.053 |
| | 20% | −0.028 | | −1.238 | 0.216 |
| | 30% | 0.035 | | 1.544 | 0.123 |
| | 40% | 0.043 | | | |
| Payback years | 5 | 0.133 | *** | 5.701 | 0.000 |
| | 10 | 0.012 | | 0.535 | 0.593 |
| | 15 | −0.053 | *** | −2.303 | 0.021 |
| | 20 | −0.092 | | | |
| Function | Cooling and heating | 0.007 | | 0.556 | 0.579 |
| | Heating | −0.007 | | −0.556 | 0.579 |
| Location | City center | −0.430 | *** | −9.767 | 0.000 |
| | Suburb | −0.028 | | −1.453 | 0.146 |
| | Rural | 0.458 | | | |
| Have investment plan | Yes | 0.195 | *** | 7.184 | 0.000 |
| | No | −0.195 | *** | −7.184 | 0.000 |
| Have building ownership | Own | 0.066 | *** | 2.729 | 0.006 |
| | Rent | −0.066 | *** | −2.729 | 0.006 |
| Number of employees | <5 | −0.349 | *** | −7.314 | 0.000 |
| | 5–20 | 0.110 | *** | 5.093 | 0.000 |
| | ≥20 | 0.239 | | | |
| Business scope | Food | −0.017 | | −0.730 | 0.466 |
| | Non-food | 0.034 | | 1.199 | 0.231 |
| | Other | −0.017 | | | |
| Endogenous effects | solar panel | 0.181 | *** | 3.380 | 0.001 |
| | Ventilation | −0.105 | * | −1.804 | 0.071 |
| R-Squared | | 0.635 | | | |

Notes: "*", "**", "***" indicate significance at the 90%, 95% and 99% confidence level, respectively.

The payback period at a five-year level has a positive sign, which suggests the shorter payback period has positive impacts on the acceptance of heat pumps. When the payback time is fifteen years or longer, the probability of adopting a heat pump decreases. In addition, results show that cost reduction increases the acceptance of heat pumps. The coefficient of energy reduction increases obviously from 40% reduction (−0.081) to 60% reduction (0.062), which means the heat pump will become more acceptable when it can save nearly half of the energy expense.

Regarding government incentives, the coefficient for no incentive (−0.067) is negative and statistically significant, which indicates incentive also has a significant impact on the acceptance of heat pumps. The result is in line with previous studies about heat pump diffusion [34]. The acceptance of heat pumps increases near-linearly if the incentive increases from 0 to 30%, which is different from the impact on solar panels. This is perhaps because heat pumps are relatively new, and many EU cities are implementing incentives to promote their installation. In contrast, the incentives for solar panels in some cities have already diminished or closed.

In addition, the coefficient of emission reduction related to heat pumps is insignificant, indicating that the environmental benefits are not the primary motivation for companies to install heat pumps. The cooling function has no significant impact on the acceptance of heat pumps. This is, however, dependent on the microclimate. In this study area, the need for cooling is not strong. The values and changes in heat pump attributes are presented in Figure 4.

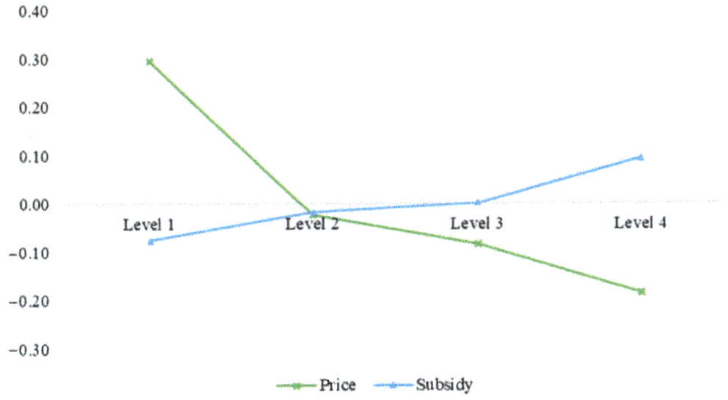


**Figure 4.** Marginal effects of ventilation attributes.

As to the company's characteristics, the number of employees has a positive impact, which implies the acceptance of heat pumps increases with the company size. This is similar to the case of solar panels. In general, smaller companies are less likely to invest in expensive energy facilities. Moreover, companies in the city center are also less likely to install a heat pump than companies in suburb and rural areas. Furthermore, building ownership shows a positive impact on investing in heat pumps in the sense that companies who own the building are more likely to adopt heat pumps. Apart from that, no significant impact was found for business scope on the adoption of heat pumps. Furthermore, the reduction in $CO_2$ emissions is found insignificant, which is different from the effect on solar panels. This means the environmental benefits are not yet a motivation for companies to adopt heat pumps; however, this conclusion may deserve further analysis because, in the choice experiment, we did not technically discriminate heat pumps according to natural resources or not.

### 4.3. Results of Ventilation

Table 7 lists the parameter results related to ventilation. As shown in Table 7, the parameter for a price of €1000 is 0.294, while the parameter for €10,000 is −0.185. Similar to solar panels and heat pumps, the installation price of ventilation has a significantly negative impact on its acceptance. The probability decreases obviously when the price of ventilation increases from €1000 to €3000. The coefficient decreases by 0.318 when the price increases from €1000 (0.294) to €3000 (−0.024), while the coefficient only decreases 0.1 when the price increases from €3000 to €10,000. These results imply that companies are more sensitive to the price when it is between €1000 and €3000.

**Table 7.** Results of ventilation.

| | | Estimate | | t-Value | p-Value |
|---|---|---|---|---|---|
| | Intercept | 1.496 | *** | 7.588 | 0.000 |
| Price | €1000 | 0.294 | *** | 11.766 | 0.000 |
| | €3000 | −0.024 | | −0.997 | 0.319 |
| | €5000 | −0.085 | *** | −3.470 | 0.001 |
| | €10,000 | −0.185 | | | |
| Incentive | €0% | −0.077 | *** | −3.145 | 0.002 |
| | €10% | −0.019 | | −0.782 | 0.434 |
| | €20% | 0.002 | | 0.082 | 0.935 |
| | €30% | 0.094 | | | |
| Demand control | Yes | −0.040 | *** | −2.877 | 0.004 |
| | No | 0.040 | *** | 2.877 | 0.004 |
| Can recover heat? | Yes | 0.062 | *** | 4.368 | 0.000 |
| | No | −0.062 | *** | −4.368 | 0.000 |
| Can filter out pollutants? | Yes | 0.103 | *** | 7.242 | 0.000 |
| | No | −0.103 | *** | −7.242 | 0.000 |
| Location | City center | −0.631 | *** | −15.405 | 0.000 |
| | Suburb | −0.035 | * | −1.704 | 0.089 |
| | Rural | 0.666 | | | |
| Have investment plan | Yes | 0.124 | *** | 4.166 | 0.000 |
| | No | −0.124 | *** | −4.166 | 0.000 |
| Building ownership | Own | −0.329 | *** | −17.410 | 0.000 |
| | Rent | 0.329 | *** | 17.410 | 0.000 |
| Number of employees | ≤5 | −0.764 | *** | −23.743 | 0.000 |
| | 5–20 | 0.150 | *** | 6.467 | 0.000 |
| | ≥20 | 0.614 | | | |
| Business scope | Food | 0.066 | *** | 2.662 | 0.008 |
| | Non-food | −0.005 | | −0.170 | 0.865 |
| | Other | −0.061 | | | |
| Endogenous effects | solar panel | 0.095 | | 1.510 | 0.131 |
| | Heat pump | −0.049 | | −0.537 | 0.591 |
| R-Squared | | 0.751 | | | |

Notes: "*", "***" indicate significance at the 90%, 95% and 99% confidence level, respectively.

The estimated parameter for incentives set at 0 is −0.077, while the estimated parameter for 30% is 0.094. Government incentives are found to have a significant and positive effect on the intention to install ventilation. Similar to the effects on heat pumps, the probability of installing ventilation increases linearly when the incentive increases from 10% to 30%. The effect of the demand control function is negative, which means companies in general are reluctant to the advanced control features. In addition, positive impacts were found for the function of filtering out pollutants and heat recovery, which suggests the added functions can increase the probability of adopting ventilation. The filtering function has the largest effect among the three functions, which means people weigh the filtering function more than the other two. The values and changes in the estimated coefficients are presented in Figure 4. The two marginal effect lines in Figure 4 indicate that the marginal effects of price and tax incentives are not linear.

In addition, company location, number of employees, and building ownership are statistically significant. Similar to the cases of heat pumps and solar panels, companies in the city center are less likely to install a ventilation system. Regarding the company size, acceptance of ventilation increases with the number of employees. The coefficient of having a food-related business scope is statistically significant and positive, which means food-related companies are more likely to adopt a ventilation system than other companies. This is understandable because food-related businesses may have a relatively higher requirement for air quality.

### 4.4. Mutual Dependency of Innovative Energy Facilities

The potential mutual dependency between different facilities is captured through the endogenous variables in the simultaneous equation model. The results are shown in Tables 5–7. It is found that the adoption of heat pumps and the adoption of ventilation both have a positive and significant effect on the adoption of solar panels. Keeping everything else constant, one unit increase in the degree of accepting heat pump and ventilation increases the level of acceptance of solar panels by 0.450 and 0.152, respectively. The mutual dependency between the acceptance of heat pumps and solar panels is stronger than that between ventilation and solar panels.

In the case of the heat pump, the effect of the adoption of the heat pump is significant and positive. This means the adoption of solar panels increases the adoption of heat pumps. This is probably because the combination of solar panels and heat pumps can largely improve overall energy efficiency; however, the acceptance of ventilation is found to have a negative and significant impact on the acceptance of heat pumps. This is probably because some advanced ventilation already integrates part of the heating and cooling, which overlaps the functionality of the heat pump. Last, in the case of ventilation, no significant impacts of the adoption of solar panels and heat pumps on the acceptance of ventilation are found.

To summarize, the results reveal a bi-directional positive relationship between the acceptance of solar panels and heat pumps and a unidirectional relationship between ventilation and solar panels (Figure 5). More precisely, it implies that companies who prefer one of the two energy facilities (solar panels and heat pumps) are more likely to install the other one. Companies that choose ventilation are more likely to install solar panels but less likely to install heat pumps.

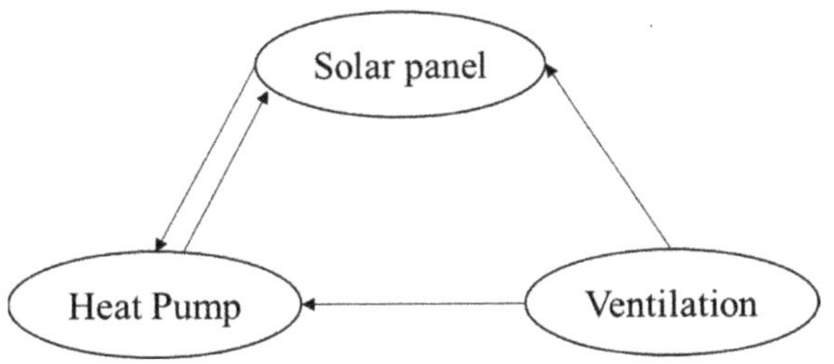


**Figure 5.** Mutual dependency on the adoption of innovative energy facilities.

## 5. Discussion

This study makes significant contributions to understanding the adoption of innovative energy facilities by small and medium-sized enterprises (SMEs) in the tertiary sector. It fills a critical gap in the literature, which has largely focused on residential energy adoption, by examining the specific decision-making processes of SME owners in service industries, such as butchers and restaurants. This contribution is crucial, as SMEs constitute a substantial portion of economic activity and energy consumption.

In comparing our findings with previous studies, we observe several similarities and differences. Similar to research on larger enterprises [35] and households [36], this study reveals that financial factors, such as price and installation costs, significantly influence the installation intentions of SMEs. Furthermore, government tax reductions and subsidy policies have a positive impact on adoption willingness, supporting the findings of earlier studies [37–39].

However, a notable difference emerges in the realm of environmental awareness. While some studies [38,40,41] suggest that the application of eco-technologies does not significantly affect investment intentions in renewable energy, our findings indicate that the technological performance of renewable energy facilities in reducing carbon emissions enhances the investment willingness of SMEs significantly. This divergence may be attributable to the differing priorities between the tertiary sector enterprises and other sectors, with the latter placing greater emphasis on how environmental performance can positively influence their social image. Additionally, SMEs in the Netherlands show a stronger focus on eco-labels compared with their counterparts in developing Asian countries [41,42], reflecting regional differences in the perception of sustainability.

Second, this study emphasizes the interdependencies between different energy facilities, a topic that has been insufficiently explored in the existing literature. Our findings indicate that the decision to adopt one facility can positively influence the acceptance of others. For instance, the adoption of solar panels may reduce electricity costs for heat pumps, creating a synergistic effect that encourages more comprehensive energy solutions. This interdependency aspect is particularly important for policymakers and technology developers, as it suggests that promoting one type of facility could enhance the overall adoption of multiple technologies, leading to greater energy efficiency and environmental benefits.

Notably, our results corroborate the widely recognized benefits of synergies among energy facilities in terms of energy savings [43,44] and economic viability [45,46], indicating that such collaborative effects are acceptable to SME managers; however, a key distinction from existing research on synergies lies in the directionality of these influences. Existing research typically asserts that combinations of these facilities enhance acceptance of them [47,48], but it often fails to analyze the internal interactions between individual devices separately. Specifically, we identified a bi-directional relationship between the adoption of solar panels and heat pumps, indicating that the acceptance of one positively influences the other. In contrast, we found a unidirectional relationship between ventilation and solar panels, where the adoption of ventilation positively influences solar panel adoption but not vice versa. This variation in directionality highlights the complexities of interdependencies among energy technologies, suggesting that the impacts of different facility combinations can differ significantly and require tailored strategies for promotion.

This study has several limitations that should be acknowledged. First, our research is focused on the Netherlands, which may limit the generalizability of the findings to other regions with different socio-economic and regulatory contexts. Factors influencing the adoption of innovative energy facilities can vary significantly across countries due to differing energy policies and cultural attitudes. While our insights are valuable for understanding adoption behaviors within the Netherlands context, they may not be directly applicable elsewhere. Additionally, the sample comprises small and medium-sized enterprises in the tertiary sector, which may not represent all industries.

## 6. Conclusions and Policy Implications

### 6.1. Conclusions

Companies in commercial buildings are a crucial sector of energy consumption and have great potential to reduce carbon emissions through the installation of innovative or renewable energy facilities. In the context of accelerated electrification and renewable energy development, a faster popularization of new energy facilities is needed and timely in the sense that companies may be expected to take a larger role in reducing carbon emissions through upgrading the current facilities; however, companies have a specific decision-making process that is different from households and has different obstacles and limitation. Identifying the decision-making mechanism of companies to adopt innovative energy facilities not only contributes to the rare literature in this perspective but also has important practical relevance in terms of policy decision-making.

In this study, we tackle three popular innovative energy facilities, including solar panels, heat pumps, and ventilation. By designing a dedicated stated choice experiment, we systematically examined companies' adoption decisions and the potential mutual dependency between different energy facilities. We found price and payback period are the two most important factors influencing the acceptance of solar panels and heat pumps, whereas payback period has an even greater effect than the price. The environmental benefits, such as the reduction in carbon emissions, are not yet a motivation for companies to adopt solar panels, heat pumps, and ventilation. In addition, companies' adoption behavior differs according to their profiles, e.g., companies located in the city center or a rented building are less likely to install energy facilities, but building ownership offers a mixed message. Furthermore, government incentive is found to play a role in promoting the adoption of innovative energy facilities, especially solar panels. In the case of the potential mutual effects, we observe a bi-directional relationship between the acceptance of solar panels and heat pumps and a unidirectional relationship between ventilation and solar panels, e.g., the intention to install ventilation increases the acceptance of solar panels.

### 6.2. Policy Implications

Practically, our findings have implications for policymakers and industry stakeholders. By identifying key drivers of adoption, this study can inform targeted policies that enhance the uptake of renewable energy technologies among SMEs. Moreover, our insights emphasize the need for increased awareness of the economic and environmental benefits of innovative energy facilities, guiding effective outreach and promotion efforts.

According to our results, policy recommendations were provided for stakeholders. The results indicate that cost-related attributes significantly influence the decision-making processes of company owners. By increasing grants, tax credits, and subsidies, government entities can effectively lower financial barriers for businesses, thereby promoting the uptake of sustainable energy solutions and advancing national environmental objectives. Local governments should establish initiatives that promote community awareness of innovative energy solutions, complemented by pilot programs demonstrating their effectiveness. By engaging communities and showcasing successful examples of energy facility adoption, local governments can create supportive environments for SMEs.

As for the industry associations, they should focus on developing and disseminating educational programs and resources that underscore the benefits and interdependencies of various energy facilities, supplemented by case studies and best practices. These organizations play a vital role in facilitating knowledge transfer and fostering collaboration among SMEs. By enhancing an understanding of the relationships between energy facilities, businesses can make informed decisions regarding their energy investments. This approach is expected to lead to increased adoption rates and improved environmental performance across the sector.

Manufacturers of energy facilities should consider collaborating with stakeholders to create bundled offerings that integrate solar panels, heat pumps, and ventilation systems, along with attractive financing options. This study highlights a bi-directional relationship between the adoption of solar panels and heat pumps. By promoting integrated solutions that emphasize these synergies, manufacturers can enhance market appeal and drive sales. This strategy not only contributes to higher sales figures but also fosters a more sustainable energy landscape.

### 6.3. Future Research

Although the present study provides a modeling approach and a perspective to investigate companies' intention to adopt innovative energy facilities, several directions deserve further elaboration. First, the present study observed the adoption intention towards each energy facility separately, and the mutual dependency between alternatives is captured by the endogenous effects. As an alternative, a direct choice of the combination of two or more energy facilities will compensate and may provide more information on the synergy of different facilities. Second, the potential constraints of companies in adopting innovative energy facilities are worthy of further investigation. In the current study, the building ownership was found with mixed effects across different facilities, which could be attributable to the difficulty in imaging the actual situation out of the choice experiment. In this regard, a larger revealed dataset would be more suitable to identify the constraints. Third, this study models the acceptance of energy facilities at linear scales. Assuming linearity may have a limitation as the ordered responses were collected. Future studies can develop a generalized simultaneous model incorporating non-linearity in the responses.

**Author Contributions:** Conceptualization, G.G. and R.H.; methodology, G.G.; validation, G.G. and R.H.; formal analysis, G.G. and R.H.; investigation, G.G. and R.H.; resources, G.G.; data curation, R.H.; writing—original draft preparation, R.H.; writing—review and editing, G.G. and R.H.; visualization, R.H.; supervision, G.G. and R.H.; project administration, G.G.; funding acquisition, G.G. All authors have read and agreed to the published version of the manuscript.

**Funding:** This research was funded by the China Postdoctoral Science Foundation, grant number 2023 M740465, and the Fundamental Research Funds from the Liaoning Education Department, Grant Number JYTMS20230168.

**Informed Consent Statement:** Informed consent was obtained from all subjects involved in this study.

**Data Availability Statement:** The datasets used during the current study are available from the corresponding author on reasonable request.

**Conflicts of Interest:** The authors declare no conflicts of interest.

## References

1. González-Torres, M.; Pérez-Lombard, L.; Coronel, J.F.; Maestre, I.R.; Yan, D. A Review on Buildings Energy Information: Trends, End-Uses, Fuels and Drivers. *Energy Rep.* **2022**, *8*, 626–637. [CrossRef]
2. Fu, H.; Baltazar, J.-C.; Claridge, D.E. Review of Developments in Whole-Building Statistical Energy Consumption Models for Commercial Buildings. *Renew. Sustain. Energy Rev.* **2021**, *147*, 111248. [CrossRef]
3. Wu, W.; Skye, H.M. Residential Net-Zero Energy Buildings: Review and Perspective. *Renew. Sustain. Energy Rev.* **2021**, *142*, 110859. [CrossRef] [PubMed]
4. Lu, X.; Pang, Z.; Fu, Y.; O'Neill, Z. Advances in Research and Applications of $CO_2$-Based Demand-Controlled Ventilation in Commercial Buildings: A Critical Review of Control Strategies and Performance Evaluation. *Build. Environ.* **2022**, *223*, 109455. [CrossRef]
5. Impram, S.; Varbak Nese, S.; Oral, B. Challenges of Renewable Energy Penetration on Power System Flexibility: A Survey. *Energy Strategy Rev.* **2020**, *31*, 100539. [CrossRef]
6. Alam, M.S.; Al-Ismail, F.S.; Salem, A.; Abido, M.A. High-Level Penetration of Renewable Energy Sources into Grid Utility: Challenges and Solutions. *IEEE Access* **2020**, *8*, 190277–190299. [CrossRef]
7. Li, S.; Gou, Z. Accepting Solar Photovoltaic Panels in Rural Landscapes: The Tangle among Nostalgia, Morality, and Economic Stakes. *Land* **2023**, *12*, 1956. [CrossRef]
8. Shi, M.; Lu, X.; Craig, M.T. Climate Change Will Impact the Value and Optimal Adoption of Residential Rooftop Solar. *Nat. Clim. Change* **2024**, *14*, 482–489. [CrossRef]
9. Chayjan, M.R.; Dehghanian, F.; Kakhki, M.D. Modeling Residential Photovoltaic Adoption: A System Dynamics Approach for Solar Energy Expansion. *Energy Policy* **2024**, *189*, 114133. [CrossRef]
10. Molnár, G.; Cabeza, L.F.; Chatterjee, S.; Ürge-Vorsatz, D. Modelling the Building-Related Photovoltaic Power Production Potential in the Light of the EU's Solar Rooftop Initiative. *Appl. Energy* **2024**, *360*, 122708. [CrossRef]
11. Sommerfeld, J.; Buys, L.; Vine, D. Residential Consumers' Experiences in the Adoption and Use of Solar PV. *Energy Policy* **2017**, *105*, 10–16. [CrossRef]
12. Inês, C.; Guilherme, P.L.; Esther, M.-G.; Swantje, G.; Stephen, H.; Lars, H. Regulatory Challenges and Opportunities for Collective Renewable Energy Prosumers in the EU. *Energy Policy* **2020**, *138*, 111212. [CrossRef]
13. Batel, S. Research on the Social Acceptance of Renewable Energy Technologies: Past, Present and Future. *Energy Res. Soc. Sci.* **2020**, *68*, 101544. [CrossRef]
14. Henriques, J.; Catarino, J. Motivating towards Energy Efficiency in Small and Medium Enterprises. *J. Clean. Prod.* **2016**, *139*, 42–50. [CrossRef]
15. Trianni, A.; Cagno, E. Dealing with Barriers to Energy Efficiency and SMEs: Some Empirical Evidences. *Energy* **2012**, *37*, 494–504. [CrossRef]
16. Peñaloza, D.; Mata, É.; Fransson, N.; Fridén, H.; Samperio, Á.; Quijano, A.; Cuneo, A. Social and Market Acceptance of Photovoltaic Panels and Heat Pumps in Europe: A Literature Review and Survey. *Renew. Sustain. Energy Rev.* **2022**, *155*, 111867. [CrossRef]
17. Abreu, J.; Wingartz, N.; Hardy, N. New Trends in Solar: A Comparative Study Assessing the Attitudes towards the Adoption of Rooftop PV. *Energy Policy* **2019**, *128*, 347–363. [CrossRef]
18. Bondio, S.; Shahnazari, M.; McHugh, A. The Technology of the Middle Class: Understanding the Fulfilment of Adoption Intentions in Queensland's Rapid Uptake Residential Solar Photovoltaics Market. *Renew. Sustain. Energy Rev.* **2018**, *93*, 642–651. [CrossRef]

19. Karytsas, S. An Empirical Analysis on Awareness and Intention Adoption of Residential Ground Source Heat Pump Systems in Greece. *Energy Policy* **2018**, *123*, 167–179. [CrossRef]
20. Wang, L.; Ma, G.; Zhou, F.; Liu, Y.; Tian, T. Multicriteria Decision-Making Approach for Selecting Ventilation Heat Recovery Devices Based on the Attributes of Buildings and the Preferences of Decision Makers. *Sustain. Cities Soc.* **2019**, *51*, 101753. [CrossRef]
21. Liu, G.; Xiao, M.; Zhang, X.; Gal, C.; Chen, X.; Liu, L.; Pan, S.; Wu, J.; Tang, L.; Clements-Croome, D. A Review of Air Filtration Technologies for Sustainable and Healthy Building Ventilation. *Sustain. Cities Soc.* **2017**, *32*, 375–396. [CrossRef]
22. De Groote, O.; Pepermans, G.; Verboven, F. Heterogeneity in the Adoption of Photovoltaic Systems in Flanders. *Energy Econ.* **2016**, *59*, 45–57. [CrossRef]
23. Li, P.-H.; Keppo, I.; Strachan, N. Incorporating Homeowners' Preferences of Heating Technologies in the UK TIMES Model. *Energy* **2018**, *148*, 716–727. [CrossRef]
24. Best, R.; Trück, S. Capital and Policy Impacts on Australian Small-Scale Solar Installations. *Energy Policy* **2020**, *136*, 111082. [CrossRef]
25. Nurunnabi, M.; Esquer, J.; Munguia, N.; Zepeda, D.; Perez, R.; Velazquez, L. Reaching the Sustainable Development Goals 2030: Energy Efficiency as an Approach to Corporate Social Responsibility (CSR). *GeoJournal* **2020**, *85*, 363–374. [CrossRef]
26. Chen, J.F.; Dai, Y.J.; Wang, R.Z. Experimental and Theoretical Study on a Solar Assisted $CO_2$ Heat Pump for Space Heating. *Renew. Energy* **2016**, *89*, 295–304. [CrossRef]
27. Wang, X.; Xia, L.; Bales, C.; Zhang, X.; Copertaro, B.; Pan, S.; Wu, J. A Systematic Review of Recent Air Source Heat Pump (ASHP) Systems Assisted by Solar Thermal, Photovoltaic and Photovoltaic/Thermal Sources. *Renew. Energy* **2020**, *146*, 2472–2487. [CrossRef]
28. Lowry, G. Day-Ahead Forecasting of Grid Carbon Intensity in Support of Heating, Ventilation and Air-Conditioning Plant Demand Response Decision-Making to Reduce Carbon Emissions. *Build. Serv. Eng. Res. Technol.* **2018**, *39*, 749–760. [CrossRef]
29. Al-tuwaijri, S.A.; Christensen, T.E.; Ii, K.E.H. The Relations among Environmental Disclosure, Environmental Performance, and Economic Performance: A Simultaneous Equations Approach. *Account. Organ. Soc.* **2004**, *29*, 447–471. [CrossRef]
30. Islam, T. Household Level Innovation Diffusion Model of Photo-Voltaic (PV) Solar Cells from Stated Preference Data. *Energy Policy* **2014**, *65*, 340–350. [CrossRef]
31. Gu, G.; Feng, T. Heterogeneous Choice of Home Renewable Energy Equipment Conditioning on the Choice of Electric Vehicles. *Renew. Energy* **2020**, *154*, 394–403. [CrossRef]
32. Mah, D.N.; Wang, G.; Lo, K.; Leung, M.K.H.; Hills, P.; Lo, A.Y. Barriers and Policy Enablers for Solar Photovoltaics (PV) in Cities: Perspectives of Potential Adopters in Hong Kong. *Renew. Sustain. Energy Rev.* **2018**, *92*, 921–936. [CrossRef]
33. Sun, P.C.; Wang, H.M.; Huang, H.L.; Ho, C.W. Consumer Attitude and Purchase Intention toward Rooftop Photovoltaic Installation: The Roles of Personal Trait, Psychological Benefit, and Government Incentives. *Energy Environ.* **2020**, *31*, 21–39. [CrossRef]
34. McElroy, D.J.; Rosenow, J. Policy Implications for the Performance Gap of Low-Carbon Building Technologies. *Build. Res. Inf.* **2019**, *47*, 611–623. [CrossRef]
35. Obeng-Darko, N.A. Policy Trends on Renewable Energy for Decentralised Electrification as a Catalyst for Achieving Goal Seven of the Sustainable Development Goals in Sub-Saharan Africa. *Renew. Energy Law Policy Rev.* **2018**, *8*, 12–24. Available online: https://www.jstor.org/stable/26638282 (accessed on 4 November 2024). [CrossRef]
36. Li, L.; Dai, C. Internal and External Factors Influencing Rural Households' Investment Intentions in Building Photovoltaic Integration Projects. *Energies* **2024**, *17*, 1071. [CrossRef]
37. Aboelmaged, M.; Hashem, G. Absorptive Capacity and Green Innovation Adoption in SMEs: The Mediating Effects of Sustainable Organisational Capabilities. *J. Clean. Prod.* **2019**, *220*, 853–863. [CrossRef]
38. De Jesus Pacheco, D.A.; ten Caten, C.S.; Jung, C.F.; Ribeiro, J.L.D.; Navas, H.V.G.; Cruz-Machado, V.A. Eco-Innovation Determinants in Manufacturing SMEs: Systematic Review and Research Directions. *J. Clean. Prod.* **2017**, *142*, 2277–2287. [CrossRef]
39. Daiyabu, Y.A.; Manaf, N.A.A.; Mohamad Hsbollah, H. Extending the Theory of Planned Behaviour with Application to Renewable Energy Investment: The Moderating Effect of Tax Incentives. *Int. J. Energy Sect. Manag.* **2023**, *17*, 333–351. [CrossRef]
40. Appiah, M.K.; Anderson Akolaa, R.; Ayisi-Addo, A.K. Modeling the Impact of Macroenvironmental Forces on Investment in Renewable Energy Technologies in Ghana: The Moderating Role of Entrepreneurship Orientation Dimensions. *Cogent Econ. Financ.* **2022**, *10*, 2071387. [CrossRef]
41. Atchike, D.W.; Zhenyu, Z.; Ali, T.; Weishang, G.; Jabeen, G. Towards Sustainable Energy: Factors Affecting Solar Power System Adoption by Small and Medium-Sized Businesses. *Front. Environ. Sci.* **2022**, *10*, 967284. [CrossRef]
42. Qamar, S.; Ahmad, M.; Oryani, B.; Zhang, Q. Solar Energy Technology Adoption and Diffusion by Micro, Small, and Medium Enterprises: Sustainable Energy for Climate Change Mitigation. *Environ. Sci Pollut. Res.* **2022**, *29*, 49385–49403. [CrossRef] [PubMed]
43. Vallati, A.; Di Matteo, M.; Sundararajan, M.; Muzi, F.; Fiorini, C.V. Development and Optimization of an Energy Saving Strategy for Social Housing Applications by Water Source-Heat Pump Integrating Photovoltaic-Thermal Panels. *Energy* **2024**, *301*, 131531. [CrossRef]
44. Franco, A.; Fantozzi, F. Experimental Analysis of a Self Consumption Strategy for Residential Building: The Integration of PV System and Geothermal Heat Pump. *Renew. Energy* **2016**, *86*, 1075–1085. [CrossRef]

45. Niekurzak, M.; Lewicki, W.; Drożdż, W.; Miązek, P. Measures for Assessing the Effectiveness of Investments for Electricity and Heat Generation from the Hybrid Cooperation of a Photovoltaic Installation with a Heat Pump on the Example of a Household. *Energies* **2022**, *15*, 6089. [CrossRef]
46. Kim, H.; Junghans, L. Economic Feasibility of Achieving Net-Zero Emission Building (NZEB) by Applying Solar and Geothermal Energy Sources to Heat Pump Systems: A Case in the United States Residential Sector. *J. Clean. Prod.* **2023**, *416*, 137822. [CrossRef]
47. Palomba, V.; Dino, G.E.; Frazzica, A. Analysis of the Potential of Solar-Assisted Heat Pumps: Technical, Market, and Social Acceptance Aspects. *Sol. RRL* **2022**, *6*, 2200037. [CrossRef]
48. Social Acceptance of Renewable Energy Technologies for Buildings in the Helsinki Metropolitan Area of Finland. *Renew. Energy* **2016**, *99*, 813–824. [CrossRef]

MDPI AG
Grosspeteranlage 5
4052 Basel
Switzerland
Tel.: +41 61 683 77 34

*Buildings* Editorial Office
E-mail: buildings@mdpi.com
www.mdpi.com/journal/buildings

www.ingramcontent.com/pod-product-compliance
Lightning Source LLC
Chambersburg PA
CBHW061831210926
51850CB00006B/860
*9783725869589*